Social Sensing and Big Data Computing for Disaster Management

Social Sensing and Big Data Computing for Disaster Management captures recent advancements in leveraging social sensing and big data computing for supporting disaster management. Specifically, analysed within this book are some of the promises and pitfalls of social sensing data for disaster relevant information extraction, impact area assessment, population mapping, occurrence patterns, geographical disparities in social media use, and inclusion in larger decision support systems.

Traditional data collection methods such as remote sensing and field surveying often fail to offer timely information during or immediately following disaster events. Social sensing enables all citizens to become part of a large sensor network which is low cost, more comprehensive, and always broadcasting situational awareness information. However, data collected with social sensing is often massive, heterogeneous, noisy, and unreliable in some aspects. It comes in continuous streams, and often lacks geospatial reference information. Together, these issues represent a grand challenge toward fully leveraging social sensing for emergency management decision making under extreme duress. Meanwhile, big data computing methods and technologies such as high-performance computing, deep learning, and multi-source data fusion become critical components of using social sensing to understand the impact of and response to the disaster events in a timely fashion.

This book was originally published as a special issue of the *International Journal of Digital Earth.*

Zhenlong Li is Associate Professor in the Department of Geography at the University of South Carolina, USA where he established and leads the Geoinformation and Big Data Research Laboratory. His primary research focuses on geospatial big data analytics, spatiotemporal analysis/modelling, and CyberGIS/GeoAI. By synthesizing advanced computing technologies, geospatial methods, and spatiotemporal principles, his research aims to advance knowledge discovery and decision making to support domain applications including disaster management, climate change, human mobilities, and public health.

Qunying Huang is Associate Professor in the Department of Geography at the University of Wisconsin–Madison, USA. Her fields of expertise include spatial computing, spatial data mining, and spatial data analytics. Dr. Huang's research bridges the gap between computer and information science (CIScience) and GIScience by generating new computational algorithms and methods to make sense of complex big spatial datasets obtained from both the physical sensing (e.g. remote sensing) and social (e.g. social media) sensing networks. The problem domains of her research are related to natural hazards and human mobility.

Christopher T. Emrich is Endowed Associate Professor of Environmental Science and Public Administration within the School of Public Administration and a founding member of the newly formed National Center for Integrated Coastal Research at the University of Central Florida (UCF Coastal), USA. His research/practical service includes applying geospatial technologies to emergency management planning and practice, long-term disaster recovery, and the intersection of social vulnerability and community resilience in the face of catastrophe.

Social Sensing and Big Data Computing for Disaster Management

Edited by

Zhenlong Li, Qunying Huang, and Christopher T. Emrich

Routledge
Taylor & Francis Group
LONDON AND NEW YORK

First published 2021
by Routledge
2 Park Square, Milton Park, Abingdon, Oxon OX14 4RN

and by Routledge
52 Vanderbilt Avenue, New York, NY 10017

Routledge is an imprint of the Taylor & Francis Group, an informa business

British Library Cataloguing in Publication Data
A catalogue record for this book is available from the British Library

ISBN 13: 978-0-367-61765-3

Typeset in Minion Pro
by Newgen Publishing UK

Publisher's Note
The publisher accepts responsibility for any inconsistencies that may have arisen during the conversion of this book from journal articles to book chapters, namely the inclusion of journal terminology.

Contents

Citation Information

The chapters in this book were originally published in the *International Journal of Digital Earth*, volume 12, issue 11 (December 2019). When citing this material, please use the original page numbering for each article, as follows:

Chapter 1

Introduction to social sensing and big data computing for disaster management
Zhenlong Li, Qunying Huang and Christopher T. Emrich
International Journal of Digital Earth, volume 12, issue 11 (December 2019), pp. 1198–1204

Chapter 2

Identifying disaster-related tweets and their semantic, spatial and temporal context using deep learning, natural language processing and spatial analysis: a case study of Hurricane Irma
Muhammed Ali Sit, Caglar Koylu and Ibrahim Demir
International Journal of Digital Earth, volume 12, issue 11 (December 2019), pp. 1205–1229

Chapter 3

Deep learning for real-time social media text classification for situation awareness – using Hurricanes Sandy, Harvey, and Irma as case studies
Manzhu Yu, Qunying Huang, Han Qin, Chris Scheele and Chaowei Yang
International Journal of Digital Earth, volume 12, issue 11 (December 2019), pp. 1230–1247

Chapter 4

A visual–textual fused approach to automated tagging of flood-related tweets during a flood event
Xiao Huang, Cuizhen Wang, Zhenlong Li and Huan Ning
International Journal of Digital Earth, volume 12, issue 11 (December 2019), pp. 1248–1264

Chapter 5

Rapid estimation of an earthquake impact area using a spatial logistic growth model based on social media data
Yandong Wang, Shisi Ruan, Teng Wang and Mengling Qiao
International Journal of Digital Earth, volume 12, issue 11 (December 2019), pp. 1265–1284

Chapter 6

Mapping near-real-time power outages from social media
Huina Mao, Gautam Thakur, Kevin Sparks, Jibonananda Sanyal and Budhendra Bhaduri
International Journal of Digital Earth, volume 12, issue 11 (December 2019), pp. 1285–1299

Chapter 7

Social and geographical disparities in Twitter use during Hurricane Harvey
Lei Zou, Nina S. N. Lam, Shayan Shams, Heng Cai, Michelle A. Meyer, Seungwon Yang, Kisung Lee, Seung-Jong Park and Margaret A. Reams
International Journal of Digital Earth, volume 12, issue 11 (December 2019), pp. 1300–1318

Chapter 8

Population distribution modelling at fine spatio-temporal scale based on mobile phone data
Petr Kubíček, Milan Konečný, Zdeněk Stachoň, Jie Shen, Lukáš Herman, Tomáš Řezník, Karel Staněk, Radim Štampach and Šimon Leitgeb
International Journal of Digital Earth, volume 12, issue 11 (December 2019), pp. 1319–1340

Chapter 9

Discovering the relationship of disasters from big scholar and social media news datasets
Liang Zheng, Fei Wang, Xiaocui Zheng and Binbin Liu
International Journal of Digital Earth, volume 12, issue 11 (December 2019), pp. 1341–1363

Chapter 10

A cyberGIS-enabled multi-criteria spatial decision support system: A case study on flood emergency management
Zhe Zhang, Hao Hu, Dandong Yin, Shakil Kashem, Ruopu Li, Heng Cai, Dylan Perkins and Shaowen Wang
International Journal of Digital Earth, volume 12, issue 11 (December 2019), pp. 1364–1381

Notes on Contributors

Budhendra Bhaduri, Oak Ridge National Laboratory, Urban Dynamics Institute, Oak Ridge, TN, USA.

Heng Cai, Department of Environmental Sciences, Louisiana State University, Baton Rouge, LA, USA.

Ibrahim Demir, Department of Civil and Environmental Engineering, University of Iowa, Iowa City, IA, USA.

Christopher T. Emrich, School of Public Administration and National Center for Integrated Coastal Research, University of Central Florida, Orlando, FL, USA.

Lukáš Herman, Department of Geography, Faculty of Science, Masaryk University, Brno, Czech Republic.

Hao Hu, CyberGIS Center for Advanced Digital and Spatial Studies, University of Illinois, Urbana-Champaign, IL, USA.

Qunying Huang, Department of Geography, University of Wisconsin–Madison, Madison, WI, USA.

Xiao Huang, Department of Geography, University of South Carolina, Columbia, SC, USA.

Shakil Kashem, Department of Geography and Geographic Information Science, University of Illinois, Urbana-Champaign, IL, USA.

Milan Konečný, Department of Geography, Faculty of Science, Masaryk University, Brno, Czech Republic.

Petr Kubíček, Department of Geography, Faculty of Science, Masaryk University, Brno, Czech Republic.

Caglar Koylu, Department of Geographical and Sustainability Sciences, University of Iowa, Iowa City, IA, USA.

Nina S. N. Lam, Department of Environmental Sciences, Louisiana State University, Baton Rouge, LA, USA.

Kisung Lee, Division of Computer Science and Engineering, Center for Computation and Technology, Louisiana State University, Baton Rouge, LA, USA.

Šimon Leitgeb, Department of Geography, Faculty of Science, Masaryk University, Brno, Czech Republic.

Ruopu Li, Department of Geography and Environmental Resources, Southern Illinois University, Carbondale, IL, USA.

Zhenlong Li, Geoinformation and Big Data Research Laboratory, Department of Geography, University of South Carolina, Columbia, SC, USA.

Binbin Liu, Institute of Safety Science and Technology, Department of Engineering Physics, Shenzhen Graduate School, Tsinghua University, Shenzhen, People's Republic of China.

Huina Mao, Oak Ridge National Laboratory, Urban Dynamics Institute, Oak Ridge, TN, USA.

Michelle A. Meyer, Department of Sociology, Louisiana State University, Baton Rouge, LA, USA and Department of Landscape Architecture and Urban Planning, Texas A&M University, College Station, TX, USA.

Huan Ning, Department of Geography, University of South Carolina, Columbia, SC, USA.

Seung-Jong Park, Division of Computer Science and Engineering, Center for Computation and Technology, Louisiana State University, Baton Rouge, LA, USA.

Dylan Perkins, Advanced Research Computing Center, University of Wyoming, Laramie, WY, USA.

Mengling Qiao, State Key Laboratory of Information Engineering in Surveying, Mapping and Remote Sensing, Wuhan University, Wuhan, China.

Han Qin, Department of Geography and Geoinformation Science, George Mason University, Fairfax, VA, USA.

Margaret A. Reams, Department of Environmental Sciences, Louisiana State University, Baton Rouge, LA, USA.

Tomáš Řezník, Department of Geography, Faculty of Science, Masaryk University, Brno, Czech Republic.

Shisi Ruan, State Key Laboratory of Information Engineering in Surveying, Mapping and Remote Sensing, Wuhan University, Wuhan, China.

Jibonananda Sanyal, Oak Ridge National Laboratory, Urban Dynamics Institute, Oak Ridge, TN, USA.

Chris Scheele, Department of Geography, University of Wisconsin–Madison, Madison, WI, USA.

Shayan Shams, Division of Computer Science and Engineering, Center for Computation and Technology, Louisiana State University, Baton Rouge, LA, USA.

Jie Shen, Key Laboratory of Virtual Geographic Environment of Ministry of Education, Nanjing Normal University, Nanjing, China; Jiangsu Center for Collaborative Innovation in Geographical Information Resource Development and Application, Nanjing, China.

Muhammed Ali Sit, Department of Computer Science, University of Iowa, Iowa City, IA, USA.

Kevin Sparks, Oak Ridge National Laboratory, Urban Dynamics Institute, Oak Ridge, TN, USA.

Zdeněk Stachoň, Department of Geography, Faculty of Science, Masaryk University, Brno, Czech Republic.

Radim Štampach, Department of Geography, Faculty of Science, Masaryk University, Brno, Czech Republic.

Karel Staněk, Department of Geography, Faculty of Science, Masaryk University, Brno, Czech Republic.

Gautam Thakur, Oak Ridge National Laboratory, Urban Dynamics Institute, Oak Ridge, TN, USA.

Cuizhen Wang, Department of Geography, University of South Carolina, Columbia, SC, USA.

Fei Wang, Institute of Safety Science and Technology, Department of Engineering Physics, Shenzhen Graduate School, Tsinghua University, Shenzhen, People's Republic of China.

Shaowen Wang, CyberInfrastructure and Geospatial Information Laboratory, Department of Geography and Geographic Information Science, University of Illinois, Urbana-Champaign, IL, USA and CyberGIS Center for Advanced Digital and Spatial Studies, University of Illinois, Urbana-Champaign, IL, USA.

Teng Wang, State Key Laboratory of Information Engineering in Surveying, Mapping and Remote Sensing, Wuhan University, Wuhan, China.

Yandong Wang, State Key Laboratory of Information Engineering in Surveying, Mapping and Remote Sensing, Wuhan University, Wuhan, China and Collaborative Innovation Center for Geospatial Information Technology, Wuhan, China.

Chaowei Yang, Department of Geography and Geoinformation Science, George Mason University, Fairfax, VA, USA.

Seungwon Yang, School of Library and Information Science, Center for Computation and Technology, Louisiana State University, Baton Rouge, LA, USA.

Dandong Yin, CyberInfrastructure and Geospatial Information Laboratory, Department of Geography and Geographic Information Science, University of Illinois, Urbana-Champaign, IL, USA and CyberGIS Center for Advanced Digital and Spatial Studies, University of Illinois, Urbana-Champaign, IL, USA.

Manzhu Yu, Department of Geography and Geoinformation Science, George Mason University, Fairfax, VA, USA.

Zhe Zhang, CyberInfrastructure and Geospatial Information Laboratory, Department of Geography and Geographic Information Science, University of Illinois, Urbana-Champaign, IL, USA and CyberGIS Center for Advanced Digital and Spatial Studies, University of Illinois, Urbana-Champaign, IL, USA.

Liang Zheng, Institute of Safety Science and Technology, Department of Engineering Physics, Shenzhen Graduate School, Tsinghua University, Shenzhen, People's Republic of China.

Xiaocui Zheng, Institute of Safety Science and Technology, Department of Engineering Physics, Shenzhen Graduate School, Tsinghua University, Shenzhen, People's Republic of China.

Lei Zou, Department of Geography, Texas A&M University, College Station, TX, USA and Department of Environmental Sciences, Louisiana State University, Baton Rouge, LA, USA.

Introduction to social sensing and big data computing for disaster management

Zhenlong Li, Qunying Huang and Christopher T. Emrich

ABSTRACT
Traditional data collection methods such as remote sensing and field surveying often fail to offer timely information during or immediately following disaster events. Social sensing enables all citizens to become part of a large sensor network, which is low cost, more comprehensive, and always broadcasting situational awareness information. However, data collected with social sensing is often massive, heterogeneous, noisy, unreliable from some aspects, comes in continuous streams, and often lacks geospatial reference information. Together, these issues represent a grand challenge toward fully leveraging social sensing for emergency management decision making under extreme duress. Meanwhile, big data computing methods and technologies such as high-performance computing, deep learning, and multi-source data fusion become critical components of using social sensing to understand the impact of and response to the disaster events in a timely fashion. This special issue captures recent advancements in leveraging social sensing and big data computing for supporting disaster management. Specifically analyzed within these papers are some of the promises and pitfalls of social sensing data for disaster relevant information extraction, impact area assessment, population mapping, occurrence patterns, geographical disparities in social media use, and inclusion in larger decision support systems.

1. Introduction

Rapid onset disasters, often difficult to prepare for and respond to, make disaster management a challenging task worldwide. Disaster and emergency management effectiveness depends heavily on making good decisions in near-real time under extreme duress. These key, often life-saving, decisions are possible only with real-time situational awareness (SA) describing 'the idealized state of understanding what is happening in an event with many actors and other moving parts, especially with respect to the needs of command and control operations' (Vieweg et al. 2010). However, detailed levels of SA are mainly depended on real-time data availability and a capacity to collect, process, synthesize, and analyze these multi-sourced data in a timely fashion.

Traditional data collection practices such as in-situ sensing, field surveying and remote sensing often fail to offer timely information during or directly following a disaster event. For example, stream gauges positioned in rivers (in-situ sensing) are only useful while the stations are functioning properly and before they are overtopped by floodwaters and rendered inoperable. Field surveying is often costly, labor intensive, and suffers significant delays making it a less optimal method for obtaining information in affected areas during and immediately following disasters as lifesaving missions take place. Remote sensing renders a synoptic view in a large geographic area, significantly contributing to a holistic understanding of disaster situation. Remotely sensed imagery generally provides substantial visual evidence of impact and damage areas and has widely been utilized for monitoring pre-disaster conditions, evaluating impacts, and assessing damages following a disaster. However, some inherent restrictions within remote sensing imagery hamper its utility for disaster management. For example, the coarse temporal resolution (e.g. long revisit cycle) and extreme weather conditions (e.g. cloud cover during a storm) prevent remote sensors from acquiring timely good-quality images during critical disaster response phases (Li et al. 2018).

As a new data collection approach, social sensing first appearance in the literature (2006) where it was viewed as one paradigm of the citizen-initiated sensing in personal, social, and urban settings enabled by embedded sensing technology and internet infrastructure (Srivastava et al. 2006). It was later discussed in Campbell et al. (2008) where social sensing was considered as a type of people-centric sensing along with private sensing and public sensing, 'in which information is shared within social and special interest groups (p.13)'. Meanwhile, geographer Goodchild (2007) coined he term 'voluntary geographic information (VGI)' when considering human as sensors in the world of volunteered geography. VGI draws upon the concept of collaborative user-generated content through crowdsourcing, by which many users with varying levels of expertise contribute geographic data via the web (Goodchild and Glennon 2010). A more recent study defines social sensing as 'a category of spatiotemporally tagged big data that provide an observatory for human behavior, as well as the methods and applications based on such big data.' (Liu et al. 2015, 525). In that study, VGI is differentiated from social sensing as an active data collection approach while social sensing is more passive (such as data collected from the social media platforms). We consider social sensing inclusive of both active human sensors (e.g. those participating in a crowdsourcing project) and passive human sensors (e.g. social media users), where they sense the environment and broadcast such sensed data via social media or dedicated crowdsourcing apps.

In the disaster management context, social sensing enables all citizens to become part of a large sensor network and a homegrown disaster response team by sharing information such as texts, images, and videos through social media platforms (e.g. Twitter, Facebook, Instagram) as well as dedicated crowdsourcing applications such as the USGS 'Did You Feel It' project (Atkinson and Wald 2007). Compared to traditional physical sensors, such a citizen-sensor network is low cost, more spatially comprehensive, and always broadcasting SA information. Social sensing enables wide-scale interaction where primary data becomes 'collectively resourceful, self-policing, and generative of information that cannot otherwise be easily obtained' in a useful timeframe for disaster areas (Sutton, Palen, and Shklovski 2008, 7). For example, context-rich micro-level disaster information (e.g. site specific damage and flood water depth) can be captured in real-time through social media platforms such as Twitter, enabling rapid assessment of evolving disaster situations.

While social sensing offers a new data source in disaster management, the data itself is often massive, heterogeneous, noisy, unreliable, comes in continuous streams, and mostly lacks geospatial reference information. Millions of microblog posts from different social media platforms can be generated in a short time during or right after an impactful disaster. For example, during the 2011 East Coast earthquake impacting regions from North Carolina, USA to Quebec, Canada, over 5,000 earthquake-related tweets were being sent per second according to Twitter (DHS 2013). Effectively synthesizing and processing such social sensing data exhibit the typical five-V challenges of big data, including volume (millions of posts), variety (different data formats such as textual, multimedia), veracity (noisy and biased), velocity (rapid data streams), and value (transformation from data to knowledge). Hence, big data computing methods and technologies such as high-performance computing, distributed geoinformation processing, deep learning, spatial statistics/modeling, and multi-source data fusion become critical components when using social sensing to understand impacts from and response to disaster events in a timely fashion. We believe that social sensing coupled with innovative big data computing methods offers enormous opportunities for disaster management by examining the physical infrastructure (e.g. physical damage), environment (e.g. disaster specific extents), and nature-human interaction (e.g. evacuation) from spatial, temporal, and social dimensions.

Along these lines, this special issue captures and highlights recent advancements in leveraging social sensing and big data computing for enabling decision making in one or more disaster phases (mitigation, preparedness, response, and recovery). The articles included in the issue makes contribution to the research of social sensing and big data computing in the context of disaster management by presenting new methodologies and techniques for the examination, extraction, modeling, mapping and/or assessment of disaster relevant information (Sit, Koylu, and Demir 2019; Huang et al. 2018; Yu et al. 2019), impact area (Wang et al. 2018), geographic location information (Mao et al. 2019), population (Kubíček et al. 2019), occurrence patterns (Zheng et al. 2019), and geographical disparities in social media use (Zou et al. 2019), as well as the development of the decision making system (Zhang et al. 2018).

2. Overview of the articles featured in this special issue

Extracting SA information from massive 'citizen sensed' social media datasets is quite challenging as only a small portion of such datasets are disaster relevant and useful. In addition, it is difficult to further classify disaster relevant datasets into finer 'actionable' categories (e.g. caution, donation, damage) after discarding noise (Huang and Xiao 2015). To address this issue, the first paper by Sit, Koylu, and Demir (2019) introduces a framework integrating deep learning and natural language processing for identifying disaster-relevant information and specific information categories. Within this framework, a deep artificial neural network architecture is first used for classifying social media messages (e.g. tweets in this case) about the disaster context. Topic modeling is leveraged to identify various categories of disaster relevant tweets such as affected individuals, donations and support, advice, caution and alerts. Finally, temporal, spatial, and spatiotemporal aspects of different types of disaster relevant information are analyzed. The results with hurricane Irma revealed the progression of the disaster including the preparedness, response and recovery efforts, and pinpoint the potential areas of damage and affected individuals.

Similarly, Huang et al. (2018) also intend to address the challenge of extracting disaster relevant tweets from a massive twitter data pool. This paper introduces a flood relevant tweet extraction approach leveraging both visual and textual information contained within a tweet. A convolutional neural network (CNN) model classifies the visual content of flood pictures, followed a sensitivity test and duplication test to create a tweet pool for a flood event. The results indicated that coupling CNN classification results with flood sensitive words in tweets improves classification performance. The elimination of tweets containing duplicative photos greatly contributes to higher spatiotemporal relevance to the flood. Using 2017's Houston Flood as a study case, the proposed approach to tagging the flood tweets from massive social media data is demonstrated. This work enables the selection of tweets with flood-related visual information and flood-sensitive textual information. The automatically derived geotagged flood pictures could seed a wide range of flood studies for improved SA and rapid flood response.

Yu et al. (2019) further refine the utility of disaster relevant information in real time, by looking back at past disasters in preparation for future threats. Each disaster event evolves differently through time, leading to a diverse set of topics discussed throughout social media platforms. However disparate an individual disaster event appears, looking across multiple disasters will produce common themes. Once identified, a classification model trained on datasets generated from historic events should be capable of automatically categorizing the text messages for future events. Unfortunately, existing models based on machine learning techniques, such as support vector machine (SVM), are mostly event specific and difficult to handle data generated during a new disaster event. To tackle this challenge, Yu et al. (2019) examines the capability of a CNN model in cross-event message classification based on three geotagged twitter datasets collected during Hurricanes Sandy, Harvey, and Irma. Experiment results indicate that CNN models, comparing to traditional machine learning methods, achieved a consistently better performance for both single event and cross-event evaluation scenarios. The study also reveals that the CNN model has the capability of pre-training Twitter data from past events to classify for an upcoming event for SA.

As noted earlier, SA is crucial for emergency management decision making in time-critical situations. Having access to the most reliable and accurate information can save lives and protect property. The fourth paper by Wang et al. (2018) utilizes social media (geotagged tweets) to quickly estimate the impact area of a large earthquake. At its core, positive results from studies such as these can help the emergency management community gain access to better SA information much more quickly than traditional impact assessment techniques. This paper combines social media and other ancillary data using a spatial logistic growth model to estimate earthquake impact areas. Results and confirmatory analytics indicate that social media data containing geospatial references representing real-time citizen sensor data, can be used to substantially enhance emergency management by rapidly mapping the impact area of a disaster.

Mao et al. (2019) take a slightly different approach to understanding disasters from twitter information recognizing that although geo-spatially enabled social networking data provides detailed locational information the fact remains that geotagged tweets, among other platforms, account for a very small portion of social media's big data. Because location is an important aspect of disaster impacts and damage estimation we must attempt to identify, develop, and deploy novel approaches that can utilize a broader set of social media data. The paper by Mao et al. (2019) explores a new deep learning method to extract outage locations from social media messages to map near real-time power outages rather than relying on locational information attached to such messages. Promising results point to strong positive linear correlations between socially derived outage maps and national outage indicators developed by leading national research labs. This novel approach to extracting locations directly from messages helps overcome data scarcity challenges with current low 'buy in' to geotagged tweets and highlights the value add provided by free and public socially sensed data.

Kubíček et al. (2019) provide a case study of cellular phone information for SA. Cellular phones, perhaps the most ubiquitous personal technology available to society, provide a unique platform from which people can communicate not only verbally, but also through social media. Although cellular phone data is not easy to access because of privacy concerns, this data contains a wealth of useful information for understanding the site and situation of users. During disaster scenarios specifically, linking locational information with hazard zone information can quickly inform decision makers about the number of at-risk people. This paper devised a novel approach to model and map population distribution at fine spatiotemporal scale based on big mobile phone data. A number of research questions drive the focus of this study on the utility of cellular data to better understand population movement and locational patterns in the context of a disaster. The general, and specific emergency management utility of cellular based location data is proven through a series of geostatistical tests. Furthermore, use of cellular data identified deficits in current census level demographics for understanding how many people are at a given place and time.

Zou et al. (2019) move from data coverage toward examining coverage disparities during disasters. Knowing where at risk and vulnerable populations might not have adequate access to information is a key component for emergency management planning and response. Zou et al. (2019) examined the questions related to geographical disparities in twitter use before, during, and after a major disaster (Hurricane Harvey) in Texas and Louisiana using ratio and sentiment analysis. This article explores twitter use patterns under the presumption that better socioeconomic conditions would also have higher twitter usage. A novel twitter data mining framework developed by the authors captured disaster-related tweets in a Hadoop computing environment searching on hurricane related key words. One of two approaches provided a geolocation for each tweet, enabling geospatial analysis across the study area. Ratio and sentiment analysis, derived for multiple geographic and time scales, provided specific twitter datasets for each phase of the emergency management cycle. Summary statistics, correlations, and linear regressions revealed significant social and geographical disparities in Twitter usage.

Zheng et al. (2019) build from the premise that a single hazard can cascade into multiple disaster events to develop a better understanding of how socially sensed data can help us understand these linkages. Many historical disaster events prove that disasters are typically not isolated or static. Rather, many cases have shown that secondary and derivative disasters will occur along with the primary event, resulting in a chain of cascading disasters. For example, a blizzard could trigger a series of hazards, such as blocked roads, power grids collapse, air travel delays, and flash floods. Interlocking disaster-causal factors, including natural, biological, physical, chemical, and human characteristics/interactions further contribute to often complex chains of disasters. Understanding the regular patterns of the occurrence and development of disasters helps control and reduce their negative impacts of disasters. Zheng et al. (2019) proposed an approach to extract the disaster chain of a specific disaster type by collecting the big scholar and social news data with disaster-related keywords, analyzing the strengths of their relationships with the co-word analysis method, and constructing a complex network of all defined disaster types. Within their work, the scholar and social news data are collected from Google Scholar, Baidu Scholar, and Sina News search engines, and the obtained disaster chins are compared with each other to demonstrate the feasibility of the proposed approach. The achieved disaster chains are also compared with the ones concluded from traditional research methods and the very reasonable result is demonstrated. The proposed method can assist in the emergency manager to better prepare in advance and minimum the losses by eliminating the key triggered factors and cutting off the chain.

With the availability of timely and critical information from various data sources including social sensing, the next step is to develop a responsive decision-making process that can integrate this multi-sourced information to support disaster management in the event of disasters. The paper by Zhang et al. (2018) proposes a cyberGIS-enabled multi-criteria spatial decision support system for informing rapid decision-making during emergencies. The system consists of a high-performance computing environment, and multi-criteria decision analysis models, including Weighted Sum Model (WSM) and Technique for Order Preference by Similarity to Ideal Solution Model (TOPSIS), along with various types of social vulnerability indicators to solve decision problems with conflicting evaluation criteria during a flood disaster. Using two decision goals generated based on a historic flood event as a case study, the results indicate that WSM generates more diverse values and higher output category estimations than the TOPSIS model. This spatial decision support system enables collaborative problem solving and efficient knowledge transformation between decision makers, where different responders can formulate their decision objectives, select relevant evaluation criteria, and perform interactive weighting and sensitivity analyses.

3. Conclusion and innovation opportunities

Emergency and crisis emergency management remains a challenging task worldwide. Real-time data sources and the ability to timely collect, process, synthesize, and analyze these data is paramount for enabling better decisions in near-real time under extreme duress. The research articles featured in this issue shed light on some of the opportunities, challenges and solutions of leveraging social sensing and big data computing for supporting disaster management. As evidenced in this special issue, social sensing coupled with innovative analytical and modeling methods offers enormous possibilities for disaster management, including mapping near real-time power outages during a disaster, rapidly estimation of an earthquake impact, modeling population distribution at a fine spatiotemporal scale, and uncovering the relationships of multiple disasters. Meanwhile, innovative big data computing methods based on deep learning and high-performance computing have proven to be effective and efficient in extracting relevant and actionable information, and identify useful patterns from massive unstructured social sensing data in a timely manner. Lastly, the cyberGIS-enabled spatial decision support system demonstrates the utility of integrating social sensing and big data computing in supporting flood emergency management.

Moving forward, research opportunities related to social sensing and big data computing for disaster management lie in four focus areas: (1) theory innovations: how to develop new theoretical frameworks for effectively fusing multi-sourced social sensing datasets by leveraging state-of-the-art big data computing infrastructures and models; (2) application innovations: how to innovatively use social sensing, a new data collection approach, for disaster management activities including event detection, early warning, flood mapping and damage assessment; (3) computing innovations: how to effectively and efficiently analyze big social sensing data for rapid information extraction and knowledge discovery to better understand the physical, environmental, and social dynamics of a disaster; and (4) engineering innovations: how to develop practical systems by integrating big data computing and social sensing along with other data sources for rapid decision making. From the GIScience perspective, advancements in these innovation areas demand a cross-disciplinary and collaborative approach. Application innovations require domain knowledge from disaster experts, computing innovations benefit from collaborating with computer scientists, and engineering innovation demands collaborations with the first responders.

Acknowledgement

We would like to thank the authors who contributed to this special issue and the reviewers for their helpful suggestions and criticisms that improved the papers in the special issue. We also want to extend our gratitude to Professor Changlin Wang and Dr. Linlin Guan for their editorial support and guidance in preparing the special issue.

Disclosure statement

No potential conflict of interest was reported by the authors.

References

Atkinson, G. M., and D. J. Wald. 2007. "Did You Feel It?" Intensity Data: A Surprisingly Good Measure of Earthquake Ground Motion." *Seismological Research Letters* 78 (3): 362–368.

Campbell, A. T., S. B. Eisenman, N. D. Lane, E. Miluzzo, R. A. Peterson, H. Lu, X. Zheng, M. Musolesi, K. Fodor, and G-S. Ahn. 2008. "The Rise of People-Centric Sensing." *IEEE Internet Computing* 12 (4): 12–21.

DHS. 2013. Innovative Uses of Social Media in Emergency Management. Last accessed on 09/11/2019. https://www.dhs.gov/sites/default/files/publications/Social-Media-EM_0913-508_0.pdf.

Goodchild, M. F. 2007. "Citizens as Sensors: the World of Volunteered Geography." *GeoJournal* 69 (4): 211–221.

Goodchild, M. F., and J. A. Glennon. 2010. "Crowdsourcing Geographic Information for Disaster Response: A Research Frontier." *International Journal of Digital Earth* 3 (3): 231–241.

Huang, X., C. Wang, Z. Li, and H. Ning. 2018. "A Visual–Textual Fused Approach to Automated Tagging of Flood-Related Tweets During a Flood Event." *International Journal of Digital Earth* 12 (11): 1248–1264.

Huang, Q., and Y. Xiao. 2015. "Geographic Situational Awareness: Mining Tweets for Disaster Preparedness, Emergency Response, Impact, and Recovery." *ISPRS International Journal of Geo-Information* 4 (3): 1549–1568.

Kubíček, P., M. Konečný, Z. Stachoň, J. Shen, L. Herman, T. Řezník, K. Staněk, R. Štampach, and Š Leitgeb. 2019. "Population Distribution Modelling at Fine Spatio-Temporal Scale Based on Mobile Phone Data." *International Journal of Digital Earth* 12 (11): 1319–1340.

Li, Z., C. Wang, C. T. Emrich, and D. Guo. 2018. "A Novel Approach to Leveraging Social Media for Rapid Flood Mapping: A Case Study of the 2015 South Carolina Floods." *Cartography and Geographic Information Science* 45 (2): 97–110.

Liu, Y., X. Liu, S. Gao, L. Gong, C. Kang, Y. Zhi, G. Chi, and L. Shi. 2015. "Social Sensing: A New Approach to Understanding our Socioeconomic Environments." *Annals of the Association of American Geographers* 105 (3): 512–530.

Mao, H., G. Thakur, K. Sparks, J. Sanyal, and B. Bhaduri. 2019. "Mapping Near-Real-Time Power Outages from Social Media." *International Journal of Digital Earth* 12 (11): 1285–1299.

Sit, M. A., C. Koylu, and I. Demir. 2019. "Identifying Disaster-Related Tweets and Their Semantic, Spatial and Temporal Context Using Deep Learning, Natural Language Processing and Spatial Analysis: A Case Study of Hurricane Irma." *International Journal of Digital Earth* 12 (11): 1205–1229.

Srivastava, M., M. Hansen, J. Burke, A. Parker, S. Reddy, G. Saurabh, and D. Estrin. 2006. Wireless Urban Sensing Systems. https://escholarship.org/uc/item/6sj003r4.

Sutton, J. N., L. Palen, and I. Shklovski. 2008. "Backchannels on the Front Lines: Emergency Uses of Social Media in the 2007 Southern California Wildfires." Proceedings of the 5th International ISCRAM Conference, Washington, DC, USA, May 2008.

Vieweg, S., A. L. Hughes, K. Starbird, and L. Palen. 2010. "Microblogging During Two Natural Hazards Events: What Twitter May Contribute to Situational Awareness." In *Proceedings of the 2010 SIGCHI Conference on human factors in computing systems*, Atlanta, GA, USA, April 2010.

Wang, Y., S. Ruan, T. Wang, and M. Qiao. 2018. "Rapid Estimation of an Earthquake Impact Area Using a Spatial Logistic Growth Model Based on Social Media Data." *International Journal of Digital Earth* 12 (11): 1265–1284.

Yu, M., Q. Huang, H. Qin, C. Scheele, and C. Yang. 2019. "Deep Learning for Real-Time Social Media Text Classification for Situation Awareness – Using Hurricanes Sandy, Harvey, and Irma as Case Studies." *International Journal of Digital Earth* 12 (11): 1230–1247.

Zhang, Z., H. Hu, D. Yin, S. Kashem, R. Li, H. Cai, and S. Wang. 2018. "A cyberGIS-Enabled Multi-Criteria Spatial Decision Support System: A Case Study on Flood Emergency Management." *International Journal of Digital Earth* 12 (11): 1364–1381.

Zheng, L., F. Wang, X. Zheng, and B. Liu. 2019. "Discovering the Relationship of Disasters from Big Scholar and Social Media News Datasets." *International Journal of Digital Earth* 12 (11): 1341–1363.

Zou, L., N. S. N. Lam, S. Shams, H. Cai, M. A. Meyer, S. Yang, K. Lee, S-J. Park, and M. A. Reams. 2019. "Social and Geographical Disparities in Twitter use During Hurricane Harvey." *International Journal of Digital Earth* 12 (11): 1300–1318.

Identifying disaster-related tweets and their semantic, spatial and temporal context using deep learning, natural language processing and spatial analysis: a case study of Hurricane Irma

Muhammed Ali Sit, Caglar Koylu and Ibrahim Demir

ABSTRACT

We introduce an analytical framework for analyzing tweets to (1) identify and categorize fine-grained details about a disaster such as affected individuals, damaged infrastructure and disrupted services; (2) distinguish impact areas and time periods, and relative prominence of each category of disaster-related information across space and time. We first identify disaster-related tweets by generating a human-labeled training dataset and experimenting a series of deep learning and machine learning methods for a binary classification of disaster-relatedness. We employ LSTM (Long Short-Term Memory) networks for the classification task because LSTM networks outperform other methods by considering the whole text structure using long-term semantic word and feature dependencies. Second, we employ an unsupervised multi-label classification of tweets using Latent Dirichlet Allocation (LDA), and identify latent categories of tweets such as affected individuals and disrupted services. Third, we employ spatially-adaptive kernel smoothing and density-based spatial clustering to identify the relative prominence and impact areas for each information category, respectively. Using Hurricane Irma as a case study, we analyze over 500 million keyword-based and geo-located collection of tweets before, during and after the disaster. Our results highlight potential areas with high density of affected individuals and infrastructure damage throughout the temporal progression of the disaster.

1. Introduction

Recent advancements in the spatio-temporal analysis of big data and the extensive use of social media data during crisis situations enable extraction of site-specific information for environmental monitoring (Demir et al. 2015) and disaster management (Li et al. 2018). Such information is invaluable in every stage of disaster management including preparedness, response and recovery (Krajewski et al. 2017). Locations of emergency situations and damage can be captured through social sensors or individuals who generate and diffuse eyewitness information and enable situational awareness using social media platforms. Social sensing and multi-sourced analysis of disaster information have been successfully employed to evaluate the impact of disasters by many researchers (Cervone et al. 2016; De Albuquerque et al. 2015; Feng and Sester 2018; Huang, Cervone, and Zhang 2017; Li et al. 2018; Restrepo-Estrada et al. 2018). Due to the noise in social media data, and the locational and contextual uncertainty of shared information, it is challenging to extract useful and actionable information generated by citizen sensors in real-time or in the aftermath of an event when an analysis is done retrospectively to coordinate recovery efforts.

Despite the significant advancements in deep learning and machine learning methods, previous researchers (Burel et al. 2017; Nguyen et al. 2016; Olteanu et al. 2014) have reported lower performance for classifying fine-grained details of information such as affected individuals, donations and support, and sympathy and prayers from social media data. The challenge for classifying social media messages have often been attributed to the noise in the data, conflicting annotation done by online workers and volunteers, and the multi-label problem, which a tweet can belong to multiple categories (Nguyen et al. 2016). Moreover, there is a critical need for improving the classification of fine-grained details on disaster-relevant information by taking into account the whereabouts and timing of messages sent through social media.

In this article, we introduce an analytical framework that integrates a two-step classification approach with spatial analysis to (1) identify and categorize fine-grained details about a disaster such as affected individuals, damaged infrastructure and disrupted services from tweets (2) distinguish impact areas and time periods, and relative prominence of each category of disaster-related information across space and time. Our analytical framework consists of the following steps. Using Hurricane Irma as a case study, we first develop a context-specific training dataset by manually labeling 20,000 (5%) keyword-based and geo-located corpus of tweets generated before, during and after Hurricane Irma in its impact area. After generating the training dataset, we conduct a classification experiment using a series of classification methods including linear classifiers such as Logistic Regression, Linear Support Vector Machines (SVM) and Ridge, deep learning and machine learning methods such as Convolutional Neural Networks (CNNs) and Long Short-Term Memory (LSTM) Networks. We employ LSTM networks for the classification task because LSTM networks outperform other methods by taking into account the order of words and the whole text structure using long-term semantic word and feature dependencies. Our training data and the experiments on classification methods provide valuable input for future studies that aim to classify tweets about hurricanes in real-time and retrospectively. Different from the supervised methods that classify the type of tweets using predetermined categories, we employ Latent Dirichlet Allocation (LDA) after the binary classification of disaster-related tweets to extract the fine-grained categories of information whose target audience is diverse. For example, we identify affected individuals and infrastructure damage category, which targets emergency responders, rescue missions and utility companies; donations and support category, which targets government organizations and non-profit institutions for planning recovery and relief efforts; affected individuals and pets category, which targets animal control and shelters; and advice, warning and alerts category, which targets local governments and FEMA for building and enhancing situational awareness.

After identifying disaster-related tweets with their information types, we compute a smoothed rate for each type of information using a spatially-adaptive kernel smoothing method. Ultimately, smoothed rates allow us to pinpoint areas where there is significant impact such as damaged infrastructure, affected individuals and disrupted services, as well as areas and time periods where other types of information such as advice and caution, donations and support, and sympathy and prayers are more prominent. In addition to using adaptive kernel smoothing to capture the relative prominence of each category of tweets across geographic space, we employ density-based spatial clustering with noise (DBSCAN) to identify areas where there is high density of tweets for each category. Different from the smoothed surface of probability of each category, DBSCAN allows us to identify the areas in which the absolute density of tweets that belong to a category is greater than a defined threshold. Finally, we overlay the spatial and temporal patterns of tweets with rainfall and flood data to pinpoint potential areas for emergency response and recovery efforts and identify potential geographic biases in social media usage during the hurricane.

2. Related work

Previous work analyzing micro-blogging behavior during a crisis situation commonly reported that shared information varies substantially depending on the type and stage of crisis situations (Kanhabua and Nejdl 2013; Munro and Manning 2012; Olteanu et al. 2014). On the other hand, Olteanu, Vieweg, and Castillo (2015) found consistencies and commonalities about how people share information during different crisis situations after investigating a variety of natural hazards and human-induced disasters. One of the most common findings is that public information seeking increases during crisis situations (Hagen et al. 2017; Nelson, Spence, and Lachlan 2009), and people utilize social media as a resource to ask for help and coordinate efforts through sharing locational and contextual information (Stefanidis et al. 2013). Social media plays an important role in crisis communication by enabling information dissemination to a wider audience, sending and receiving emergency alerts, control information diffusion, collaborate with emergency responders and create situational awareness (Lachlan et al. 2016). The utilization of social media in public information seeking has been evaluated in a variety of disaster events such as the spread of the Zika virus (Hagen et al. 2017), collapsed bridge in Minnesota (Nelson, Spence, and Lachlan 2009), the Mexican drug wars (Monroy-Hernández et al. 2013), the Arab Spring (Kumar et al. 2013), the Haiti earthquake (Caragea et al. 2011) and numerous flood events (Kongthon et al. 2012; Kwon and Kang 2016; Maantay, Maroko, and Culp 2010; Restrepo-Estrada et al. 2018).

Although there have been significant advancements in automated event detection in crisis situations, extracting fine-grained event-related information such as affected individuals and damaged infrastructure remains to be challenging (Burel et al. 2017). In order to improve the event detection, a number of strategies have been introduced. For example, Zhang, Szabo, and Sheng (2016) analyzed not only the content and type of information but also the source of information by combining lexical analysis with user profiling. Pekar et al. (2016) found that lexical features help to achieve better precision, while semantic and stylistic features derived from metadata help improve recall. Table 1 summarizes the previous work by three sections: (1) ontologies for classifying disaster-relevant information from social sensors, (2) classification methods to distinguish disaster-related tweets

Table 1. Typologies for disaster-relevant tweets, classification methods and overlaying with disaster data.

Ontologies for information type	Classification methods	Overlaying with disaster-specific data
Affected individuals, post-impact responses (Liu et al. 2008); caution and advice, causalities and damage, donations and support, affected individuals, and information source (Imran et al. 2013)	Content analysis and human-coding (Kwon and Kang 2016; Lachlan et al. 2016); crowdsourcing (Caragea et al. 2011)	Impact areas of disasters (Acar and Muraki 2011; Cervone et al. 2016; Huang, Cervone, and Zhang 2017); proximity to disaster extent (Kryvasheyeu et al. 2016; Vieweg et al. 2010)
Informative and not informative; infrastructure and utilities, sympathy and emotional support, the criticism of government responses, government outreach efforts, and political discussions (Olteanu, Vieweg, and Castillo 2015).	The Bag-of-Words (Bow) approach, feature abstraction, feature selection, and the topic-words output of Latent Dirichlet Allocation (LDA) as alternative inputs into training Support Vector Machine (SVM) for content classification (Caragea et al. 2011)	Hydrological data, digital elevation models and satellite imagery (De Albuquerque et al. 2015; Herfort et al. 2014; Li et al. 2018; Schnebele 2013; Smith et al. 2015)
Linguistic features such as crisis sensitive features, personal or impersonal expressions, time-awareness, reciprocal conversational discourse and word embedding to identify eye-witness events (Doggett and Cantarero 2016; Fang et al. 2016; Li et al. 2018; Zhang, Szabo, and Sheng 2016); response and recovery efforts, weather predictions and videos from eye-witnesses (De Choudhury, Diakopoulos, and Naaman 2012; Kumar et al. 2013)	Binary classification (e.g. informative or not informative) and multi-label classification (e.g. type of information such as affected individuals, damaged infrastructure and etc.) using Convolutional Neural Networks (CNNs) (Burel et al. 2017; Caragea, Silvescu, and Tapia 2016; Nguyen et al. 2016).	Food, shelter and evacuation, the locations of power outages, evacuation plans, general risk (Spielhofer et al. 2016); warning, preparatory activity, hazard location, flood level, weather, wind, visibility, road conditions and advice, evacuation information, volunteer information and damage and injury reports (Vieweg et al. 2010)

and information categories, (3) studies that integrate location-specific information with social media data to guide disaster management.

3. Case study: Hurricane Irma

In September 2017, Hurricane Irma, a Category 5 storm, devastated Caribbean Islands and Florida with damaging winds, heavy precipitation and flooding. Hurricane region suffered from extensive power outages, loss of lives and infrastructure damage including airports, hospitals and schools. Many areas have become inhabitable with likely damage of billions of dollars to rebuild the infrastructure. Massive evacuation orders were given affecting millions of people living in the coastal areas of Florida. Social media was used extensively during the event to disseminate information and coordinate disaster response and recovery efforts. Agencies used Twitter to post updates about the trajectory of the hurricane and evacuation information. Social media became an effective component of relief efforts, extracting eye-witness information, and help develop situational awareness. As compared to other disasters such as earthquakes, wildfires and landslides, predictability of hurricanes allows collecting social sensing data prior to and during the event, which is invaluable for planning the preparation and emergency response efforts.

4. Methodology

In this article, we introduce an analytical framework to identify, categorize and analyze disaster-related tweets generated before, during and aftermath of a hurricane. Our overall objective is rather holistic to capture any useful information that maybe related to the progression of a hurricane including the preparedness, response and recovery stages. Figure 1 illustrates our analytical

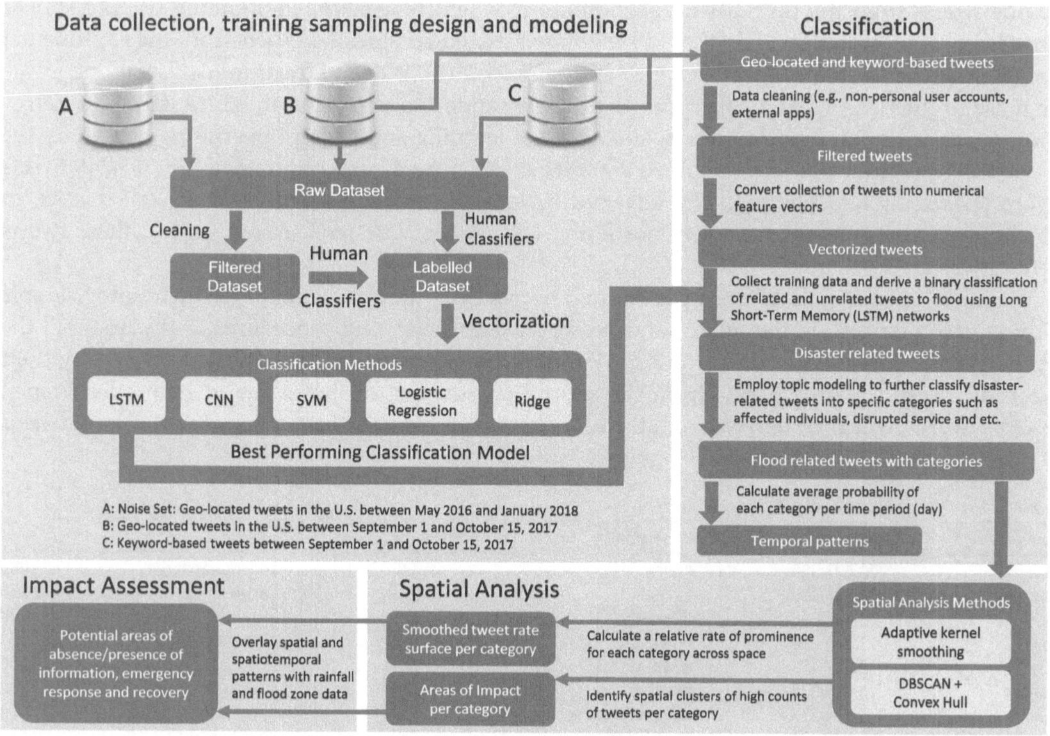

Figure 1. Analytical framework for identifying disaster-related information and impact areas from tweets.

framework. The first phase includes the data collection, sampling design and procedures for training and test data collection, and classification experiments on the training and test sets to evaluate a variety of classification methods. The second phase includes a binary classification of disaster-related tweets using LSTM, and a multi-label classification of detailed information. By splitting the training data into training and test sets, we first perform an experiment on the performance of classification methods Logistic Regression, Linear SVM, Ridge, CNN and LSTM on the binary classification of disaster-related information. Based on the evaluation results and LSTM's conceptual fit to the sequential order of textual data, we choose LSTM to classify the rest of the dataset into related and unrelated tweets. Next, we extract detailed categories of information on disaster-relevant tweets such as affected individuals, advice and warnings, and prayers and support from tweets using an unsupervised topic modeling approach. We then employ spatially-adaptive kernel smoothing and density-based spatial clustering to identify the relative prominence and impact areas for each information category, respectively. Finally, we overlay the spatial and temporal patterns of categorized tweets with rainfall and flood data to pinpoint potential areas for emergency response and recovery efforts and identify geographic biases in social media usage during the hurricane.

4.1. Data collection, training sampling design and modeling

In this study, we used three collections of Twitter data collected by Twitter's Streaming API. The first dataset was collected using 11 hurricane-related keywords: 'flood', 'irma', 'hurricaneirma', 'irmahurricane', 'irma2017', 'hurricane', 'damage', 'storm', 'rain', 'disaster', 'emergency' in order to derive any tweet related to the disaster event. Although we did not use a statistical method to identify keywords that were mentioned statistically more during the disaster (Huang et al. 2018), we continuously updated our keywords using the most frequent hashtags throughout the disaster. We collected the second dataset using a spatial bounding box that covers the impact area of Irma in order to obtain geo-located tweets before, during and aftermath of the disaster. Both the first and second dataset include tweets from the time period beginning on 1 September 2017 and ending on 15 October 2017. Finally, we used a third dataset of randomly selected geo-located tweets from the Continental United States between the dates May 2017 and January 2018. We used the third dataset to incorporate noise in the training data collection. We picked random tweets from the third dataset to introduce the classification models noise to enhance the learning and prevent overfitting caused by the tweets from the study period. Combined database of the three datasets contained 756,913,748 tweets before processing and filtering. The keyword-based collection consisted of 64,266,068 tweets, in which 99% of these tweets have geo-location at least at the state level. About 64% of these tweets were retweets.

We included randomly selected geo-located tweets from the third dataset in the training sample without any filtering. On the other hand, we applied the following procedures to the first and the second dataset prior to the classification. We removed (1) retweets; (2) duplicated tweets by their tweet ids; (3) geo-located tweets from the second dataset that do not contain the predetermined set of keywords, which we derived from the list of keywords from the first dataset (Table 2); (4) tweets

Table 2. Top 26 keyword counts in the raw dataset.

Keyword	Frequency	Hashtag	Frequency		
rain	20,491,588	#rain	47,235	#hurricaneharvey	7883
hurricane	19,897,133	#irma	37,918	#disaster	5458
flood	13,489,668	#flood	20,821	irmahurricane	4298
storm	11,340,708	hurricaneirma	20,149	#irmahurricane	4180
irma	10,127,435	#emergency	19,972	#irma2017	2586
disaster	5,890,874	#storm	19,289	#damage	1932
emergency	5,514,237	#hurricaneirma	15,638	harveystorm	424
damage	5,179,521	hurricaneharvey	10,718	#harveystorm	416
#hurricane	56,389	irma2017	8727		

from non-personal user accounts such as weather apps and bots using the tweet source metadata. Table 3 lists the top sources for tweets, and whether we filtered them or not; (5) tweets that fall outside the time interval from 29 August 2017 to 16 October 2017; (6) tweets that do not contain information on geographic location, or tweets that are located outside the impact area of Hurricane Irma's path. After the filtering, the data were reduced to 557,541 tweets that spanned a time period of one and a half months.

We designed a web-based interface to collect the training data for the binary classification of tweets: whether tweets are related to a hurricane, flooding or an extreme weather event (Figure 2 (b)). We formed a team that consisted of the authors, undergraduate and graduate students for labeling the tweets. We trained our team based on the tasks using examples of both related and unrelated tweets. We considered any tweet whether it is informative or not, or it is about eye-witness information, or news story or warning, or critiques of the government and political discussions about the extreme weather events, flooding and the hurricanes as relevant. The web interface displays the tweet in its original form, and includes 'Yes' and 'No' buttons to answer the question. We included a 'Not Sure' button to let users skip the tweet if they are not sure about whether the tweet is relevant or not. The web interface starts with instructions on how to label the tweets with examples (Figure 2(a)). We specifically kept the question generic in contrast to being specific to Hurricane Irma, because the timespan of Irma overlapped with the other hurricanes such as Harvey, Maria and Jose, and some of the tweets in our dataset were related to those other hurricanes. This allowed us to generate a generic training dataset that can potentially be used for classification tasks for these multiple events. We collected approximately 20,000 labeled tweets with our team of human coders and the web-based interface.

4.2. Binary classification of tweets

Using our training data, we performed classification experiments using Logistic Regression, Support Vector Machines (SVM), Convolutional Neural Networks (CNNs) and Long Short-Term Memory (LSTM) Networks. We describe each of these models and parameters in the following subsections.

4.2.1. Baseline classifier

We selected two different types of machine learning classifiers for comparison: (1) SVM with the Gaussian kernel to obtain a baseline classifier (2) a model with Principal Component Analysis (PCA) summarization step pipelined prior to the Gaussian SVM. We used the second model to

Table 3. Filtering with the most common sources from Twitter metadata.

Source	% of tweets (and number)	Included
(Empty)	79.42% (601,179,529)	Yes
Twitter Web Client	6.79% (51,435,265)	Yes
Instagram	4.14% (31,389,138)	No
TweetMyJOBS	2.79% (21,186,456)	No
iPad	1.05% (7,991,578)	Yes
Foursquare	0.54% (4,130,888)	No
Tweetbot for iOS	0.54% (4,092,581)	Yes
SafeTweet by TweetMyJOBS	0.43% (3,312,923)	No
Sandaysoft Cumulus	0.35% (2,715,771)	No
IFTTT	0.35% (2,674,435)	No
Twitter Lite	0.33% (2,521,005)	Yes
TweetDeck	0.29% (2,223,845)	Yes
dlvr.it	0.22% (1,711,766)	No
Cities	0.16% (1,238,873)	No
Facebook	0.12% (956,507)	Yes
Tweetbot for Mac	0.11% (895,958)	Yes
Hootsuite	0.10% (820,211)	No
Untappd	0.10% (813,913)	No

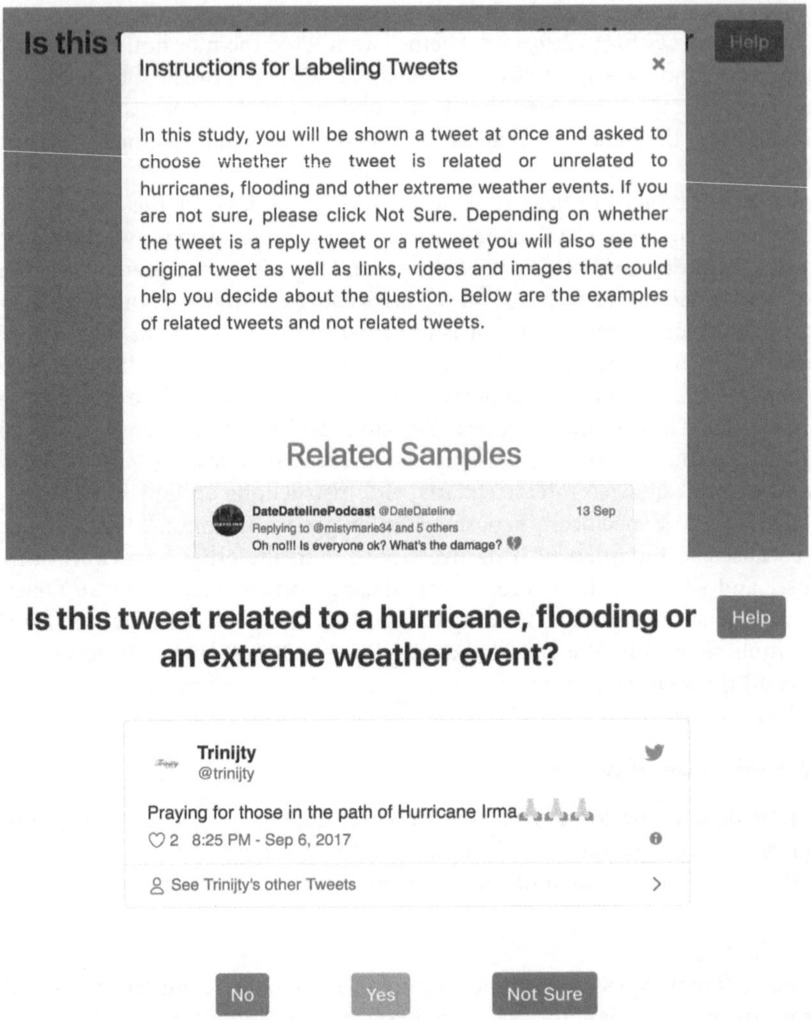

Figure 2. (a) Instructions for the labeling task (b) Training data collection interface. 'No' and 'Yes' buttons to allow the user to label the presented tweet as 'Related' or 'Unrelated', or 'Not Sure' button is used to skip the tweet.

evaluate the PCA's effect on the classification task. We also evaluated results for several linear classifiers including Logistic Regression, Linear SVM and Ridge. These linear classifiers are expected to provide similar results due to the fixed linearity of the problem. The Gaussian SVMs use the radial basis function (RBF) as the similarity function while Linear SVMs use a linear similarity function. Thus, their difference in similarity functions makes them converge on different points. We evaluated alternative parameters and optimization settings for these models to improve the classification accuracy, and we report the experiment results for all five models in the Results section.

4.2.2. Convolutional Neural Networks (CNNs)

Artificial Neural Networks (ANNs) is a deep learning method that have been widely used for text classification tasks. ANNs allow learning the features of a phenomenon (e.g. words, combinations of words related to disasters) by progressively learning and evolving relevant characteristics from the learning material (training set) that they process. ANNs do not require any prior knowledge about the phenomenon. As a category of ANNs, Convolutional Neural Networks (CNNs) have been commonly used in text classification tasks. By eliminating the full connectivity of nodes in regular neural network layers, CNNs speed up the training process while sustaining the ability of creating non-linear models. Since CNNs were developed with the idea of the input as an image, CNNs utilize 3D neuron structures that have width, height and depth which commonly represent the width and height of an image as well as the color channels of the image respectively. While CNNs are mostly used in tasks that involve images, they are also capable of producing efficient and effective models in text classification (Burel, Saif, and Alani 2017).

Since our classification task is binary, we used the sigmoid (Equation (1)) function, that maps output values into *[0, 1]* set, as the final activation of the output layer. The output of the sigmoid function can then be converted into a class of 0 or 1 by using a threshold of *0.5.* We used binary cross-entropy (Equation (2)) as the loss function in the CNN structure (Figure 1). Binary cross-entropy function measures the distance of a prediction made by the model from its actual value. The neural network is then updated in order to minimize this cost.

Sigmoid function, the activation function for neural network architectures

$$\sigma(x) = \frac{1}{1+ e^{-x}}. \quad (1)$$

Binary cross-entropy function that is used as the loss function

$$\mathcal{L}(y_i, \hat{y}_i) = -\frac{1}{N}\sum_{i=1}^{N}[y_i \log(\hat{y}_i) + (1-y_i)\log(1-\hat{y}_i)]. \quad (2)$$

Figure 3 illustrates the architecture of our proposed CNN model which consists of a convolutional layer followed by two fully connected layers. Input is embedded before feeding it into the convolutional part. Dropouts and an additional pooling layer are applied to the outputs of some layers in order to prevent and control over-fitting.

4.2.3. Long Short-Term Memory (LSTM) Networks

Although CNNs may create robust classifiers (Burel et al. 2017; Nguyen et al. 2016), they are limited in memorizing features throughout the training process. Recurrent Neural Networks (RNNs) incorporate a pair of memory vectors that are trained at each step, which are then moved to next nodes or steps of training. For instance, in a word sequence, correlation of previous words with the intended output may affect the meaning of and possible inferences that can be made from latter words. This correlation of former words with the output should be carried on to the next nodes in order to

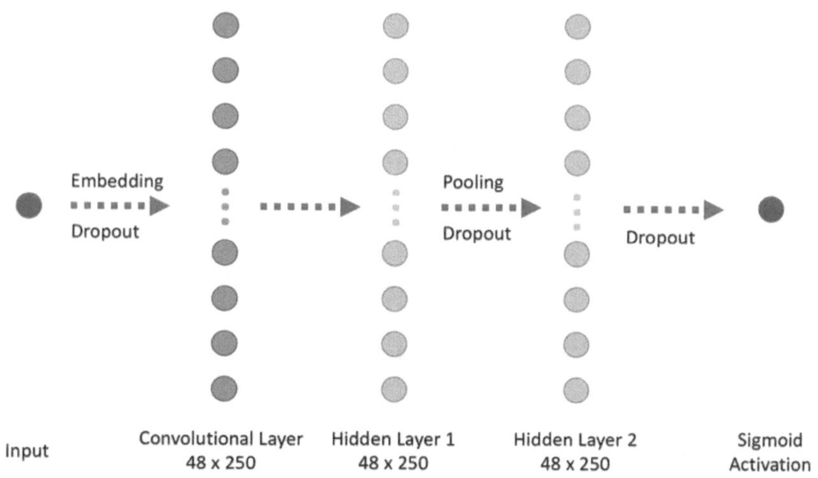

Figure 3. Convolutional Neural Network Architecture.

capture the sequential nature of sentences. In a 'vanilla' RNN, the information that is being learned and kept in a feature vector (hidden state) changes irregularly. Due to the vanishing gradient problem, the weight of the neural network receives a vanishingly small proportion of the gradient of the loss which prohibits the learning process. As a result, the process of learning weights may be slow or even stop. Thus, a simple and modest RNN is not capable of remembering the features learned for long times. This mostly impacts the overall accuracy of text classification tasks.

A Long Short-Term Memory Neural Network (Hochreiter and Schmidhuber 1997) extends the lifespan of the short-term memory to inherit more features into the deep neural network model. LSTMs and their variations are widely used in classification tasks that need longer memories, such as image classification (Byeon et al. 2015), text classification (Liu, Qiu, and Huang 2016) and time-series analysis (Xingjian et al. 2015). Figure 4 and the Equations (4)–(9) provide a brief summary of an LSTM node and mathematical background of an LSTM Node. The LSTM node is applied to tensors x_t and h_{t-1} in which x_t represents the input, and h_{t-1} represents the hidden state of the previous LSTM Node in the layer. Sigmoid function is represented by $\sigma(.)$ in Equation (1), and $\tanh(x)$ denotes hyperbolic tangent function (Equation (3)).

Hyperbolic tangent function

$$\tanh(x) = \frac{e^x - e^{-x}}{e^x + e^{-x}}. \tag{3}$$

The weight matrices W and U transform input and hidden state vectors and forms i_t, f_t, o_t and $\tilde{c}_t$ values, namely input gate (Equation (4)), forget gate (Equation (5)), output gate (Equation (6)) and cell update gate (Equation (7)) respectively. By virtue of these gates, LSTMs provide a forgetting mechanism to remove the features that are learned in previous nodes but are now obsolete. The $\circ$ operator represents an element-wise product, and the output of an LSTM Node is calculated by using these gates. While keeping the information in cell state (Equation (8)), LSTM nodes remember the information that will be used immediately in the hidden state (Equation (9)) which is similar to the hidden layer in RNNs. Finally, the hidden state and cell state are conveyed to the next LSTM Node, and the hidden state is conveyed to the next layer in the neural network architecture (Figure 5).

Input gate of an LSTM Node

$$i_t = \sigma(W^{(i)}x_t + U^{(i)}h_{t-1}). \tag{4}$$

Forget gate of an LSTM Node

$$f_t = \sigma(W^{(f)}x_t + U^{(f)}h_{t-1}). \tag{5}$$

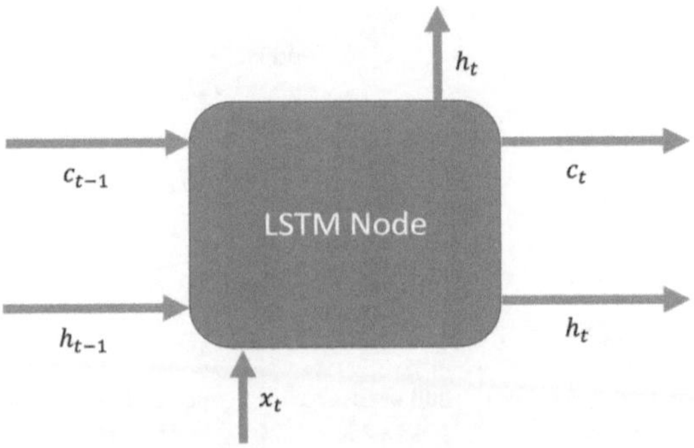

Figure 4. An LSTM Node and its connection formulations.

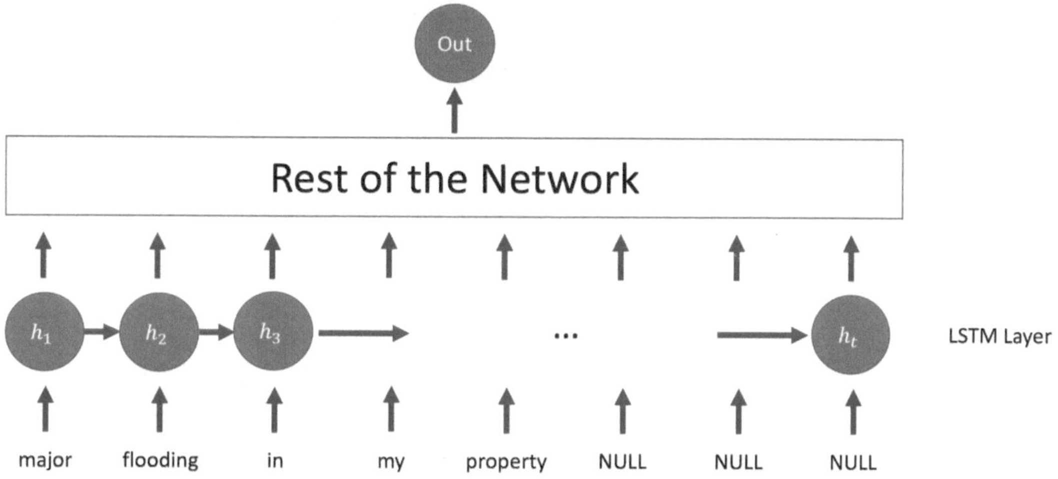

Figure 5. An LSTM layer and connections between LSTM nodes.

Output gate of an LSTM Node

$$o_t = \sigma(W^{(o)}x_t + U^{(o)}h_{t-1}). \tag{6}$$

Cell update gate of an LSTM Node

$$\tilde{c}_t = \tanh(W^{(c)}x_t + U^{(c)}h_{t-1}). \tag{7}$$

Cell state of an LSTM Node

$$c_t = f_t \circ c_{t-1} + i_t \circ \tilde{c}_{t-1}. \tag{8}$$

Hidden state of an LSTM Node

$$h_t = o_t \circ \tanh(c_t). \tag{9}$$

The deep neural network proposed here depends on LSTM networks along with other layers such as Embedding and Dropout layers (Figure 6). Similar to the proposed CNN model, we

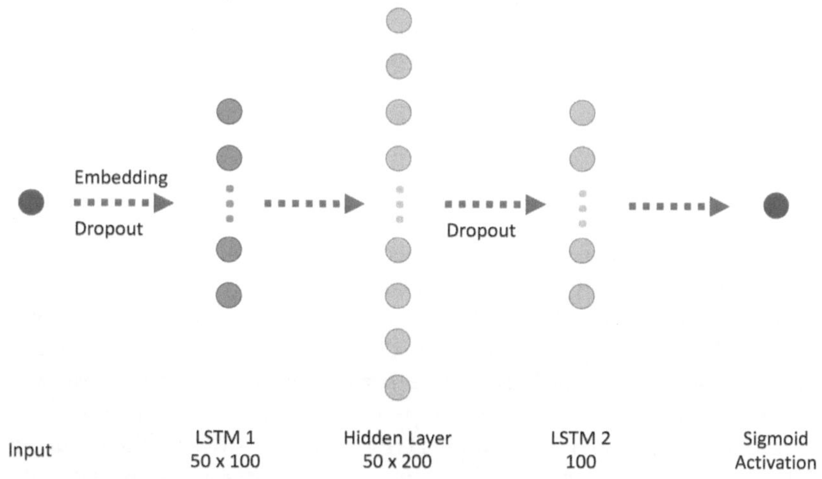

Figure 6. Proposed Long Short-term Memory (LSTM) Architecture.

employed the sigmoid (Equation (1)) and binary cross-entropy (Equation (2)) functions in LSTM. We also evaluated the categorical cross-entropy function which did produce a better performance than the binary cross-entropy function. Thus, we do not report the results of the categorical cross-entropy function.

4.2.4. Vectorization of the dataset

We vectorized the tweets to create an array that contains the tweet text, date and time stamp, and the binary label assigned by human classifiers during training dataset collection. An example tweet entry is illustrated below.

$$D_x = [\text{Tweet Text, Date and Time, Label}],$$

$$D_x = [\text{"Major flooding at my backyard. https : //t.co/UJn3s"}, \text{ "Sep 12,2017, 22.13 ET"}, 1].$$

To feed the information within the dataset into any kind of probabilistic classifier, the dataset needs to be represented by vectors. We performed pre-processing to filtering-out noisy tweets prior to converting the data into vectors. We first cleaned any kind of punctuation and URLs and transformed all characters to lower-case. We then split the text into a sequence of words in which sequences will be restructured as vectors. At the end of the pre-processing step, we derived a triple (tweet content, date and time stamp, and the label) of each tweet for the next step of tokenization.

$$D_x = [[\text{"major", "flooding", "at", "my", "backyard"}], \text{"Sep 12, 2017, 22.13 ET"}, 1].$$

After the sequencing step, we transformed each tweet to a vector using a tokenization centric approach, which is based on each word's number of occurrences in the whole dataset. The more a word occurs in the dataset, the less it has an effect as a feature in the initialization of the network. We sorted the words in the dataset by their occurrence, and assigned indices starting from 1 to replace each word with an index. For instance, the word 'flood' has the highest frequency in the whole dataset, and therefore its index (replacement) is 1. Below is the transformed triple in which words were replaced by their integer indices.

$$D_x = [[515, 252, 34, 17, 4200], \text{"Sep 12, 2017, 22.13 } ET\text{"}, 1].$$

Since each tweet has different length, it's crucial to pad the tweets to construct vectors with the same length and feed them into the neural network. We empirically tested different sizes for the vectorization process: 25, 30, 35, 40, 45, 50, 55 and 60. Classification test accuracy for the vectorized dataset with the length of 50 performed the best among all vector length options on the proposed LSTM architecture. While one column was used for the tweet post-date, the rest of the columns were used for the tweet tokens. Therefore, we chose 50 as the optimum number of vector size (49 word tokens + 1 date column) to derive semantics from a tweet. Following this assumption, we padded all vectors to generate a vector of zeros with a length of 49 words for each tweet.

$$D_x = [[0, \ldots 0, 515, 252, 34, 17, 4200], \text{"Sep 12, 2017, 22.13 ET"}, 1].$$

We incorporated the tweet post-date in tweet vectors in order to reveal the relationship between a tweet's post-date and its relevancy with flooding and extreme weather events. Date-time tagging could be done for different temporal granularities such as by minute, hour, day or week. Although floods may occur suddenly, their impact, i.e. flood inundation often lasts days, weeks and in some instances longer time periods. Because our purpose was to capture the progression of the disaster over an extended period of time that include during and aftermath of Hurricane Irma, we chose day as the temporal resolution considering the occurrence and duration of the flood event and its impact in the aftermath of the hurricane. As a result, we mapped the dates between 29 August 2017 and 17 October 2017 on to an integer scale between 1 and 50, and we appended the integers into the feature vector. Finally, we derived a vector of length 50 for each tweet in the dataset, and the final vector of the example tweet is shown below:

$$D_x = [[0, \ldots 0, 515, 252, 34, 17, 4200], 15, 1].$$

4.3. Thematic classification of disaster-relevant information

After classifying tweets related to flooding, hurricanes and extreme weather events, we employed an unsupervised topic modeling approach, Latent Dirichlet Allocation (LDA), to extract detailed information such as affected individuals, donations and support, caution and advice from tweet content. LDA is a Bayesian probabilistic model of documents that allows discovering a collection of latent topics defined as a multinomial distribution over words (Blei, Ng, and Jordan 2003). LDA assigns each document a mixture of topics.

$$P(Z|W, D) = \frac{W_{Z+\beta w}}{\text{total tokens in } Z + \beta} * D_{Z+\propto}.$$

For each topic Z, $P(Z \mid W, D)$, the probability that word W came from topic Z, is calculated by multiplying the normalized frequency of W in Z with the number of other words in document D that already belong to Z. Hyper parameters such as β and βw are used to incorporate the probability that word W belongs to topic Z even if it is nowhere else associated with Z (Blei, Ng, and Jordan 2003). LDA iteratively goes through the collection of documents word by word and reassigns each word to a topic. After each iteration, the model becomes more consistent as topics with specific words and documents and reaches an equilibrium that is as consistent as the collection allows (Goldstone and Underwood 2012). The words become more common in topics when their frequencies are higher, and topics become more common in documents where they occur more often in documents. We used each tweet as a document to train the topic model. We first tokenized the tweets and converted them into lower case, and used stop words (e.g. commonly used words such as 'the', 'of', 'am') from 28 languages prior to training the model using the Mallet Toolkit (McCallum 2002). We did not use stemming, because stemming introduces noise to the results, and does not improve the interpretation of topic models (Schofield and Mimno 2016).

4.4. Spatially-adaptive kernel smoothing

Once each tweet is classified into a set of topics, one can calculate the average topical probabilities per unit area. For example, among 1000 disaster-related tweets in Sarasota County in Florida, one can calculate the average probability of a topic such as affected individuals, by simply dividing the sum of the probabilities of tweets that belong to affected individuals category to the total number of disaster-related tweets generated within the county. However, such an approach may suffer from the modifiable areal unit problem (MAUP) due to the mismatch between the impact of the disaster and the arbitrarily assigned administrative boundaries. Moreover, rates calculated for different geographic areas would have unequal reliability due to different population sizes, in our case, the total number of tweets, across the spatial units. This problem is exacerbated for small areas or in other words areas with low density of observations. Therefore, rates for areas with small population values become less reliable. To address the problem, spatially-adaptive kernel smoothing (Tiwari and Rushton 2005) can be applied to estimate a reliable density value of a feature by taking into account the nearby observations. Different from fixed-distance kernel smoothing, adaptive kernel smoothing defines a threshold for the denominator of the rate using the k-nearest observations (e.g. tweets or users). Using the number of users for determining the k-nearest neighbors help remove the user contribution bias, in which a few active users produce most of the content. However, in this study, we use number of tweets to determine the k-nearest neighbors because of two reasons. First, locational information is critical for the disaster context although it may be produced by only a few number of users. Second, large portion of tweets share the exact coordinates because the significant portion of tweets is place-tagged.

Definitions, the equations and steps of the adaptive kernel smoothing are defined below. In Step 1, the area is divided into a grid of 2.5 km resolution (G), which covers Florida and the nearby affected areas of Hurricane Irma. There is a substantial amount of place-tagged tweets with the same coarse coordinates at city or neighborhood level. In Step 2, we aggregate the tweets that share the exact coordinates into distinct tweet locations prior to determining the k-nearest neighbors in order to prevent the potential bias. Therefore, we obtain a list of distinct tweet locations that include information on the total tweet count, the list of tweets and their assigned topical probabilities. We define k as the minimum number of distinct tweet locations to calculate a reliable estimate of the rate for each category of tweets. In Step 3, we employ a Sort-Tile-Recursive Tree algorithm to compute an index of the k-nearest distinct tweet locations for each grid cell in order to improve the computational efficiency. Once the neighborhood reaches the defined threshold k, we determine the list of tweets, T_i, bandwidth $h\ (G_i, k)$, and the weights of distinct tweet locations for each grid cell in Step 3. K is the kernel function, and h is the bandwidth for smoothing. Kernel functions determine the weight of each observation within a kernel, and the choice of function often does not have substantial impact on the result. The most commonly used kernel functions are Uniform, Epanechnikov, Triangular and Gaussian. In this study, we employ the Gaussian kernel function to determine the weight of each tweet within the kernel of each grid cell. Given the list of tweets, the spatial weights and the probability of topics for each tweet, we compute a weighted average probability of each topic for each grid cell using the formula defined in Step 4.

Definitions:

G: The study area defined by the total set of grid cells.
G_i: Grid cell i. $G_i \in G$.
k: Adaptive filter (neighborhood) threshold based on the total number of tweets.
T_i: The list of tweets within the neighborhood of G_i.
$h\ (G_i, k)$: The bandwidth of the k-size Neighborhood of the grid cell G_i is defined as the smallest KNN (G_i, k) that has a total count of distinct tweet locations that is greater than or equal to k: $\sum_i^t \geq k$.
K: Kernel function. Gaussian function is used in this study.

Steps:

(1) Compute G, the grid of the study area given a resolution r. In this study $r = 2.5$ km is used.
(2) Aggregate tweet statistics such as tweet count and topic probabilities of tweets for each distinct tweet location.
(3) Construct a Sort-tile-recursive (STR) tree for finding the smallest k-nearest neighboring distinct tweet locations of G_i that meets the size constraint, k.
(4) Compute T_i, $h\ (G_i, k)$ and the weights of observations (tweets) for each grid-cell using the adaptive kernel estimation

$$f_h(P_z(G_i|\theta)) = \frac{1}{nh} \sum_{j=1}^{x \in T_i} K\left(\frac{x - x_j}{h}\right) p_z(t_i),$$

where $P_z(G_i|\theta)$ is the weighted average probability of topic z given all the topical probabilities within the neighborhood of G_i, n is the total number of points (tweets), h is the bandwidth, $x_1, x_2, \ldots, x_{n}$- $\in T_i$ is a set of tweet locations within the neighborhood of G_i, x is the location of estimation, and $p_z(t_j)$ is the probability of topic z in tweet t_j.

4.5. Spatial clustering

In addition to using adaptive kernel smoothing to capture the relative prominence of each category of tweets across geographic space, we employ density-based spatial clustering with noise (DBSCAN) to identify areas where there is a high density of tweets for each category. Different from the smoothed surface of probability of each category, DBSCAN allows us to identify the areas in which the absolute density of tweets is greater than a defined threshold (Ester et al. 1996). DBSCAN requires two parameters: a search radius (*Eps*) and the minimum number of points within the search radius (*MinPts*). Unlike other clustering methods such as K-Means or K-Medoids, DBSCAN does not require the number of clusters as an input, and can capture clusters with arbitrary shapes. While a larger search radius (*Eps*) results in larger areas of spatial clusters, a smaller search radius results in smaller areas of interest. On the other hand, a larger *MinPts* results in more reliable detection of clusters, whereas a smaller *MinPts* results in many more clusters with increased noise and outlier points that are not assigned to any clusters. Finally, we calculate the smallest convex polygon that encloses each of the spatial clusters in order to identify areas in which high counts of tweets that belong to an information category (e.g. affected individuals) is spatially clustered. Ultimately, this allows us to identify areas where tweets from a particular information category are spatially clustered.

5. Results and evaluation

5.1. Experiments on binary classification

Using the vectorized version of the manually labeled dataset, we trained a series of binary classifiers. We included both related and unrelated data manually classified by human coders in our training set in order to increase the accuracy of models for predicting unrelated tweets. The dataset before any sampling consisted of 20,413 total tweets in which 7801 of them were manually classified as related and 12,612 of them were classified as unrelated.

We sub-sampled the labeled dataset to feed the model with the same amount of sequences for the binary classification, which consisted of 7801 related and 7801 unrelated tweets. Sub-sampled data were then split into the training and test sets in which the training set consisted 80% and the test set was 20%. After that, 12,481 training sequences and 3121 test sequences were used. Using the training set, we trained each disjoint model that include Logistic regression, SVM, CNN and LSTM models described in the methodology section with various parameter settings independently using the training set. Using the test set, we carried out a validation task for each model run. We optimized the model parameters to prevent overfitting and provide the best validation accuracy.

Table 4 illustrates the evaluation results of the models, which reveals that the LSTM networks and CNN-based models were significantly better than other conventional machine learning algorithms. Accuracy is measured as the fraction of true positive and true negatives over all of the classified tweets. While precision is the fraction of relevant tweets among all relevant tweets (i.e. true positives/(true positives + false positives)), recall is the fraction of relevant tweets over the total amount of tweets (true positives/(true positives + false negatives)). Finally, F1 score considers both the

Table 4. Accuracy, precision, recall and *F*1 scores for proposed classifier models.

	LSTM	CNN	CNN Kim (2014)	Logistic Regression	Gaussian SVM	Linear SVM	Gaussian SVM & PCA	Ridge
Accuracy	0.7478	0.7391	0.7250	0.5434	0.4985	0.4982	0.4915	0.5456
Precision	0.7541	0.7602	0.6999	0.5516	0.7608	0.5097	0.6250	0.5473
Recall	0.7488	0.7123	0.7826	0.5512	0.0220	0.3782	0.0031	0.6224
*F*1 Score	0.7514	0.7355	0.7389	0.5514	0.0428	0.4342	0.0062	0.5824

precision and recall given the formula:

$$\text{F1 score} = 2 * \frac{\text{precision} * \text{recall}}{\text{precision} + \text{recall}}.$$

Overall, linear models produced better performance (F1 score) than the Gaussian RBF-based models. Although its difference from a random classifier is insignificant, Linear SVM produced an *F1 Score* of *0.4342*, Gaussian SVMs produced high precision and low recall scores, which indicates a large number of false negative predictions. This inclination is reflected on the *F1 Scores*, as well. The Gaussian models can be chosen over the non-Gaussian SVM if high precision and low recall values are desired. By having an accuracy of approximately *55%*, non-SVM Linear models revealed slightly better performance than a random classifier, however, these models were outperformed by the deep neural network classifiers.

We employed a variety of architectures utilizing both LSTM and CNN-based approaches in terms of the number of nodes and layers trained. We systematically evaluated the number of nodes in the input layer, which resulted in changes in padding values of the vectorization process for the training and test data. Additionally, we incorporated auxiliary neural network layers, such as pooling, drop-out et cetera layers, into the proposed architectures in order to further optimize the model. In the training of the models that involved CNNs and LSTM networks, we obtained the best accuracy scores within the first few epochs by using the same batch size of 128 for both networks. To decrease the computation time, we accelerated the training processes of detailed neural networks by utilizing NVidia Tesla K80 GPUs. In addition to the models presented in the methodology section, we also report the results for the CNN model proposed by Kim (2014) that is trained on our training data and tested on our test set for further evaluation.

While both neural network architectures we proposed and the one we adopted from Kim (2014) performed similar, our LSTM-based network produced slightly a better performance than any of the CNN-based architectures in terms of both accuracy and *F*1 score. This uplift of the performance can be related to LSTMs' ability in conveying previous information to the next nodes of the network. Although LSTMs take more time to train, time elapsed while predicting the class of a tweet did not change significantly. We should note that since our CNN model provided a better precision score of 0.7602 in return of lower recall score, one can choose the CNN over the LSTM network if the correctness of positive predictions has crucial importance. Similarly, the CNN architecture proposed by Kim (2014) resulted in a better recall score of 0.7826. Therefore, if the rate of actual positives is more important, model selection can be made in favor of this CNN model. We continued the rest of our analysis based on the outcome prediction of the LSTM model. We made this choice based on the overall performance of the model as well as the model's ability to consider the sequential nature of text constructs. The LSTM based classifier predicted 378,474 related tweets (positives) and 179,050 unrelated tweets (negatives). We used this predicted set for thematic classification, and identifying temporal, spatio-temporal patterns of the thematic categories of disaster-related tweets.

5.2. Experiments and results of topic modeling

In this section, we present the experiments for selecting and interpreting the topic model parameters and results. In order to determine the optimum number of topics, we performed LDA using different number of topics such as 10, 20, 30, 40 and 50 topics with 2000 iterations. We evaluated the model outcomes to minimize the overlap of topics within each model, and the number of unique topics between topic models with the consecutive number of topics. In general, a topic model with a fewer number of topics generates combined topics, whereas a model with a larger number of topics results in less overlapping content (Newman et al. 2010). We used cosine similarity to measure the similarity of the models with different number of topics, and employed the following criteria to determine unique topics between two model outputs (Koylu 2018a): (1) The topic must have a probability threshold above 50% in order to eliminate noisy topics that only exist in a small number of documents. (2) The topic in the model with a larger number of topics should have a cosine similarity less than 20% with any of the topics of the model with a smaller number of k topics. Following these criteria, we compared the topic models with consecutive numbers such as 50–40, 40–30, 30–20 and 20–10 topics. The systematic evaluation of cosine similarity values both within and between models did not yield a clear result to justify the number of topics chosen for the model. However, the topic model with 30 topics produced much less within overlap than the 20-topic model, and this rate was smaller than the comparison of the 30-topic model to the 40-topic model. We manually observed the topics and individual tweets that were predominantly classified by each topic to determine the optimum number of topics. This evaluation also led us to the same number of topics, 30, as the optimum number that helped us determine the number of topics with minimal overlapping.

In order to diagnose the results of the topic model, we computed the measures of probability (P), document entropy (E) and corpus distance (CD) (Koylu 2018b). The probability of a topic represents how common a topic is across all tweets and is calculated by dividing the number of word tokens assigned to the topic by the sum of the token counts for all topics. If a topic has small probability, it means it exists only on a small number of tweets, and oftentimes there are not enough observations to examine the topic's word distribution. On the other hand, if a topic has large probability, the topic is extremely frequent across many tweets. Topics with large probability of occurrence across documents are often considered as a collection of corpus specific stop-words. Besides the low and high probability topics, more interesting topics reside within the non-extreme values. We approximately set this range to be between 0.005 and 0.5. In addition to topic probability, document entropy represents the distribution characteristics of a topic across documents. For example, if entropy is high the topic is distributed evenly over documents (tweets), whereas if entropy is low the topic is concentrated on a smaller number of tweets. Finally, we use corpus distance which is calculated by Kullback–Leibler divergence. Corpus distance represents how far a topic is from the overall distribution of words in the corpus. A greater corpus distance means that the topic is more distinct as compared to the overall distribution of words, whereas a smaller distance means that the topic is more similar to the corpus distribution.

Table 5 illustrates the top 10 coherent topics, and the 2 outlier topics at the end of the table. The table includes both the words and metrics of each topic, and the words are ranked by their probability of occurrence from the highest to the lowest. One can infer the latent topic using the combination of the words that commonly co-occur. The first and the foremost important topic represents tweets about affected individuals and infrastructure that include personal updates, injuries, deaths, missing, found, or displaced people; and buildings, roads, utilities/services that are damaged, interrupted, restored or operational. The second topic represents the caution, warning and advice category, which includes guidance, tips, warnings issued or lifted, forecasting, and alerts about emergency response situations. The third topic represents donations and support which include needs, requests, or efforts for help, offers of money blood, shelter, supplies, and/or services by volunteers or professionals. The fourth topic represents the sympathy and emotional support category which includes thoughts, prayers, gratitude and sadness. The rest of the topics illustrates news, official reports, government response, critiques, and political discussions, missing pets, disrupted services, and self-reporting of hurricane preparedness of individuals. The last two topics, Topic 11 and Topic 12, in Table 5 are outlier topics. The eleventh topic is the topic with the highest probability that includes live-tweeting about the hurricane. These tweets are often not informative, and do not contain information about eye-witness accounts, or site-specific situations. The last topic in this table represents Spanish tweets about the hurricane.

Table 5. Twelve latent topics with words and diagnostics measures.

Topic	Words	*P*	*E*	CD
Affected individuals and infrastructure damage	hurricane, power, damage, storm, rain, house, wind, trees, water, winds, tree, flood, lost, lot, safe, update, flooding, Florida, hope, area	0.021	10.11	1.33
Caution, advice and warnings	Storm, rain, flood, surge, warning, tropical, hurricane, winds, wind, heavy, flooding, flash, Florida, area, watch, country, south, issued, tornado, water	0.015	9.72	1.68
Donations and volunteers	hurricane, relief, victims, donate, flood, Houston, money fund, disaster, support, Texas, efforts, donated, donations, benefit, hand, helping, Red, Cross, donating	0.019	10.04	1.73
Sympathy and emotional support	hurricane, safe, Florida, stay, storm, hope, prayers, family praying, path, friends, thoughts, coming, hoping, safety, irmahurricane, damage, south, heart	0.030	10.45	1.30
News and official reports	hurricane, Florida, shoot, police, looters, Miami, guns, Richard, island, Branson, storm, news, video, looting, caught, warn, sign, private, foot, rescue	0.050	8.76	2.40
Government response, critiques and political discussions	hurricane, relief, disaster, victims, aid, Trump, tax, FEMA, Florida, Texas, house, recovery, voted, emergency, efforts, response, funding, money, flood, cuts	0.011	9.46	1.85
Government response, critiques and political discussions	hurricane, disaster, natural, damage, Trump, storm, flood, victims, disasters, lives, relief, country, Texas, North, Korea, America, Houston, worst, season, president	0.021	10.12	1.31
Affected individuals and missing pets	hurricane, flood, victims, pets, animals, dogs, left, dog, rescue, Houston, abandoned, cats, Texas, key, owners, storm, west, Florida, family, shelter	0.009	9.25	2.07
Disrupted service and tips for recovery	hurricane, power, storm, service, phone, internet, free, Florida, app, wifi, data, lost, work, cell, watch, lose, emergency, week, working, wait	0.010	9.40	1.80
Preparedness	hurricane, water, food, snacks, eat, gas, easting, supplies, prep, ready, storm, house, buy, power, coming, store, party, beer, open, bought	0.019	10.01	1.60
Chatting about the hurricane	hurricane, coming, Florida, y'all, storm, house, season, hope, scared, leave, category, party, bout, mom, jose, hits, making, work, Miami, cat	0.065	11.23	0.94
Spanish tweets	huracan, Florida, Cuba, paso, vea, Miami, categoria, caribe, video, vivo, fotos, llegada, isla, minute, mertos, danos, mar, imagenes, vientos, cayos	0.008	9.29	2.99

Notes: *P*: Probability, *E*: Entropy, CD: Corpus Distance.

5.3. Temporal, spatial and spatio-temporal patterns

Figure 7 illustrates the temporal probability distribution of the 10 coherent topics shown in Table 5. The probability of donations and support, and sympathy and prayers categories increased prior to the disaster, decreased during the disaster, and increased after Hurricane Irma dissipated. As an exception, both of these two categories were high on September 3, which coincided with the immediate aftermath of Hurricane Harvey that hit Texas and the surrounding areas prior to Hurricane Irma. Similarly, sympathy and prayers increased prior to the disaster, decreased during the disaster, and started to increase after Hurricane Irma dissipated. Advice, warning and alerts also had a similar increasing trend before the disaster, and in the aftermath of the disaster's impact in Florida. Government critiques also had an increasing trend after the hurricane dissipated. Another notably interesting pattern is that tweets about affected individuals started to increase right after September 6 when Irma reached its peak with wind speed and made the first landfall in Florida on September 10. We provide a random sample of original tweet content with their assigned topics for affected individuals and infrastructure category in Table 6, and for all other categories in Table 7 in the Appendix section.

In this article, we only report the semantic, spatial and spatio-temporal characteristics of affected individuals and damaged infrastructure category, which we will refer to this category as affected individuals in the rest of the article. In order to calculate the smoothed ratio of affected individual tweets, we used $k = 10$ as the minimum number of distinct tweet locations. In addition to the smoothed rate

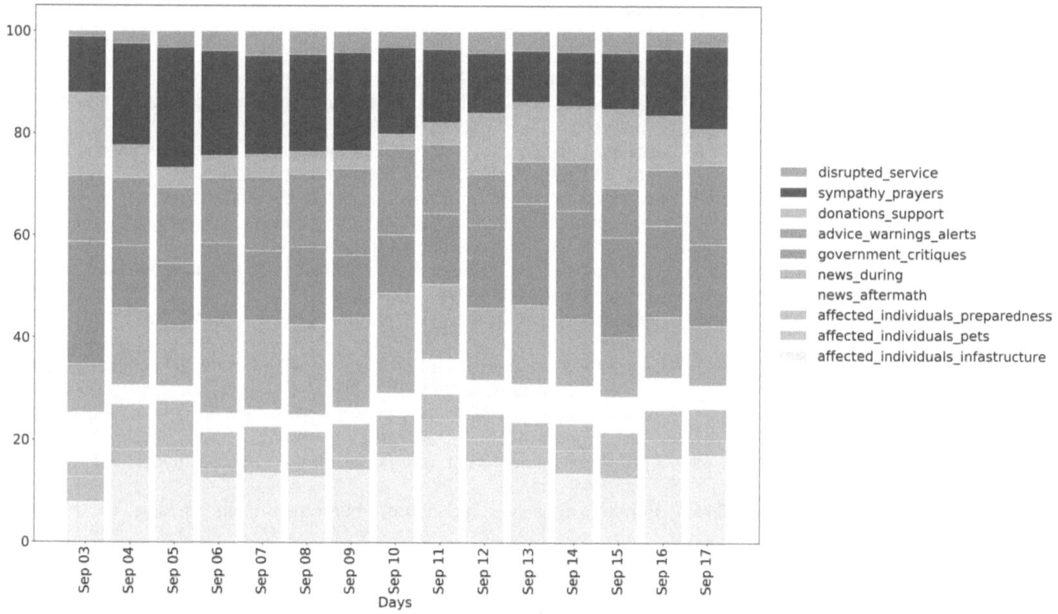

Figure 7. Temporal distribution of top 10 coherent categories of disaster-related tweets.

that illustrates the prominence of each category relative to all categories, we also extracted areas where there is a high density of tweets that belong to each category using the DBSCAN algorithm. We experimented with a range of *Eps* values: 10, 20, 30, 40, and 50 km; and a range of *MinPts* such as 5, 10, 15 and 20 distinct tweet locations. Higher *Eps* and lower *MinPts* values resulted in clusters that are quite large, while lower *Eps* and higher *MinPts* values resulted in many clusters, and outlier tweet locations that were not classified part of the cluster. Values of 30 km for Eps and 10 distinct tweet locations for MinPts produced 32 clusters which were consistent with the impacted areas of the hurricane. Also, this combination minimized the number of outliers in the clustering process. Figure 8 illustrates the ratio of affected individual tweets to tweets from all categories (i.e. the results of Section 4.4), the areas where there is a high density of tweets for affected individual tweets (i.e. the results of Section 4.5) as well as the temporal histogram of frequency for this category. There were a total of 14,724 tweets that had at least 50% probability for belonging to the affected individuals and infrastructure category. Because of the high presence of place-tagged tweets, the number of distinct tweet locations was 7709, which corresponds to approximately 2% of all tweets in the study area. Among these tweets, 501 tweets were place-tagged at the state level (Florida), therefore, these tweets were excluded prior to computing the smoothed rate.

In Figure 8, the temporal frequency of tweets about affected individuals was higher between September 5 and September 13, when the storm made the landfall, and areas in Florida were inundated. The smoothed rate surface illustrates the relative prominence of affected individual and infrastructure tweets, while the convex hulls of tweet clusters helps distinguish the areas where there are higher count of tweets from the selected category. This figure confirms some of the highly damaged areas and areas with disrupted services such as the urban areas of Tallahassee, Jacksonville, Orlando and Tampa, Florida Keys, and the coastal areas in the east such as Palm Bay and Rockledge.

In order to better assess the potential damage and affected individuals, we overlayed the tweets and potential impact areas with flood zone (Figure 9(a)) and rainfall data (Figure 9(b)). The total rainfall is spatially correlated with clusters of affected individual tweets in some impact areas such as the north-east of Cape Coral, the impact zones around Jacksonville, Orland and Port St. Lucie,

Table 6. Sample of tweets for each category of information.

C	EST 2017	Tweet	*P*
Affected individuals and preparedness	Tue Sep 05 19:21	Everybody stocking up for the hurricane	0.83
	Wed Sep 06 10:35	Prepping for Hurricane Irma but I'm never too busy to get down on some delicious.	0.68
	Wed Sep 06 11:18	It's hard to prepare for a storm when there are no supplies to buy and no gas to leave I'm all good but friends not so much!!	0.60
	Wed Sep 06 21:18	Im gonna eat all the hurricane food before the actual hurricane	0.92
	Thu Sep 07 20:00	Super disappointed to see my local gas station in Clermont; Florida selling water for $22 per case during hurricane prep.	0.97
	Sat Sep 09 18:56	I'd try to be a bit more helpful; but still in the middle of hurricane prep atm	0.31
	Sat Sep 09 20:28	Just spent over $40 on gas; thanks Irma	0.83
	Sun Sep 10 14:50	Well I Just Ate My Last Bag Of Hot Cheetos And Them Was My Hurricane Snacks	0.94
	Sun Sep 10 19:33	I know for a fact everyone is bored out of their mind and has eaten a good 50% of their hurricane food	0.92
	Mon Sep 25 21:33	Why are people so dumb. Why does nobody keep a 3 month emergency kit? 3 months food and emergency supplies; put it in a big bin. Done.	0.83
Affected individuals and infrastructure	Thu Sep 07 12:40	I just realized I might not have power for the comeback because of Hurricane Irma!!	0.19
	Sun Sep 10 10:45	Hunkered down in a block house w/hurricane shutters. 4 people; 2 dogs; 2 cats; and 4 fish. Come at us	0.48
	Mon Sep 11 15:56	I ain't think irma hit this bad but there's trees in the streets and all	0.83
	Mon Sep 11 17:51	Hurricane Irma left a mark on my yard in Grayson …	0.92
	Tue Sep 12 11:56	Savannah may have some cleaning up to do. Lots of folks in S. Georgia are out of power due to Irma	0.94
	Tue Sep 12 12:23	Too bad Irma took my power	0.42
	Wed Sep 13 21:32	People driving ruthlessly during Irma power outage. Crossing over double line; not looking before turning; speeding in parking lots.	0.96
	Fri Sep 15 15:35	I'm very sorry we are still recuperating from hurricane Irma and we still have no power. Very sorry that we will not be opened this weekend.	0.74
	Fri Sep 15 17:27	From inside the gazebo! Hurricane Irma took out the garden … we need chainsaws and bucket trucks!!!	0.62
	Mon Oct 09 17:36	I live in FL I lost power in Irma don't recall relying on fed's to get our power. PR needs to do something	0.48

Notes: *C* (Category), *P* (Probability).

while spatial clusters of affected individual tweets do not correlate with rainfall in most other areas (Figure 9(b)). This could potentially be explained by the damage from other weather events such as high winds and tornadoes. In Figure 9(a), we illustrate the area of East Central Florida, in which five tornadoes caused significant damage during the hurricane.

In order to reveal the temporal progression of the affected individual tweets, we interpolated the date and time of tweets using inverse-distance weighting (IDW) interpolation (Figure 10). Three color hues are used to illustrate the temporal progression of affected individual tweets for the study area before (blue), during (orange) and after (green) Irma. An interesting finding of this map is that there are a large number of areas with large variation (small patches of different color hues) in terms of the time of tweets, as people continued to tweet before, during and after the hurricane. On the other hand, green areas illustrate highly impacted zones which required greater intensity of response and recovery efforts in the aftermath of the hurricane. People in these areas did not tweet as much before or during the storm, which could be attributed to the potential loss of power and phone services.

6. Discussion and conclusion

Knowing what information is available through social media helps emergency responders, humanitarian organizations, governments and utility companies prepare and respond to disasters by effective allocation of their limited resources. Identifying disaster-related social media posts have recently been the focus of many studies that aim to understand the extent and magnitude

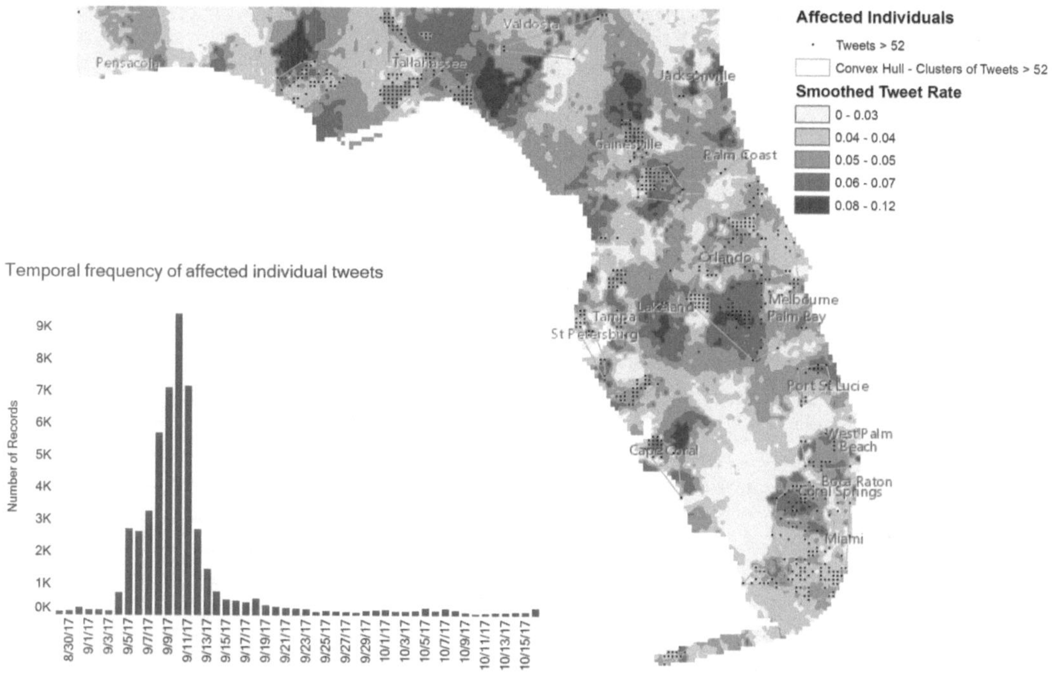

Figure 8. Spatial and temporal distribution of tweets about affected individuals and infrastructure.

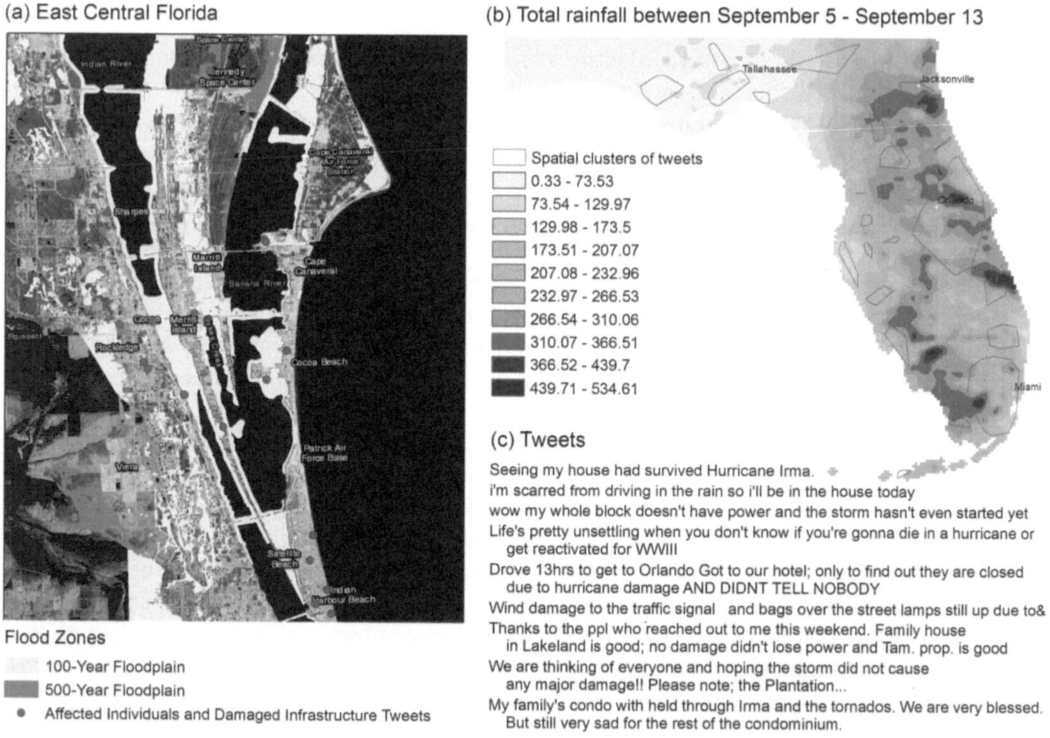

Figure 9. Overlaying of tweets with total rain in inches and flood zones between September 5 and September 12.

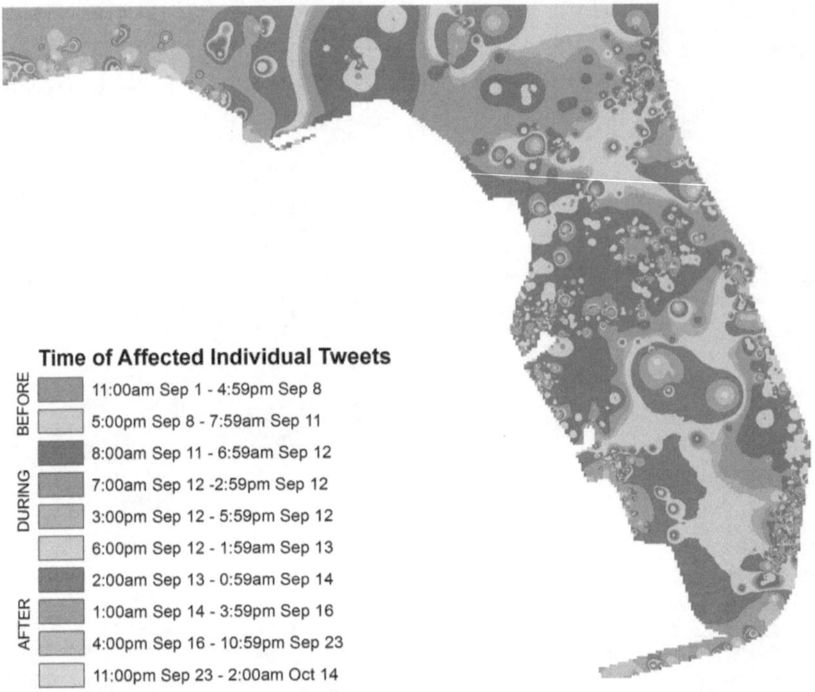

Figure 10. Temporal progression of affected individuals and infrastructure tweets by area. On average, Blue areas tweeted before, orange areas tweeted during, and green areas tweeted after Hurricane Irma.

of damage, and improve the ways we identify affected individuals, develop situational awareness and coordinate recovery efforts. Our study contributes to these studies by introducing a methodology that integrates deep learning, natural language processing and spatio-temporal analysis to identify content, whereabouts and temporal context of disaster-related social media posts.

In order to define our specific contributions to the studies that focus on the classification of crisis-related information from social media data, here we compare and contrast our methodology and results with the previous work done by Burel, Saif, and Alani (2017), Nguyen et al. (2016), Olteanu, Vieweg, and Castillo (2015) and others. First, unlike the previous work which focuses on identifying events and targeting immediate response to disasters, our overall objective is rather holistic to capture any useful information that may be related to the progression of a hurricane including the preparedness, response and recovery stages. Using a two-step classification approach, we identified relevant tweets to a hurricane with specific categories of information which have different target audience such as emergency responders, rescue missions and utility companies, animal control and shelters; government organizations and non-profit institutions that organize recovery and relief efforts; and local governments and FEMA for building and enhancing situational awareness. Second, we collapsed the three categories: related and informative, related but not informative, and not related introduced by Olteanu, Vieweg, and Castillo (2015) into a binary category of related and not related. We used the binary classification for two reasons. Human classifiers have shown significant variation in the labeling task especially when the task is complex and ambiguous (Nguyen et al. 2016). Moreover, we argue that labeling tweets as informative and not informative by simply evaluating the text of the tweet may often be inadequate without examining the content, time and location of tweets.

Burel et al. (2017) employed a three-tiered classification system for extracting (1) Crisis vs non-crisis (2) Type of crisis (flood, fire, etc.) (3) Type of information (affected individuals, donations, etc.). There are many differences between our study and Burel and colleagues. Data used by Burel and colleagues consist of 28,000 tweets. In our study, our training data consist of approximately 20,000 labeled data, and we classified 557,524 tweets using our training set and the classification methods. We chose the LSTM model for the binary classification task because (1) LSTMs are conceptually a better fit for the text classification task, since LSTMs consider the ordering of words and the whole text structure by using long-term semantic word and feature dependencies. (2) While both neural network architectures performed similarly, LSTMs produced slightly a better performance than CNN-based architecture in terms of both accuracy and $F1$ score. While Burel achieves high performance ($F1$ scores) for detecting crisis and non-crisis situations and event types, the performance of their model for identifying fine-grained event information such as affected individuals, damaged infrastructure is low (60%). Also, their classification model cannot handle multiple labels where a tweet can belong to two categories such as affected individual and advice and caution. Similarly, Nguyen and colleagues reported lower classification performance for information types, and they attributed the lower performance to the conflicting annotation done by online workers and volunteers. In contrast to these studies, we employed LDA as an unsupervised approach to extract the fine-grained details of information about the crisis situation, which also allowed us to classify each tweet with multiple labels such as affected individual and caution and advice.

Another unique aspect of our study is that we obtained information before, during and aftermath of the disaster as compared to retrospective studies which often lack the data during the first few hours of the event. Since the hazard type is a hurricane we were able to start collecting the data using a set of keywords related to the forecasted hurricane as well as geo-located tweets using the study area. We continued to collect the data a month after Hurricane Irma was formed, and we continuously updated the keywords based on the most common hashtags used throughout the hurricane. For future studies, we would like to compare the impact of the hurricanes Harvey, Irma and Maria.

The methodology introduced in this article has great potential to be used during a disaster for real-time identification of damage and emergency information, and making informed decisions by assessing the situation in impacted areas. Due to the limited space, we did not provide an in-depth analysis of the site-specific information we gained through analysis, however, we plan to conduct future work to evaluate damage, extract site and individual-specific information from various categories of disaster-relevant tweets.

Disclosure statement

No potential conflict of interest was reported by the authors.

ORCID

Muhammed Ali Sit http://orcid.org/0000-0002-2707-6692
Caglar Koylu http://orcid.org/0000-0001-6619-6366
Ibrahim Demir http://orcid.org/0000-0002-0461-1242

References

Acar, A., and Y. Muraki. 2011. "Twitter for Crisis Communication: Lessons Learned from Japan's Tsunami Disaster." *International Journal of Web Based Communities* 7 (3): 392–402.

Blei, D. M., A. Y. Ng, and M. I. Jordan. 2003. "Latent Dirichlet Allocation." *The Journal of Machine Learning Research* 3: 993–1022.

Burel, G., H. Saif, and H. Alani. 2017. "Semantic Wide and Deep Learning for Detecting Crisis-information Categories on Social Media." Paper presented at the International Semantic Web Conference.

Burel, G., H. Saif, M. Fernandez, and H. Alani. 2017. "On Semantics and Deep Learning for Event Detection in Crisis Situations." In Workshop on Semantic Deep Learning (SemDeep), at ESWC 2017, 29 May 2017, Portoroz, Slovenia.

Byeon, W., T. M. Breuel, F. Raue, and M. Liwicki. 2015. "Scene Labeling with Lstm Recurrent Neural Networks." Paper presented at the Proceedings of the IEEE Conference on Computer Vision and Pattern Recognition.

Caragea, C., N. McNeese, A. Jaiswal, G. Traylor, H.-W. Kim, P. Mitra, and B. J. Jansen. 2011. "Classifying Text Messages for the Haiti Earthquake." Paper presented at the Proceedings of the 8th International Conference on Information Systems for Crisis Response and Management (ISCRAM2011).

Caragea, C., A. Silvescu, and A. H. Tapia. 2016. "Identifying Informative Messages in Disaster Events Using Convolutional Neural Networks." Paper presented at the International Conference on Information Systems for Crisis Response and Management.

Cervone, G., E. Sava, Q. Huang, E. Schnebele, J. Harrison, and N. Waters. 2016. "Using Twitter for Tasking Remote-sensing Data Collection and Damage Assessment: 2013 Boulder Flood Case Study." *International Journal of Remote Sensing* 37 (1): 100–124.

De Albuquerque, J. P., B. Herfort, A. Brenning, and A. Zipf. 2015. "A Geographic Approach for Combining Social Media and Authoritative Data Towards Identifying Useful Information for Disaster Management." *International Journal of Geographical Information Science* 29 (4): 667–689.

De Choudhury, M., N. Diakopoulos, and M. Naaman. 2012. "Unfolding the Event Landscape on Twitter: Classification and Exploration of User Categories." Paper presented at the Proceedings of the ACM 2012 conference on Computer Supported Cooperative Work.

Demir, I., H. Conover, W. F. Krajewski, B.-C. Seo, R. Goska, Y. He, and W. Petersen. 2015. "Data-enabled Field Experiment Planning, Management, and Research Using Cyberinfrastructure." *Journal of Hydrometeorology* 16 (3): 1155–1170.

Doggett, E. V., and A. Cantarero. 2016. "Identifying Eyewitness News-worthy Events on Twitter." Paper presented at the Conference on Empirical Methods in Natural Language Processing.

Ester, M., H.-P. Kriegel, J. Sander, and X. Xu. 1996. "A Density-based Algorithm for Discovering Clusters in Large Spatial Databases with Noise." Paper presented at the Knowledge Discovery and Data Mining (KDD), Portland, OR.

Fang, R., A. Nourbakhsh, X. Liu, S. Shah, and Q. Li. 2016. "Witness Identification in Twitter." Paper presented at the Proceedings of the Fourth International Workshop on Natural Language Processing for Social Media, Austin, TX, USA.

Feng, Y., and M. Sester. 2018. "Extraction of Pluvial Flood Relevant Volunteered Geographic Information (VGI) by Deep Learning from User Generated Texts and Photos." *ISPRS International Journal of Geo-Information* 7 (2): 39.

Goldstone, A., and T. Underwood. 2012. "What Can Topic Models of PMLA Teach us about the History of Literary Scholarship." *Journal of Digital Humanities* 2 (1): 39–48.

Hagen, L., T. Keller, S. Neely, N. DePaula, and C. Robert-Cooperman. 2017. "Crisis Communications in the Age of Social Media: A Network Analysis of Zika-related Tweets." *Social Science Computer Review* 36 (5): 523–541.

Herfort, B., J. P. de Albuquerque, S.-J. Schelhorn, and A. Zipf. 2014. "Exploring the Geographical Relations Between Social Media and Flood Phenomena to Improve Situational Awareness." In *Connecting a Digital Europe Through Location and Place*, edited by Joaquin Huerta, Sven Schade, and Carlos Granell, 55–71. Heidelberg: Springer.

Hochreiter, S., and J. Schmidhuber. 1997. "Long Short-term Memory." *Neural Computation* 9 (8): 1735–1780.

Huang, Q., G. Cervone, and G. Zhang. 2017. "A Cloud-enabled Automatic Disaster Analysis System of Multi-sourced Data Streams: An Example Synthesizing Social Media, Remote Sensing and Wikipedia Data." *Computers, Environment and Urban Systems* 66: 23–37.

Huang, X., C. Wang, Z. Li, and H. Ning. 2018. "A Visual–Textual Fused Approach to Automated Tagging of Flood-related Tweets During a Flood Event." *International Journal of Digital Earth*, 1–17. doi:10.1080/17538947.2018.1523956.

Imran, M., S. Elbassuoni, C. Castillo, F. Diaz, and P. Meier. 2013. "Extracting Information Nuggets from Disaster-related Messages in Social Media." Paper presented at the ISCRAM.

Kanhabua, N., and W. Nejdl. 2013. "Understanding the Diversity of Tweets in the Time of Outbreaks." Paper presented at the Proceedings of the 22nd International Conference on World Wide Web.

Kim, Y. 2014. "Convolutional Neural Networks for Sentence Classification." arXiv preprint arXiv:1408.5882.

Kongthon, A., C. Haruechaiyasak, J. Pailai, and S. Kongyoung. 2012. "The Role of Twitter During a Natural Disaster: Case Study of 2011 Thai Flood." Paper presented at the Technology Management for Emerging Technologies (PICMET), 2012 Proceedings of PICMET'12.

Koylu, C. 2018a. "Modeling and Visualizing Semantic and Spatio-temporal Evolution of Topics in Interpersonal Communication on Twitter." *International Journal of Geographical Information Science*: 1–28. doi:10.1080/13658816.2018.1458987.

Koylu, C. 2018b. "Uncovering Geo-social Semantics from the Twitter Mention Network: An Integrated Approach Using Spatial Network Smoothing and Topic Modeling." In *Human Dynamics Research in Smart and Connected Communities*, edited by S.-L. Shaw, and D. Sui, 163–179. Cham: Springer International.

Krajewski, W. F., D. Ceynar, I. Demir, R. Goska, A. Kruger, C. Langel, and B.-C. Seo. 2017. "Real-time Flood Forecasting and Information System for the State of Iowa." *Bulletin of the American Meteorological Society* 98 (3): 539–554.

Kryvasheyeu, Y., H. Chen, N. Obradovich, E. Moro, P. Van Hentenryck, J. Fowler, and M. Cebrian. 2016. "Rapid Assessment of Disaster Damage Using Social Media Activity." *Science Advances* 2 (3): e1500779.

Kumar, S., F. Morstatter, R. Zafarani, and H. Liu. 2013. "Whom Should I Follow?: Identifying Relevant Users During Crises." Paper presented at the Proceedings of the 24th ACM Conference on Hypertext and Social Media.

Kwon, H. Y., and Y. O. Kang. 2016. "Risk Analysis and Visualization for Detecting Signs of Flood Disaster in Twitter." *Spatial Information Research* 24 (2): 127–139. doi:10.1007/s41324-016-0014-1.

Lachlan, K. A., P. R. Spence, X. Lin, K. Najarian, and M. Del Greco. 2016. "Social Media and Crisis Management: CERC, Search Strategies, and Twitter Content." *Computers in Human Behavior* 54: 647–652.

Li, Z., C. Wang, C. T. Emrich, and D. Guo. 2018. "A Novel Approach to Leveraging Social Media for Rapid Flood Mapping: A Case Study of the 2015 South Carolina Floods." *Cartography and Geographic Information Science* 45 (2): 97–110. doi:10.1080/15230406.2016.1271356.

Liu, S. B., L. Palen, J. Sutton, A. Hughes, and S. Vieweg. 2008. "In Search of the Bigger Picture: The Emergent Role of On-line Photo Sharing in Times of Disaster." Paper presented at the Proceedings of the Information Systems for Crisis Response and Management Conference (ISCRAM).

Liu, P., X. Qiu, and X. Huang. 2016. "Recurrent Neural Network for Text Classification with Multi-task Learning." arXiv preprint arXiv:1605.05101.

Maantay, J., A. Maroko, and G. Culp. 2010. "Using Geographic Information Science to Estimate Vulnerable Urban Populations for Flood Hazard and Risk Assessment in New York City." In *Geospatial Techniques in Urban Hazard and Disaster Analysis*. Vol. 2., edited by P. S. Showalter and Y. Lu, 71–97. Dordrecht: Springer.

McCallum, A. K. 2002. MALLET: A Machine Learning for Language Toolkit [online]. Available from: http://mallet.cs.umass.edu [Accessed 28 Dec 2018]

Monroy-Hernández, A., E. Kiciman, M. De Choudhury, and S. Counts. 2013. "The New War Correspondents: The Rise of Civic Media Curation in Urban Warfare." Paper presented at the Proceedings of the 2013 Conference on Computer Supported Cooperative Work.

Munro, R., and C. D. Manning. 2012. "Short Message Communications: Users, Topics, and In-language Processing." Paper presented at the Proceedings of the 2nd ACM Symposium on Computing for Development.

Nelson, L. D., P. R. Spence, and K. A. Lachlan. 2009. "Learning from the Media in the Aftermath of a Crisis: Findings from the Minneapolis Bridge Collapse." *Electronic News* 3 (4): 176–192.

Newman, D., J. H. Lau, K. Grieser, and T. Baldwin. 2010. "Automatic Evaluation of Topic Coherence." Paper presented at the Human Language Technologies: The 2010 Annual Conference of the North American Chapter of the Association for Computational Linguistics.

Nguyen, D. T., S. Joty, M. Imran, H. Sajjad, and P. Mitra. 2016. "Applications of Online Deep Learning for Crisis Response Using Social Media Information." arXiv preprint arXiv:1610.01030.

Olteanu, A., C. Castillo, F. Diaz, and S. Vieweg. 2014. "CrisisLex: A Lexicon for Collecting and Filtering Microblogged Communications in Crises." Paper presented at the ICWSM.

Olteanu, A., S. Vieweg, and C. Castillo. 2015. "What to Expect When the Unexpected Happens: Social Media Communications Across Crises." Paper presented at the Proceedings of the 18th ACM Conference on Computer Supported Cooperative Work & Social Computing.

Pekar, V., J. Binner, H. Najafi, and C. Hale. 2016. "Selecting Classification Features for Detection of Mass Emergency Events on Social Media." Paper presented at the Proceedings of the International Conference on Security and Management (SAM).

Restrepo-Estrada, C., S. C. de Andrade, N. Abe, M. C. Fava, E. M. Mendiondo, and J. P. de Albuquerque. 2018. "Geo-social Media as a Proxy for Hydrometeorological Data for Streamflow Estimation and to Improve Flood Monitoring." *Computers & Geosciences* 111: 148–158.

Schnebele, E. 2013. "Improving Remote Sensing Flood Assessment Using Volunteered Geographical Data." *Natural Hazards and Earth System Sciences* 13 (3): 669.

Schofield, A., and D. Mimno. 2016. "Comparing Apples to Apple: The Effects of Stemmers on Topic Models." *Transactions of the Association for Computational Linguistics* 4: 287–300.

Smith, L., Q. Liang, P. James, and W. Lin. 2015. "Assessing the Utility of Social Media as a Data Source for Flood Risk Management Using a Real-time Modelling Framework." *Journal of Flood Risk Management* 10 (3): 370–380.

Spielhofer, T., R. Greenlaw, D. Markham, and A. Hahne. 2016. "Data Mining Twitter During the UK Floods: Investigating the Potential Use of Social Media in Emergency Management." Paper presented at the Information and Communication Technologies for Disaster Management (ICT-DM), 2016 3rd International Conference on.

Stefanidis, A., A. Cotnoir, A. Croitoru, A. Crooks, M. Rice, and J. Radzikowski. 2013. "Demarcating New Boundaries: Mapping Virtual Polycentric Communities Through Social Media Content." *Cartography and Geographic Information Science* 40 (2): 116–129. doi:10.1080/15230406.2013.776211.

Tiwari, C., and G. Rushton. 2005. "Using Spatially Adaptive Filters to Map Late Stage Colorectal Cancer Incidence in Iowa." In *Developments in Spatial Data Handling*, edited by F. Peter Fisher, 665–676. Berlin: Springer.

Vieweg, S., A. L. Hughes, K. Starbird, and L. Palen. 2010. "Microblogging During Two Natural Hazards Events: What Twitter May Contribute to Situational Awareness." Paper presented at the Proceedings of the SIGCHI Conference on Human Factors in Computing Systems.

Xingjian, S., Z. Chen, H. Wang, D.-Y. Yeung, W.-K. Wong, and W.-C. Woo. 2015. "Convolutional LSTM Network: A Machine Learning Approach for Precipitation Nowcasting." Paper presented at the Advances in Neural Information Processing Systems.

Zhang, Y., C. Szabo, and Q. Z. Sheng. 2016. "Improving Object and Event Monitoring on Twitter Through Lexical Analysis and User Profiling." Paper presented at the International Conference on Web Information Systems Engineering.

Deep learning for real-time social media text classification for situation awareness – using Hurricanes Sandy, Harvey, and Irma as case studies

Manzhu Yu, Qunying Huang, Han Qin, Chris Scheele and Chaowei Yang

ABSTRACT

Social media platforms have been contributing to disaster management during the past several years. Text mining solutions using traditional machine learning techniques have been developed to categorize the messages into different themes, such as caution and advice, to better understand the meaning and leverage useful information from the social media text content. However, these methods are mostly event specific and difficult to generalize for cross-event classifications. In other words, traditional classification models trained by historic datasets are not capable of categorizing social media messages from a future event. This research examines the capability of a convolutional neural network (CNN) model in cross-event Twitter topic classification based on three geo-tagged twitter datasets collected during Hurricanes Sandy, Harvey, and Irma. The performance of the CNN model is compared to two traditional machine learning methods: support vector machine (SVM) and logistic regression (LR). Experiment results showed that CNN models achieved a consistently better accuracy for both single event and cross-event evaluation scenarios whereas SVM and LR models had lower accuracy compared to their own single event accuracy results. This indicated that the CNN model has the capability of pre-training Twitter data from past events to classify for an upcoming event for situational awareness.

1. Introduction

Recently, we have unfortunately witnessed a series of deadly hurricane events (e.g. Harvey, Irma, Florence, and Michael) and Northern California wildfires. These events claim many lives, result in billions of dollars of economic loss, and significantly impact the environment. When a natural hazard occurs, managers and responders rely on timely and accurate information about the disaster situations (e.g. damages) to make effective response decisions and improve management strategies. Such information is referred as 'situational awareness' (SA), i.e. an individually as well as socially cognitive state of understanding 'the big picture' during critical situations (Vieweg et al. 2010). Since SA relevant information mostly has a spatial component (e.g. all damages occur at a specific location), it has been referred as geographically grounded situational awareness (MacEachren et al. 2011) or geographic situational awareness (GSA) (Huang and Xiao 2015) in geospatial fields. GSA focuses on identifying the locations or areas of damaged infrastructure, affected people, evacuation zones, and the communities of great needs of resources, etc.

Because of the massive popularity of social media networks and their real-time production of data, social media offers new opportunities for natural hazards. For example, social media has contributed significantly in disaster management as a tool to communicate information during emergencies (Charalabidis et al. 2014). Social media is increasingly being used by both Non-Governmental Organizations and government emergency management agencies to determine public sentiment and reaction to an event (Emergency Alerting Platform Working Group 2016). As a multi-directional flow of communication and information platform, social media efficiently facilitates response and recovery efforts (Roche, Propeck-Zimmermann, and Mericskay 2013).

Much useful information in social networks such as Twitter is textual. For example, people post text messages related to the resource, damage, donation or aid during a disaster. From the text, disaster relevant data can be extracted to establish GSA or produce a crisis map (Huang and Xiao 2015; Vieweg et al. 2010). As the messages broadcasted and shared through the social media networks are extremely massive and varying, an effective text mining solution should be developed to separate the messages into different themes before we can understand the meaning and leverage useful information from the social media text content (Aggarwal and Wang 2011). Information about casualties and damage, donation efforts, and alerts are more likely to be used and extracted to improve SA during a time-critical event (Huang and Xiao 2015). As a result, messages are typically categorized into these categories. For example, previous work (Imran et al. 2013) extracted tweets published during a natural disaster into five categories, including caution and advice, casualty and damage, donation and offer, people, and information source to augment our understanding about disaster situation. While manually examining 5000 sampled tweets generated during the 2012 Hurricane Sandy, we noticed that a large number of informative messages were reporting the status of infrastructure (e.g. transportation) and resource (e.g. gas, power, internet, food). Therefore, we created an additional category named as infrastructure and resource in our classification schema for this study.

Traditional text mining methods mostly rely on machine learning techniques, such as support vector machine (SVM; Cortes and Vapnik 1995) and logistic regression (LR; Freedman 2009). Those methods have shown mixed results and are criticized for being unable to generalize beyond the scope of the designed study. Such retrospect models have little value for handling data generated during a new disaster event, as they yield the best result in a specific disaster situation. What has been learnt from a specific event was difficult to generalize for other events. One possible reason is that different disaster events evolve differently through time, resulting in different topics could be discussed throughout the social media platform. Another possible reason is that incidents and damages occurred within a specific disaster are different and unpredictable in nature (Lazreg, Goodwin, and Granmo 2016).

This research aims to address this challenge by examining the capability of deep learning in cross-event topic classification of Twitter data for disaster SA in different hurricane events. Recently, deep learning has produced promising results for various tasks in text mining (Collobert et al. 2011), particularly topic classification, sentiment analysis, question answering (Yu et al. 2014), and language translation (Wang, Raj, and Xing 2017). The key characteristics of deep learning is that nonlinear, high-level features are self-learned directly from unstructured data, such as images, sound, video, and text (LeCun, Bengio, and Hinton 2015). These characteristics have granted deep learning the capability of solving much more complicated artificial intelligence (AI) tasks, including speech recognition, genomic pattern discovery, and unmanned car driving (Wang, Raj, and Xing 2017). Currently, deep learning methods include autoencoder, restricted Boltzmann model, deep belief network, convolutional neural network (CNN) and recurrent neural network, and so on.

Using three geo-tagged twitter datasets collected during Hurricane Sandy, Hurricane Harvey, and Hurricane Irma, this research evaluates the performance of the CNN model for cross-event topic classification by comparing with two traditional machine learning models, including SVM and LR. Specifically, two groups of experiments are conducted, including (1) single-event experiment, which randomly selects 80% of the dataset from each disaster event to build the classifiers, and tests the performance of every classifier using the remaining 20%, and (2) cross-event validation experiment, which compares the classifiers that are trained with historical Twitter datasets to classify tweet messages generated during a later event into different topics. Experimental results reveal the proposed model based on CNN is able to establish an evolving SA knowledge base using historical Twitter datasets to facilitate the topic classification for future hurricane events with insufficient training data.

The rest of paper is organized as follows. Section 2 reviews the state-of-the-art work in social media text analysis efforts for disaster management and deep learning efforts for text mining. Section 3 describes the CNN model architecture and the underlying layers inside the network. Section 4 introduces the Twitter dataset for training and testing, the model configuration, and the experiment results of the CNN classifier compared to two well-known machine learning models, including SVM and LR. Section 5 demonstrates the experiment for cross-event validation of the CNN model, and compares it with SVM and LR. Finally, in Section 6, we conclude the research and provide future research directions.

2. Literature review

2.1. Social media for disaster analysis

Previously, methodologies such as phone calls, direct observation or personal interview are commonly used by disaster responders, relief coordinators and damage evaluators to gain SA for emergency response and relief coordination during a critical crisis. However, these data collection methods are time-consuming and laborious in processing the data. Social media data, however, can provide 'real-time' information for decision-making through multiple stages of the disaster management. Recently, much work has been done to leverage these new data sources to support various disaster relevant analysis and management, such as disaster event detection and tracking (Sakaki, Okazaki, and Matsuo 2010), disaster response and relief (Ashktorab et al. 2014; Kumar et al. 2011; Purohit et al. 2014), and damage assessment (Cervone et al. 2016).

Among the massive social media data collected during a period for a natural hazard, only a small portion of the messages are related and contribute to SA. Due to the volume of data is far beyond the capabilities of manual examination, text mining of social media messages typically starts with filtering out non-relevant messages and transforms unstructured text data sets into a structured form using various natural language processing techniques (NLP; Chae et al. 2012). Next, more advanced text mining techniques based on machine learning algorithms then can be applied to assign pre-defined topics to text documents (classification) or automatically structure document collections to find groups of similar documents (clustering; Hotho, Nürnberger, and Paaß 2005). A few attempts have been made to uncover and explain the social media message topics discussed when Twitter users communicate during mass emergencies (Huang and Xiao 2015; Imran et al. 2013; Vieweg et al. 2010). Imran et al. (2013), for instance, extracted tweets published during a natural disaster into several categories, including caution and advice, casualty and damage, donation and offer, and information source. The content categories (or topics) defined in those studies (Vieweg et al. 2010; Imran et al. 2013) are very useful to explore and extract the data involved in the disaster response and recovery phases. However, those content categories or topics (Imran et al. 2013; Imran and Castillo 2015; Vieweg et al. 2010; Vieweg 2012) mostly considered and are very useful to explore and extract the data involved in the disaster response and recovery phases. Huang and Xiao (2015) made an effort by coding social media messages into different categories within various stages of disaster management by including useful information that could be posted before or after a disaster event. Since our study focuses on the messages during the response and recovery phases, we will use the message coding schema based on Imran et al. (2013)'s work.

While classifying social media messages into different topics, previous work primarily applies traditional machine learning techniques, such as SVM (Cortes and Vapnik 1995) and LR (Freedman 2009). One drawback for traditional machine learning approaches is that these supervised classifiers learnt from a specific event do not generalize well to other events as different events may generate unique responses on social media (Li et al. 2015). Due to this drawback, when using these methods, we have to manually examine tweets for a new event and train a new classifier to identify different topics, which is labor intensive and time-consuming.

2.2. Deep learning for text mining

Deep learning technology is applied in common text mining tasks, such as sentiment analysis (Severyn and Moschitti 2015), text classification (Liu et al. 2015), text generation (Graves 2013) and social event detection (Chen et al. 2015). In this study, we examine the capability of CNN in text classification for disaster relevant tweets and facilitate the rapid identification of disaster response and relief contents. Therefore, here we review the deep learning technologies, particularly for text classification.

CNN was one of the most popular methods for pattern recognition. Kim (2014) described the basics of CNN for sentence-level classification tasks using pre-trained word vectors. Detection filters with different lengths were used to convolve text matrix to search for the presence of specific features or patterns present in the text. Then max pooling, operating extractive vectors of every filter, was employed to down-sample the input representation by reducing its dimensionality and allowing for assumptions to be made about features contained in the filtered subregions. Finally, each filter corresponded to a digit and these filters were connected to obtain a vector representing this sentence, on which the final prediction was based. Zhang and LeCun (2015) applied deep learning to text understanding using temporal convolutional networks (ConvNets; LeCun et al. 1998) and conducted experiments, including ontology classification, sentiment analysis, and text categorization, on various large-scale datasets. Results showed that temporal ConvNets can achieve satisfactory performance without the knowledge of words, phrases, sentences and any other syntactic or semantic structures with regards to a human language. Furthermore, Zhang, Zhao, and LeCun (2015) applied ConvNets only on characters using large-scale datasets, and the experiment results showed that deep ConvNets is able to classify words without the knowledge of words or the syntactic or semantic structure of a language. The aforementioned literature deals with sentence-based text data, whereas Lai et al. (2015) proposed a recurrent convolutional neural network (RCNN) and achieved promising results on document-level datasets.

More recently, deep learning methods have been applied to the field of disaster response utilizing social media datasets. Caragea, Silvescu, and Tapia (2016) and Nguyen et al. (2017) utilized CNN models to identify informative tweets for flood events and earthquake events respectively. Aipe et al. (2018) utilized a deep CNN model to extract linguistic features from tweets and utilize these features to classify tweets into different informative categories for crisis management. However, the capability of deep learning on cross-event multi-class tweet topic classification needs to be further examined, and this research serves as the starting point of establishing an evolving knowledge base for Twitter topic classification for disaster situation awareness.

3. Methodology

The general CNN architecture can be created using multiple layers, such as input, convolution, pooling, activation, and output layers. A deep CNN is defined when the architecture is composed of convolution layers and auxiliary layers (e.g. dropout and batch normalization layers). This section explains the overall architecture, the layers used in this study, and the backgrounds of each layer.

3.1. Overall architecture

Figure 1 presents the CNN architecture, which is the configuration for geo-tagged tweet theme classification. The first step is preprocessing, where each word of the tweet is represented by an integer (Section 3.2). The preprocessed tweet passes through the first layer, word embedding, which expands the word integers to a larger matrix and represents them in a more meaningful way (Section 3.3). The convolution layer then extracts features from the word embedding and transforms them through global max pooling (Sections 3.4 and 3.5). Then two fully connected layers predict the themes of each tweet (Section 3.8). Dropout layers are utilized before the convolution layer and the last fully connected layer (Section 3.6). Activation functions are used after the convolution layer and the fully connected layers (Section 3.7).

3.2. Preprocessing

Since machines understand numbers better than words, we first need to split each text into words and represent each word by a number to train our neural network with texts. The most common words are represented with lower numbers, thus we can easily train on only the *N* most common words in our dataset, and adjust *N* as necessary. In this way, we can ignore rare words, which neural networks can hardly learn from and might increase the processing time. In the experiment, we calculate the frequency of each word and select the 2,000,000 first most-common words in the dataset. Then all tweets are converted to list of integers. Finally, each list of integers is 'padded', so that all training and testing examples are in the same size (50 integers).

3.3. Embedding layer

The list of integers is further computed into known embedding, by parsing the data using pre-trained embedding, word2vec, which were trained on 100 billion words from Google News (Mikolov et al. 2013). In word2vec, words with similar semantic meanings are represented closely in the embedded vector space. The words oak, elm, and birch may cluster in one corner, while war, conflict and strife group together in another. This layer lets the network expand each token to a larger vector, allowing the network to represent words in a meaningful way (Table 1).

3.4. Convolution layer

The convolution layer performs the following three operations throughout an input array. First, it performs element-by-element multiplications (i.e. dot product) between a subarray of an input array, and a filter, which is also often called the filter, or kernel. Second, the multiplied values are

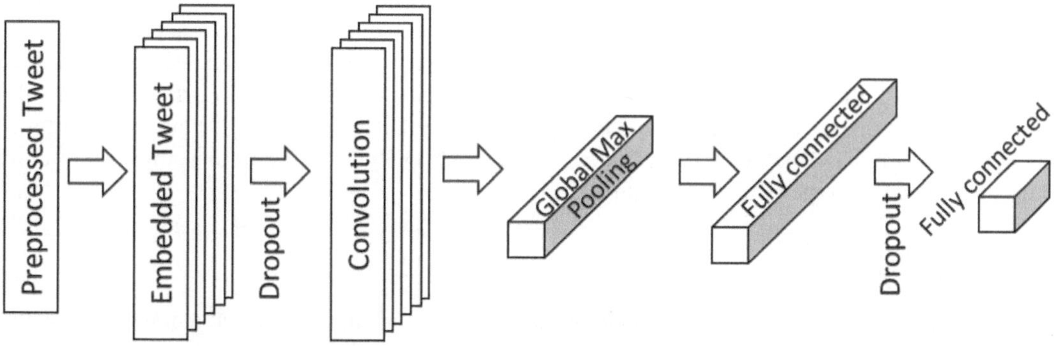

Figure 1. Overall architecture of the CNN.

Table 1. Dimensions of layers and operations.

Layer	Operator	Output height	Output width	Output depth
Input	–	1	50	–
Embedding		1	50	300
Dropout	Rate = 0.25	1	50	300
Convolution	Stride = 1, zero padding = 0, depth = 250, filter size = 3; activation = ReLU	1	50	250
Global max pooling	–	1	1	250
Fully connected	Output depth = 250; activation = ReLU	1	1	250
Dropout	Rate = 0.25	1	1	250
Fully connected	Output depth = 5; activation = Softmax	1	1	5

summed, and bias is added to the summed values. One of the advantages of the convolution is that it reduces input data size, which reduces computational cost. A convolution operation involves a weight matrix w, which is applied to a window of h words to produce a new feature. For example, a feature c_i is generated through

$$c_i = f(w * x_{i:i+h-1} + b) \tag{1}$$

Here b is a bias term and f is a nonlinear function. This operation is applied to each possible window of words in the text array $\{x_{1:h}, x_{2:h+1}, \ldots, x_{n-h+1:n}\}$and produces a feature map:

$$c = [c_1, c_2, \quad \ldots, c_{n-h+1}] \tag{2}$$

Three hyperparameters control the size of the output: the depth, padding and stride. The depth corresponds to the number of filters, and each different number will learn different information from the input. For example, the convolution layer takes the word embedding as input, and then different neurons along the depth dimension may activate in presence of various oriented edges. So, the depth column is a set of neurons looking at the same region of the input. The padding controls the size of the output volumes. The stride defines for the filter how many of columns and rows slide at a time across the input array's width and height. A larger stride size leads to fewer filter applications and a smaller output size reduces computational cost but may lose features of the input data.

The convolution layer accepts word embedding of size [50*300]. It uses neurons with filter size $F = 3$, stride $S = 1$, no padding $P = 0$, and depth $K = 250$. The output size of a convolution is [(input width – filter size)/stride + 1 = 48]. Each of the 48*250 neurons in this volume is connected to a size [1*3] region of the input. Moreover, all 250 in each depth column are connected to the same [1*3] region of the input, but with different weights.

3.5. Global max pooling layer

The global max pooling layer (Lin, Chen, and Yan 2013) reduces the size of the representation to decrease the number of parameters and computation in the network, and hence also controls overfitting. Consecutive layers of the network are activated by 'higher' or more complex features, thus strategically reducing the size of the representation will still pass on the valid information from the previous layer to the next one. The global max pooling operation performs by reducing the dimensions of the input of 1×50×250–1×1×250, and reduces each 1×50 feature map to a single number by simply taking the average of all 1×50 values.

3.6. Dropout layers

Overfitting, a long-standing issue in the field of machine learning, happens when a network classifies a training data set effectively but fails to provide satisfactory validation and testing results. This happens when a neural network is built, two or more neurons begin to detect the same feature repeatedly, i.e. co-adaptations. To address this issue, dropout layers (Srivastava et al. 2014) are used to randomly disconnect the connections between neurons of connected layers at a certain dropout rate. Accordingly, a network can generalize training examples much more efficiently by reducing these co-adaptations. In this model, the dropout rate is 0.2.

3.7. Activation layer

Activation functions are an extremely important feature of the artificial neural networks. They basically decide whether a neuron should be activated or not. Whether the information that the neuron is receiving is relevant for the given information or should it be ignored. Since all layers in a neural network are nonlinear, values for each of the neurons are calculated and put through an activation function. The most typical way to give nonlinearity in the neural network calculation is using sigmoidal functions, such as $y = \tanh(x)$. However, it has been claimed that saturating nonlinearities slows down computations (Nair and Hinton 2010).

ReLU was introduced as a nonlinear activation function (Nair and Hinton 2010). It is defined as $f(x) = \max(0, x)$. Briefly, while other nonlinear functions are bounded to output values (e.g. positive and negative ones and zeros), the ReLU has no bounded outputs except for its negative input values, which means that it does not activate all the neurons at the same time, making the network sparse but efficient and easy for computation. Intuitively, the gradients of the ReLU are always zeros and ones. These features speed up computations than those using sigmoidal functions and achieve better accuracies.

The softmax function is another type of sigmoid function. The sigmoid function can only handle two classes, whereas softmax is effective in handling multi-classification problems. The output of the softmax function can be represented as a categorical distribution. The goal of softmax is to highlight the largest values and suppress values which are significantly below the maximum value. The softmax function can be defined as

$$\sigma(z)_j = \frac{e^{z_j}}{\sum_{k=1}^{K} e^{z_k}}, \quad j = 1, \ldots, K \tag{3}$$

The softmax function is ideally used in the output layer of the classifier where we intend to obtain the probabilities to define the class of each input.

3.8. Fully connected layer

Neurons in a fully connected layer have full connections to all activations in the previous layer, as seen in regular Neural Networks. Their activations can be computed with a matrix multiplication followed by a bias offset: $Y = X.W + b$, where Y is the predictions, X is the previous layer output, W is the weights, and b is the biases.

Adding a fully connected layer is a cheaper way of learning nonlinear combinations of these features. Most of the features from convolutional and pooling layers are good for the classification task, but combinations of those features could be even better. The first fully connected layer uses ReLU as the activation function, and its output is $[0, +\infty)$.

The output from the convolution and pooling layers represent high-level features of the input text array. The purpose of the fully connected layer is to use these features for classifying the input into various classes based on the training dataset. The sum of output probability from the fully connected layer is 1, ensured by using softmax as the activation function in the output layer of the fully connected layer. The softmax function takes a vector of arbitrary real-valued scores and squashes it to a vector of values between zero and one that sums to one.

4. Building a classifier for classifying tweet theme

This section describes the dataset generation and the underlying hyperparameters assigned to train the deep learning model. Configuring and choosing adequate hyperparameter (e.g. batch size and iteration number) is tedious and no existing guidelines for those parameter optimizations are available. Thus, the optimal network architecture for the tweet theme classification must be explored via trial-and-error and guided by checking the validation set error (Goodfellow et al. 2016). The classifier is developed by the library Keras (https://github.com/keras-team/keras) with TensorFlow as the backend.

4.1. Dataset

Twitter data from three hurricane events, including the 2012 Sandy, 2017 Harvey and 2017 Irma, are collected and used as experiment datasets. For each hurricane event, around 5000–15,000 tweets were manually reviewed (Sandy: 5000, Harvey: 7342, Irma: 15,254), and around 2000–3000 informative tweets (Sandy: 1900, Harvey: 2180, Irma: 3910) were annotated into different categories.

Our classification scheme is derived and modified from existing work with five commonly used classes, including (1) Caution and Advice (CA), (2) Casualties and Damage (CD), (3) Information Sources (IS), (4) Donation and Aid (DA), and (5) People (Imran et al. 2013). Due to few samples in People and oversized sample in CA, we removed People class and extracted a portion of CA tweets and defined an additional class named infrastructure and resource (IR). IR is related to the status of infrastructure (e.g. transportation) and resource (e.g. gas, power, internet, food), which are very useful to establish GSA. For example, disaster managers and the public may want to quickly traffic status for an effective evacuation plan. The descriptions for the five classes are illustrated in Table 2. This classification scheme is suitable for the twitter data of the three hurricane events, as it produces equally distributed samples within each class without a great imbalance issue.

Figure 2 shows the temporal distribution of the five tweet topics during the hurricane events. For Hurricane Sandy, a significant increase of tweet number for 'Caution and Advice' can be observed on October 29, since the wind, rain, and flooding occurred in the city during that night. We also observe an increase for the class 'Infrastructure and Resources' on October 30, and a continuous increase of 'Casualties and Damage' during the two days of October 30 and 31. Moving forward, we observe a gradual increase for 'Donation and Aid' throughout the study time period until it reaches its peak on November 3, and decreases gradually for the rest of the time.

For Hurricane Harvey, 'Caution and Advice' also appears to be the dominating topic with a peak of ~250 tweets on August 27 and a rapid increase and decrease before and after the peak. Both 'Infrastructure and Resources' and 'Information Sources' display a different pattern between Hurricane

Table 2. Social media classification scheme.

	Class	Description	Example
1	Caution and Advice (CA)	Warning given about a related incident	Flooded neighborhoods in Norfolk and its approaching low tide
2	Casualties and Damage (CD)	Information about casualties or infrastructure damage	This tree and power lines are down at the corner of Station Road and Bethlehem Pike in Quakertown
3	Information Sources (IS)	A message from an official news source, media or government	@NYCMayorsOffice: Mayor: All @NYCSchools are closed tomorrow
4	Infrastructure and Resource (IR)	Information about IR that is not reported from an official news source and is not related to donation or aid	Two gas stations at my hotel are out of gas. Had to buy potato bread at the store … Train services are being shut down #HurricaneSandy
5	Donation and Aid (DA)	Goods or services offered or needed by victims	I do not have any money to donate but I have lots of time, where can I help/volunteer in #Hoboken? Who do I call?

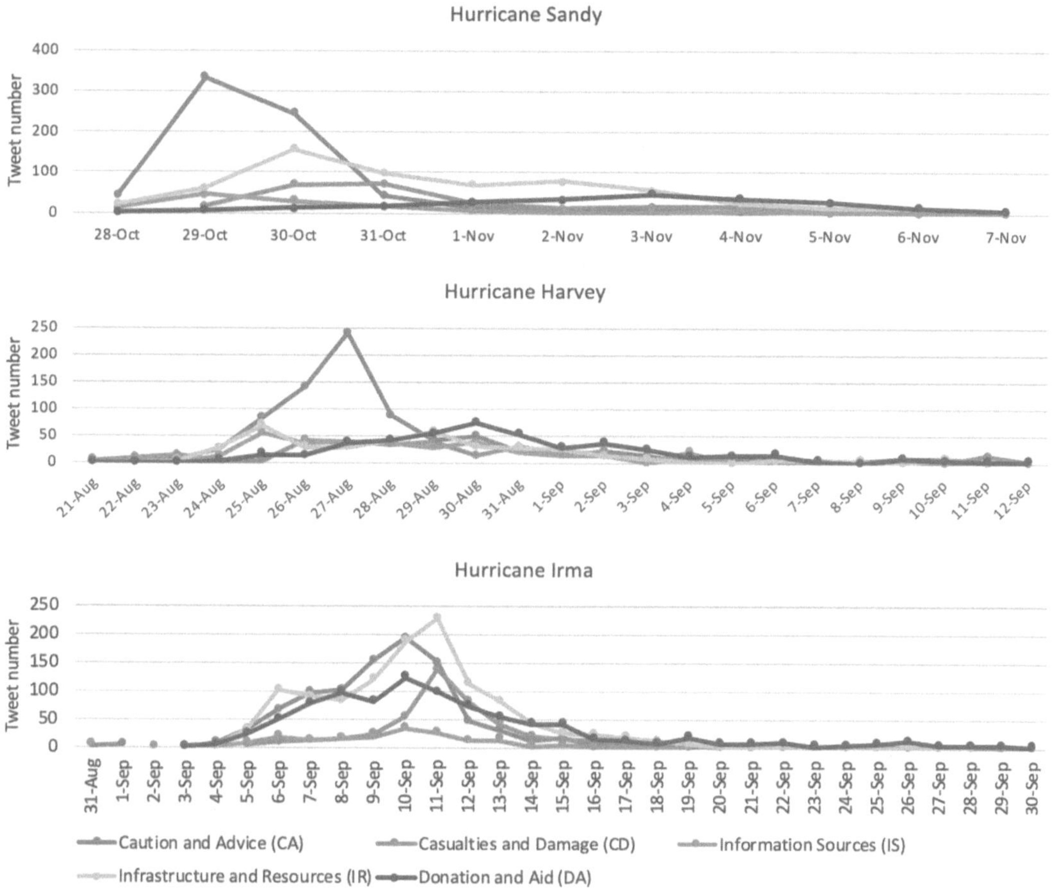

Figure 2. Tweet topic over time.

Sandy and Harvey, which has two peaks on August 25th and 29th, 2017. On 25 August 2017, Harvey hit Port Aransas and Port O'Connor near Corpus Christi with 130 mph winds. The Category 4 hurricane left 250,000 people without power. On August 29, Harvey made landfall for a third time as it hit the coastal cities of Port Arthur and Beaumont Texas on the border of Louisiana. It dumped 26 in. of rain in 24 h. It flooded Port Arthur, a city of 55,000 people. Water entered one-third of the city's building, including the shelter. Beaumont has been without drinking water since the storm. A time lag can be observed for 'Casualties and Damage' which reached its peak on August 26th, one day after Harvey made its landfall, and remained active during Harvey's peak days. 'Donation and Aid' shows a similar pattern as the one in Hurricane Sandy, which increases gradually until it reached its peak on August 30th and decreased gradually.

For Hurricane Irma, it was first developed from a tropical wave near Cape Verde on 30 August 2017. Topics about preparing for Irma started until it reached to its first peak on 6 September 2017 when it hit the Leeward Islands and left hundreds in Puerto Rice without power on 7 September 2017. Irma remained a Category 5 hurricane on September 8, downgraded to Category 3 on September 9, and regained its strength to Category 4 on September 10, when it affected Florida Keys with hugh rainfall and high storm surge. After September 11, Irma downgraded to Category 1 and then a tropical storm.

Figure 3 shows the spatial distribution of the five topics in the New York area during the Hurricane Sandy using heatmaps. It is observed that most tweets for 'Caution and Advice' are from the

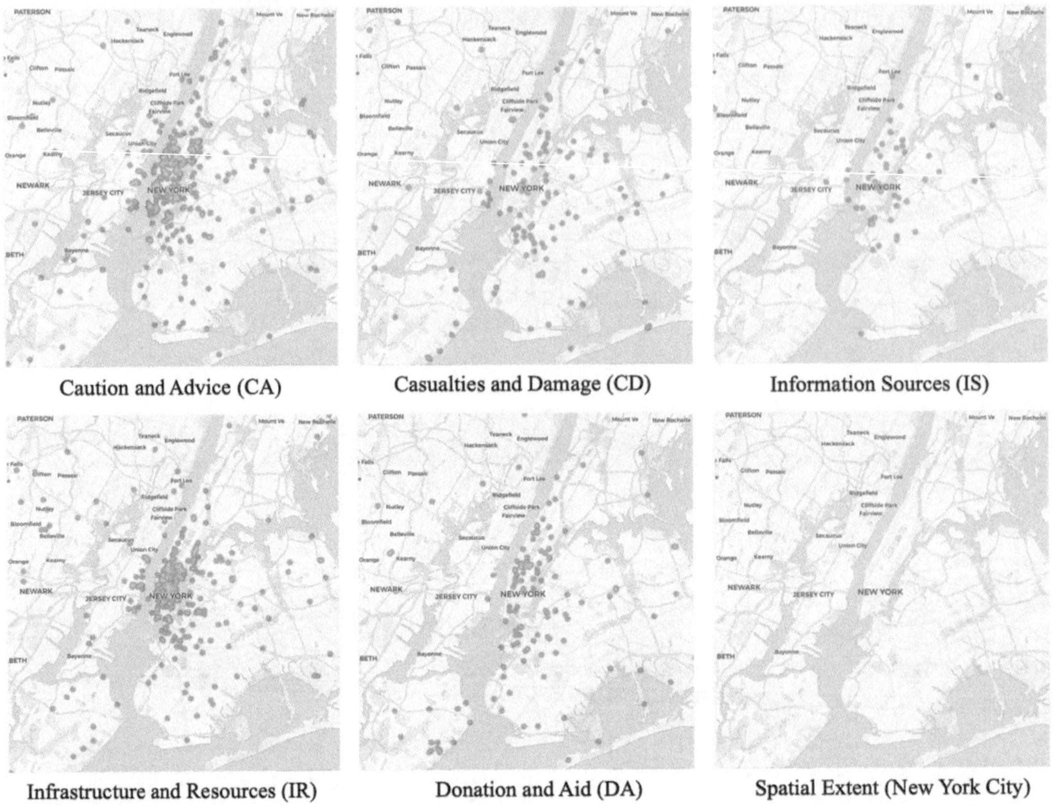

Figure 3. Spatial distribution of Hurricane Sandy tweet topic.

communities of lower Manhattan, since news reports broadcasted that this area would be impacted and drew people's attention. Similar distribution can be observed for the class 'Information Resources' due to the dense population in the same area. Tweets about 'Casualties and Damage' are more distributed in the area indicating damages of storm surge and high winds occurred throughout the area. Similar patterns can be observed for the class 'Donation and Aid' mentioning about 'red cross', 'FEMA', and 'volunteering'.

Figure 4 illustrates the spatial distribution of tweets in the Texas and Louisiana area during Hurricane Harvey. Tweets providing 'Caution and Advice' were spread in major cities in Texas, including Houston, Dallas, Austin, San Antonio and the Texas coast area. Given that Hurricane Harvey impacted most severely in Houston area where most of the population are located, most of the tweets related to 'Casualties and Damage', 'Information Sources', and 'Infrastructure and Resources' are located at the Houston area. 'Donation and Aid' are more broadly distributed and not so much dependent on destruction severity and geography area.

Different from Hurricane Harvey, Hurricane Irma had a more spatially distributed impact to Florida area regarding 'Caution and Advice', 'Casualty and Damage', 'Infrastructure and Resources', and 'Donation and Aid', but there was fewer official information than Harvey. Irma's large wind field resulted in strong winds across the state of Florida and over 70% homes lost power. Heavy rainfall – and storm surge, in some instances – caused at least 32 rivers and creeks to overflow, resulting in significant flooding, especially along the St. Johns River and its tributaries. These impacts occurred in a fairly short amount of time (September 9–11) sweeping from Southeast Florida towards Georgia, which explains the spatial distribution of tweets in the impacted area.

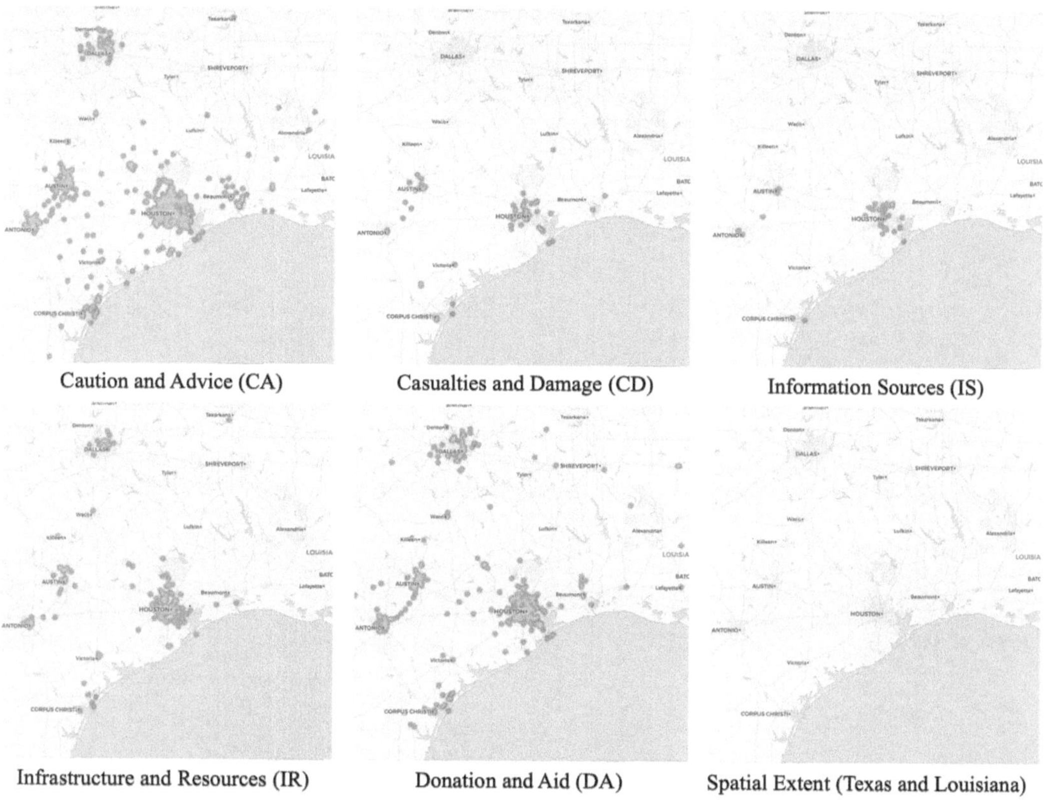

Figure 4. Spatial distribution of Hurricane Harvey tweet topic.

4.2. Experiment configuration

As the CNN model is defined and described in Section 3, it needs to be compiled for efficient training, which represents the process of finding the best set of weights to make predictions. Compiling the model requires the specification of the lost function (used to evaluate a set of weights), the optimizer (used to search through different weights for the network) and metrics (used to judge the performance of the model). The metrics are similar to a loss function, but the results from evaluating a metric are not used when training the model, whereas the loss function gets minimized by the optimizer to optimize the model during training. In this experiment, we use the categorical cross-entropy as the loss function, RMSprop (Hinton, Srivastava, and Swersky 2012) as the optimizer, and categorical accuracy as the metrics.

After being compiled for efficient computation, the model is executed on training data for the training process. The training process runs for 50 iterations, which is specified by the parameter 'epoch'. We used a batch size of 70, the number of instances that are evaluated before a weight update in the network is performed. The epoch and batch size are chosen experimentally through trial and error. Tweets are randomly selected from the dataset to generate training (80%) and validation (20%) sets.

4.3. CNN results

The classification accuracy on the training data and test data changes over time (Figure 5). The accuracy rises gradually towards 1.0, whereas the test accuracy reaches ~0.81. This indicates that our network is overfitting, i.e. the network is memorizing the training set, without understanding texts well enough to generalize to the test set. As a major problem in neural networks, overfitting is difficult to

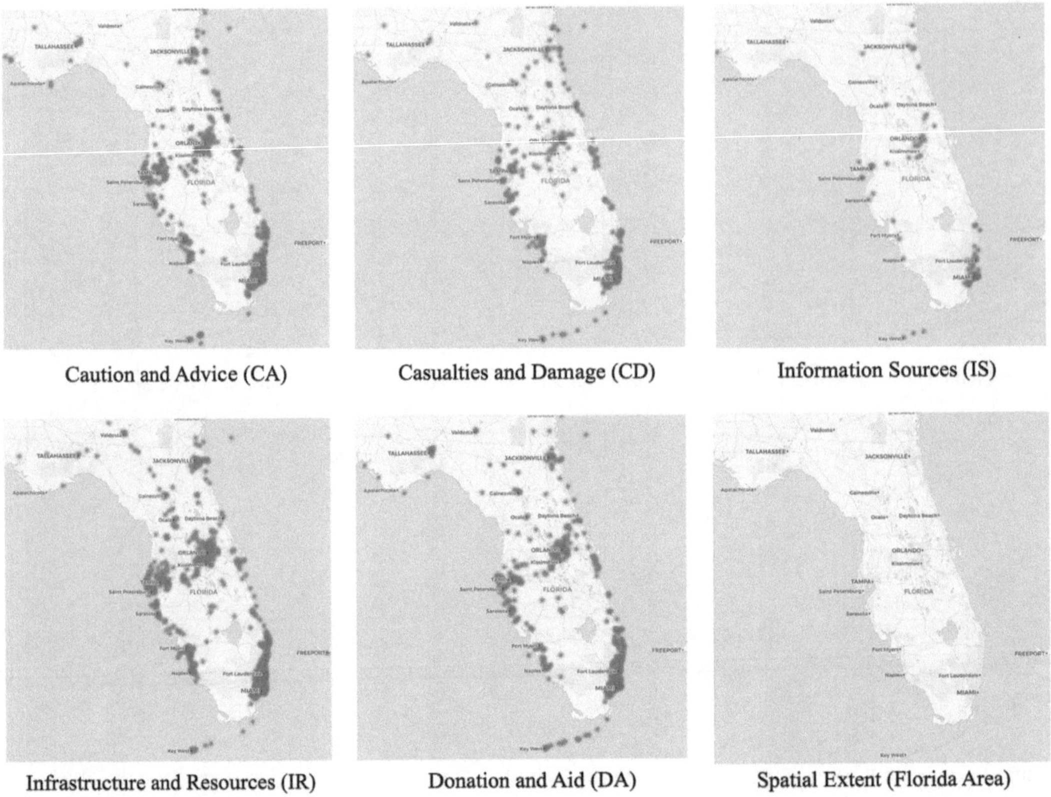

Figure 5. Spatial distribution of Hurricane Irma tweet topic.

address especially when deep learning networks often have very large numbers of weights and biases. In this case, the network has 2,138,155 parameters with 289,255 trainable parameters. Although techniques like dropout and regularization have been utilized in our network, the sign of overfitting is still not improving. The reason is that our training dataset is relatively small with ~1200 samples for each hurricane event, comparing to other benchmarking large-scale datasets, e.g. AG's news: 120,000 train samples and Amazon Review Full: 3,600,000 train samples (Zhang, Zhao, and LeCun 2015). The size of our train and test data is limited by the nature of twitter data, which was harvested real time through Twitter Streaming API. Currently, while we are not able to increase the dataset, we assume that using twitter data integrated from multiple disasters to increase the dataset will produce better performance with this CNN model (Figure 6).

5. Comparative studies

The CNN classifier is compared in terms of its capability on single-event and cross-event topic classification against the well-established text classification methods SVM and LR. For single-event experiment, we randomly select from the single-event dataset to generate training (80%) and validation (20%) sets. For cross-event experiment, we train the classifiers using the dataset from one event and test the others.

5.1. Single-event experiment among the classifiers

Three single-event experiments were conducted for using the twitter datasets for each hurricane event correspondingly. We performed a statistical analysis on the outcomes of the classification

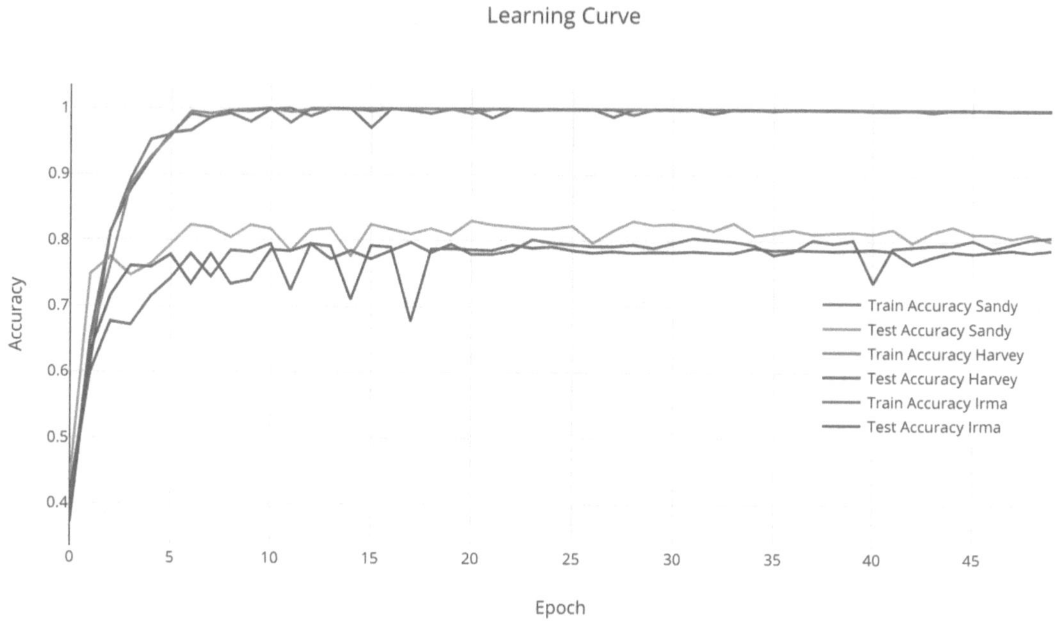

Figure 6. Learning curve of the CNN classifier.

for both methods. A confusion matrix for each experiment was created by calculating values for the five classes in main diagnostic tests including Positive predictive value (Precision), True positive rate (Recall), and F1-score. The overall or averaged accuracy scores are demonstrated in Table 3 and the scores for each topic are demonstrated in box plots in Figure 7. It is evident that the CNN model outperformed SVM and LR. With respect to Precision, the CNN model had values ~0.8 while the SVM model had 0.63–0.72 and LR had 0.44–0.6 for the three experiments. Almost similar behavior is observed in the Recall and in F1-score, where CNN achieved the best performance while SVM and LR had lower scores. These findings state clearly that CNN outperforms traditional text mining approaches for tweet classification presenting the potential for further development on tweet theme identification.

5.2. Cross-event validation

We apply the same methodology for cross-event experiments, and based on the three Hurricane events, we have designed four experiments which utilize twitter data from one or more historical events as training data to classify the data from a later event. This design ends up with four experiments: (a) Train Hurricane Sandy tweets to classify Hurricane Harvey tweets, (b) Train Hurricane Sandy tweets to classify Hurricane Irma tweets, (c) Train Hurricane Harvey tweets to classify Hurricane Irma tweets, and (d) Train Hurricane Sandy and Harvey tweets to classify Hurricane Irma tweets.

Table 3. Overall accuracy scores for single event experiments.

	(a) Hurricane Sandy			(b) Hurricane Harvey			(c) Hurricane Irma		
	CNN	SVM	LR	CNN	SVM	LR	CNN	SVM	LR
Precision	0.81	0.72	0.56	0.81	0.63	0.44	0.80	0.64	0.6
Recall	0.80	0.70	0.53	0.81	0.64	0.51	0.80	0.63	0.53
F1-score	0.80	0.70	0.50	0.81	0.63	0.40	0.79	0.63	0.51

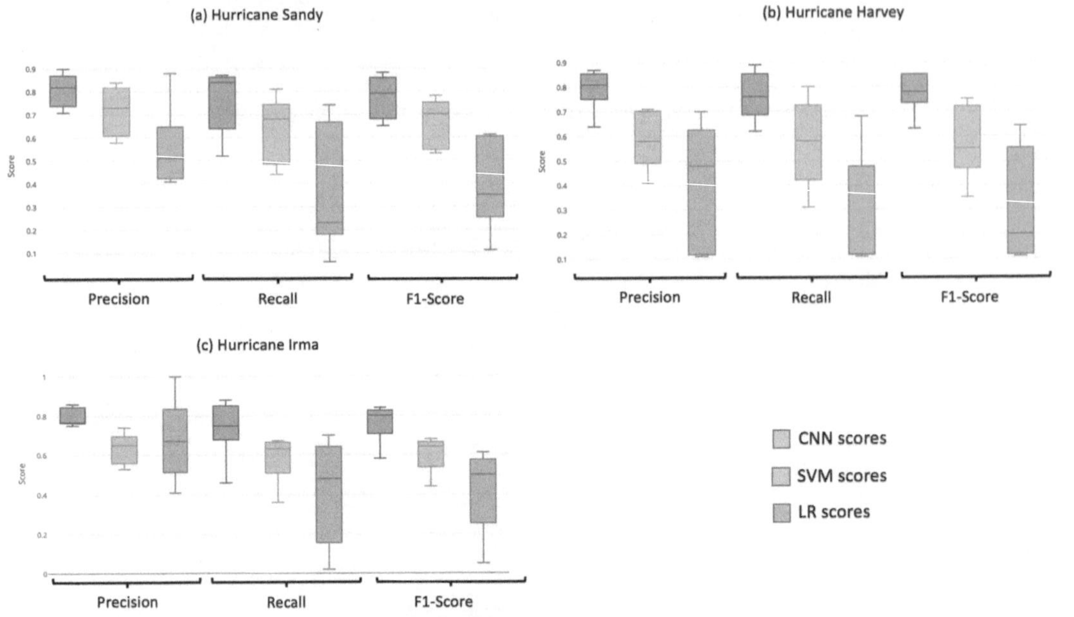

Figure 7. Performance scores comparison among CNN, SVM, and LR.

The accuracy scores are recorded in Table 4. Results showed that the CNN model achieved a consistent accuracy for the cross-event experiments with an average precision around 0.8. However, SVM and LR resulted in much lower accuracy than using a single event for training and testing. In addition, by combining the twitter data from two historical events (Sandy and Harvey) as training data to classify the topics for a later event (Irma), we achieved better accuracy results for all three classifiers (for precision, CNN: 0.83, SVM: 0.67 and LR: 0.67) than using only one event to classify another. Therefore, the CNN model outperformed SVM and LR when applying to different events; and this capability enables CNN to better predict and classify twitter topics for future events based on a pre-trained model built on the datasets generated during past events.

The boxplots (Figure 8) for each cross-event experiment reveal the limitations for SVM and LR on classifying different topic categories. It is observed that the ranges in accuracy scores are more stretched compared to the single-event experiments (Figure 7) for SVM and LR.

A notable difference is that in Hurricane Sandy twitter texts were more explicit in text messages, whereas in Hurricane Harvey and Irma twitter texts contained more links to other websites or apps, such as Instagram, news reports, and using images and videos rather than expressing in full sentences. Therefore, the messages conveyed from Hurricane Harvey and Irma tweets are less explicit than the ones from Hurricane Sandy. In future studies, we will incorporate the information discovered from images, videos, and news reports to complete the message being shared from the Twitter users to better facilitate the disaster response stage.

Table 4. Overall accuracy scores for cross-event experiments.

	(a) Train Sandy tweets to classify Harvey tweets			(b) Train Sandy tweets to classify Irma tweets			(c) Train Harvey tweets to classify Irma tweets			(d) Train Sandy and Harvey tweets to classify Irma tweets		
	CNN	SVM	LR	CNN	SVM	LR	CNN	SVM	LR	CNN	SVM	LR
Precision	0.79	0.65	0.58	0.79	0.61	0.56	0.76	0.64	0.56	0.83	0.67	0.67
Recall	0.79	0.64	0.50	0.78	0.59	0.46	0.65	0.51	0.37	0.8	0.65	0.65
F1-score	0.78	0.62	0.45	0.78	0.58	0.4	0.66	0.51	0.31	0.8	0.65	0.65

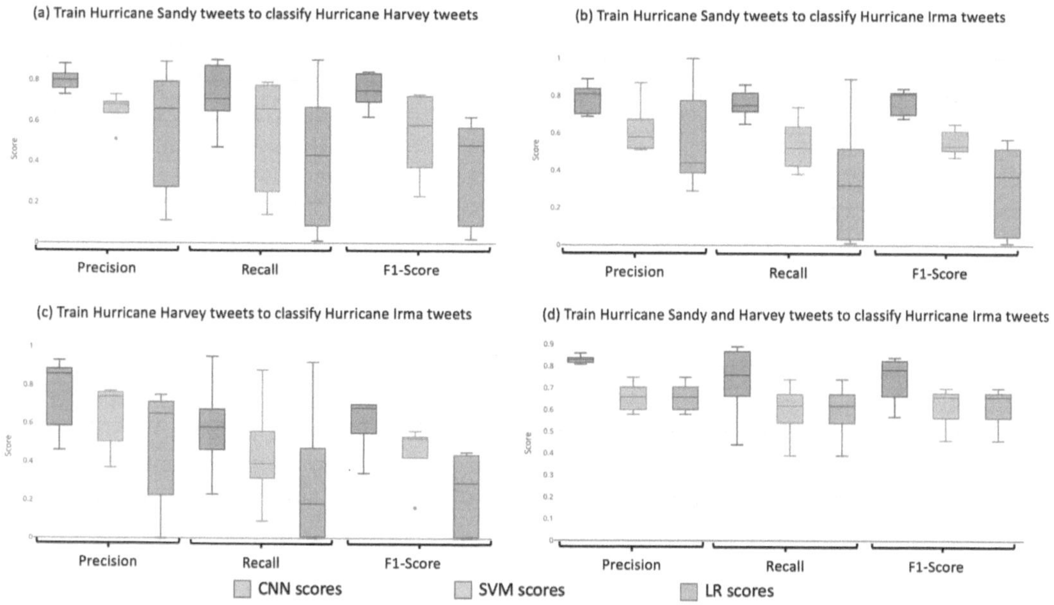

Figure 8. Cross-event classification accuracy.

5.2. Discussion

CNN outperforms SVM and LR in single-event topic classification. This is probably due to the complexity of Twitter data in nature since the tweets typically contain misspellings, abbreviations, deletions and phonetic spellings. The training set is produced through a human labeling process, which may introduce different levels of uncertainty. SVM, LR, and other machine learning methods largely depend on the correctness of the training set, whereas CNN is capable of deducting high-level abstraction from the tweets.

The capability of CNN on cross-event topic classification indicates that with existing knowledge base constructed from past hurricane events, we are able to classify twitter topics in the early hours of a new event. CNN outperforms SVM and LR in cross-event classification because it can provide a generic solution that infers similarity and dissimilarity patterns between different events. However, CNN is more time consuming than SVM and LR because there are more parameters to learn, which poses a challenge for utilizing CNN for online learning when social media data is streaming and accumulating rapidly.

6. Conclusion and future research

This research examines the capability of a deep learning model based on CNN for twitter topic classification under the background of Hurricanes Sandy, Harvey, and Irma disaster management. In the paper, we described the CNN model, and applied the classifier to three manually labeled twitter datasets, which are classified into five different topics: Caution and Advice, Casualties and Damage, Information Sources, Infrastructure and Resources, and Donation and Aid. Even though the CNN used in the experiments is a basic model and has not been further optimized to the task at hand, the results show a significantly higher performance in direct comparison to traditional approaches on the considered dataset. For single-event experiment, the CNN classifier achieves an accuracy up to 0.81 for all three datasets, and consistently outperforms two well-known classification methods SVM (0.63–0.72) and LR (0.44–0.60). The cross-event validation was conducted by using the CNN built on twitter data from historical events to classify tweet messages posted during a later event t. Experiment results showed that CNN achieved a consistent performance, whereas the accuracy of SVM and LR dropped significantly. Additionally, by combining multiple datasets from different historical events, we can achieve a higher accuracy when classifying datasets generated from a later event. This leads to the assumption that a CNN model, similar to the one evaluated in the experiments, is preferable for the task of social media data classification in a context of disaster management. In addition, the CNN model has the capability of using Twitter data from past events to classify twitter topics of an upcoming event in real time for enhanced SA.

A significant improvement can be expected using larger-scale dataset for training and testing. Future research may test the performance of CNN-based model using labeled twitter datasets from multiple disasters of the same kind (e.g. combining twitter data from multiple hurricane events, hurricanes in the past 10 years). In addition, this research is applying the CNN model to twitter text only. To improve the classification accuracy, features of each tweet can be enhanced by adding the contextual information, including its geolocation, environmental conditions (e.g. wind and precipitation), and other linked media (news, blogs, images, and videos). Finally, other deep learning technologies, such as recurrent neural networks (RNN) and long-short term memory (LSTM), may be further explored.

Acknowledgements

Dr. Manzhu Yu and Dr. Qunying Huang initiated the original idea. Dr. Manzhu Yu wrote the major part of the paper. Dr. Qunying Huang contributed on the introduction and literature review, and also revised the paper. Mr. Chris Scheele helped with the Tweeter data preprocessing and classification scheme. Dr. Han Qin helped with experiment design and implementation. Dr. Chaowei Yang revised a later version of the paper.

Disclosure statement

No potential conflict of interest was reported by the authors.

Funding

This work was supported by National Science Foundation [grant number IIP-1338925].

References

Aggarwal, C. C., and H. Wang. 2011. "Text Mining in Social Networks." In *Social Network Data Analytics*, edited by C. C. Aggarwal, 353–378. Boston, MA: Springer.

Aipe, A., N. S. Mukuntha, A. Ekbal, and S. Kurohashi. 2018. "Deep Learning Approach towards Multi-label Classification of Crisis Related Tweets." In *Proceedings of the 15th ISCRAM Conference*, edited by Kees Boersma and Brian Tomaszewski, Rochester, NY, USA May 2018.

Ashktorab, Z., C. Brown, M. Nandi, and A. Culotta. 2014. "Tweedr: Mining Twitter to Inform Disaster Response." In *Proceedings of the 11th International ISCRAM Conference*, 18–21 May 2014, University Park, PA, USA, edited by S. R. Hiltz, M. S. Pfaff, L. Plotnick, and P. C. Shih, 354–358. University Park, PA: Pennsylvania State University.

Caragea, C., A. Silvescu, and A. H. Tapia. 2016, May. "Identifying Informative Messages in Disaster Events Using Convolutional Neural Networks." International Conference on Information Systems for Crisis Response and Management, 137–147.

Cervone, G., E. Sava, Q. Huang, E. Schnebele, J. Harrison, and N. Waters. 2016. "Using Twitter for Tasking Remote-Sensing Data Collection and Damage Assessment: 2013 Boulder Flood Case Study." *International Journal of Remote Sensing* 37 (1): 100–124.

Chae, J., D. Thom, H. Bosch, Y. Jang, R. Maciejewski, D. S. Ebert, and T. Ertl. 2012. "October. Spatiotemporal Social Media Analytics for Abnormal Event Detection and Examination Using Seasonal-Trend Decomposition." In *2012 IEEE Conference on Visual Analytics Science and Technology (VAST)*, October 14–19, 2012, Seattle, WA, USA, 143–152. New York, NY: IEEE.

Charalabidis, Y., EN, Loukis, A. Androutsopoulou, V. Karkaletsis, and A. and Triantafillou. 2014. "Passive Crowdsourcing in Government Using Social Media." *Transforming Government: People, Process and Policy* 8 (2): 283–308.

Chen, Y., L. Xu, K. Liu, D. Zeng, and J. Zhao. 2015. "Event Extraction via Dynamic Multi-Pooling Convolutional Neural Networks." In *Proceedings of the 53rd Annual Meeting of the Association for Computational Linguistics and the 7th International Joint Conference on Natural Language Processing* (Vol. 1), July 26–31, 2015, Beijing, China, 167–176. Stroudsburg, PA: ACL.

Collobert, R., J. Weston, L. Bottou, M. Karlen, K. Kavukcuoglu, and P. Kuksa. 2011. "Natural Language Processing (Almost) From Scratch." *Journal of Machine Learning Research* 12 (Aug): 2493–2537.

Cortes, C., and V. Vapnik. 1995. "Support-vector Networks." *Machine Learning* 20 (3): 273–297.

Emergency Alerting Platform Working Group, 2016. Social Media & Complementary Alerting Methods – Recommended Strategies & Best Practices. Available from: https://www.atis.org/01_legal/docs/CSRIC%20 V/WG2_FinalReport_091416.docx [Accessed May 24, 2018]

Freedman, D. A. 2009. *Statistical Models: Theory and Practice*. Cambridge: Cambridge University Press.

Goodfellow, I., Y. Bengio, A. Courville, and Y. Bengio. 2016. *Deep Learning*. Vol. 1. Cambridge: MIT press.

Graves, A., 2013. Generating sequences with recurrent neural networks. arXiv preprint arXiv:1308.0850.

Hinton, G., N. Srivastava, and K. Swersky. 2012. "Rmsprop: Divide the Gradient by a Running Average of its Recent Magnitude." Neural Networks for Machine Learning, Coursera lecture 6e.

Hotho, A., A. Nürnberger, and G. Paaß. 2005. "A Brief Survey of Text Mining." *Journal for Computational Linguistics and Language Technology* 20 (1): 19–62.

Huang, Q., and Y. Xiao. 2015. "Geographic Situational Awareness: Mining Tweets for Disaster Preparedness, Emergency Response, Impact, and Recovery." *ISPRS International Journal of Geo-Information* 4 (3): 1549–1568.

Imran, M., and C. Castillo. 2015. "Towards a Data-Driven Approach to Identify Crisis-Related Topics in Social Media Streams." In *Proceedings of the 24th International Conference on World Wide Web*, May 13–17, 2013, Rio de Janeiro, Brazil, 1205–1210. New York, NY: ACM.

Imran, M., S. Elbassuoni, C. Castillo, F. Diaz, and P. Meier. 2013. "Practical Extraction of Disaster-Relevant Information From Social Media." In *Proceedings of the 22nd International Conference on World Wide Web*, May 13–17, 2013, Rio de Janeiro, Brazil, 1021–1024. New York, NY: ACM.

Kim, Y., 2014. Convolutional neural networks for sentence classification. *arXiv preprint arXiv:1408.*5882.

Kumar, S., G. Barbier, M. A. Abbasi, and H. Liu. 2011. "TweetTracker: An Analysis Tool for Humanitarian and Disaster Relief." In *Fifth International AAAI Conference on Weblogs and Social Media*, July 17–21, 2011, Barcelona, Spain, 661–662. Palo Alto, CA: AAAI Press.

Lai, S., L. Xu, K. Liu, and J. Zhao. 2015. "Recurrent Convolutional Neural Networks for Text Classification." In *Proceedings of the Twenty-Ninth AAAI Conference on Artificial Intelligence*, January 25–30, 2015, Austin, TX, USA, 2267–2273. Palo Alto, CA: AAAI Press.

Lazreg, M.B., Goodwin, M. and Granmo, O.C., 2016. "Deep learning for social media analysis in crises situations." *The 29th Annual Workshop of the Swedish Artificial Intelligence Society (SAIS)* 2–3 June 2016, Malmö, Sweden, 31.

LeCun, Y., Y. Bengio, and G. Hinton. 2015. "Deep Learning." *Nature* 521 (7553): 436.

LeCun, Y., L. Bottou, Y. Bengio, and P. Haffner. 1998. "Gradient-based Learning Applied to Document Recognition." *Proceedings of the IEEE* 86 (11): 2278–2324.

Li, H., Guevara, N., Herndon, N., Caragea, D., Neppalli, K., Caragea, C., Squicciarini, A.C. and Tapia, A.H., 2015. Twitter Mining for Disaster Response: A Domain Adaptation Approach. In *ISCRAM*, May.

Lin, M., Q. Chen, and S. Yan. 2013. Network in network. arXiv preprint arXiv:1312.4400.

Liu, P., X. Qiu, X. Chen, S. Wu, and X. Huang. 2015. "Multi-timescale Long Short-Term Memory Neural Network for Modelling Sentences and Documents." In *Proceedings of the 2015 Conference on Empirical Methods in Natural Language Processing*, September 17–21, 2015, Lisbon, Portugal, 2326–2335. Stroudsburg, PA: Association for Computational Linguistics.

MacEachren, A. M., A. Jaiswal, A. C. Robinson, S. Pezanowski, A. Savelyev, P. Mitra, X. Zhang, and J. Blanford. 2011. "Senseplace2: Geotwitter Analytics Support for Situational Awareness." In *2011 IEEE Conference on Visual Analytics Science and Technology (VAST)*, October 23–28, 2011, Providence, RI, USA, 181–190. New York, NY: IEEE.

Mikolov, T., K. Chen, G. Corrado, and J. Dean. 2013. *Efficient Estimation of Word Representations in Vector Space*. arXiv preprint arXiv:1301.3781.

Nair, V., and G. E. Hinton. 2010. "Rectified Linear Units Improve Restricted Boltzmann Machines." In *Proceedings of the 27th International Conference on Machine Learning*, June 21–24, 2010, Haifa, Israel, 807–814. Madison, WI: Omnipress.

Nguyen, D. T., K. Al-Mannai, S. R. Joty, H. Sajjad, M. Imran, and P. Mitra. 2017, May. "Robust Classification of Crisis-Related Data on Social Networks Using Convolutional Neural Networks." ICWSM, 632–635.

Purohit, H., C. Castillo, F. Diaz, A. Sheth, and P. Meier. 2014. "Emergency-relief Coordination on Social Media: Automatically Matching Resource Requests and Offers." *First Monday* 19 (1). https://uncommonculture.org/ojs/index.php/fm/article/view/4848/3809.

Roche, S., E. Propeck-Zimmermann, and B. Mericskay. 2013. "GeoWeb and Crisis Management: Issues and Perspectives of Volunteered Geographic Information." *GeoJournal* 78 (1): 21–40.

Sakaki, T., M. Okazaki, and Y. Matsuo. 2010. "Earthquake Shakes Twitter Users: Real-Time Event Detection by Social Sensors." In *Proceedings of the 19th International Conference on World Wide Web*, April 26–30, 2010, Raleigh, NC, USA, 851–860. New York, NY: ACM.

Severyn, A., and A. Moschitti. 2015. "Twitter Sentiment Analysis with Deep Convolutional Neural Networks." In *Proceedings of the 38th International ACM SIGIR Conference on Research and Development in Information Retrieval*, August 9–13, 2015, Santiago, Chile, 959–962. New York, NY: ACM.

Srivastava, N., G. Hinton, A. Krizhevsky, I. Sutskever, and R. Salakhutdinov. 2014. "Dropout: A Simple way to Prevent Neural Networks From Overfitting." *The Journal of Machine Learning Research* 15 (1): 1929–1958.

Vieweg, S. E. 2012. "Situational awareness in mass emergency: A behavioral and linguistic analysis of microblogged communications." Doctoral dissertation, University of Colorado at Boulder.

Vieweg, S., A. L. Hughes, K. Starbird, and L. Palen. 2010. "Microblogging During two Natural Hazards Events: What Twitter may Contribute to Situational Awareness." In *Proceedings of the SIGCHI Conference on Human Factors in Computing Systems*, April 10–15, 2010, Atlanta, GA, USA, 1079–1088. New York, NY: ACM.

Wang, H., B. Raj, and E. P. Xing. 2017. On the origin of deep learning. arXiv preprint arXiv:1702.07800.

Yu, L., K. M. Hermann, P. Blunsom, and S. Pulman. 2014. Deep learning for answer sentence selection. arXiv preprint arXiv:1412.1632.

Zhang, X., and Y. LeCun. 2015. Text understanding from scratch. arXiv preprint arXiv:1502.01710.

Zhang, X., J. Zhao, and Y. LeCun. 2015. "Character-level Convolutional Networks for Text Classification." In *Advances in Neural Information Processing Systems*, December 7–12, 2015, Montréal, Canada, 649–657. Red Hook, NY: Curran Associates, Inc.

A visual–textual fused approach to automated tagging of flood-related tweets during a flood event

Xiao Huang, Cuizhen Wang, Zhenlong Li and Huan Ning

ABSTRACT

In recent years, social media such as Twitter have received much attention as a new data source for rapid flood awareness. The timely response and large coverage provided by citizen sensors significantly compensate the limitations of non-timely remote sensing data and spatially isolated river gauges. However, automatic extraction of flood tweets from a massive tweets pool remains a challenge. Taking the Houston Flood in 2017 as a study case, this paper presents an automated flood tweets extraction approach by mining both visual and textual information a tweet contains. A CNN architecture was designed to classify the visual content of flood pictures during the Houston Flood. A sensitivity test was then applied to extract flood-sensitive keywords that were further used to refine the CNN classified results. A duplication test was finally performed to trim the database by removing the duplicated pictures to create the flood tweets pool for the flood event. The results indicated that coupling CNN classification results with flood-sensitive words in tweets allows a significant increase in precision while keeps the recall rate in a high level. The elimination of tweets containing duplicated pictures greatly contributes to higher spatio-temporal relevance to the flood.

1. Introduction

Floods are among the most devastating hazards on Earth. The widespread floods in Houston due to Hurricane Harvey in August 2017 have again demonstrated the power of flood and the necessity of rapid flood awareness for the first response. In recent years, crowdsourcing has become popular in collecting real-time flooding information (McDougall 2011; Triglav-Čekada and Radovan 2013). 'Citizen as sensors', firstly proposed by Goodchild (2007), describes a hypothesis of relying on the information provided voluntarily by individuals. It is highly up-to-date as eyewitnesses document their observations directly (Fohringer et al. 2015). Volunteers may contribute useful information regarding the intensity and extent of a flooding event when authoritative information is not available. Social media such as Twitter, for example, has served as an *in situ* source of flooding awareness that can largely benefit real-time flood mapping (Herfort et al. 2014; Schnebele and Waters 2014; Schnebele et al. 2014).

During a flood event, the flood-related tweets represent the textual messages with relevant keywords and pictures showing visual content of floods. If verified, both tweet sources provide real-time flood information at the tweeted locations, which is of great importance for the first responders. Especially, flood pictures provide *in situ* visual evidence of an inundated area that is superior to textual tweets in which the uncertainties need great efforts to justify (Wei et al. 2013). Additionally, good-quality flood pictures render supplemental water height information, which can feed the inundation models for real-time flood mapping and local situational awareness (Li et al. 2018). Automatic selection of flood pictures from a massive and noisy tweets pool, however, remains unsolved nowadays. Manual identification of millions of tweet pictures is labor- and time-consuming, which highly restricts its practical utilization (Feick and Roche 2013; Longueville et al. 2010; Huang, Wang, and Li 2018a).

With the development of machine learning, automatic labeling of a picture via its visual characteristics becomes possible. Common approaches include Support Vector Machines (SVMs) which learn visual features by constructing a set of hyperplanes (Suykens and Vandewalle 1999; Wu and Yap 2006); decision trees which learn features by applying a hierarchical decisive structure (Fakhari and Moghadam 2013; Jancsary et al. 2012); and Naïve Bayes which is a probabilistic learning approach by assuming independent contributions of different features to a corresponding class label (Boiman, Shechtman, and Irani 2008; McCann and Lowe 2012). Convolutional Neural Network (CNNs), a more state-of-the-art approach (Simard, Steinkraus, and Platt 2003; Simonyan and Zisserman 2014; Pinheiro and Collobert 2014), has been recently recognized as a promising technique in picture classification and labeling (Krizhevsky, Sutskever, and Hinton 2012; Wei et al. 2016). Ciresan et al. (2011) presented a flexible CNN for classifying benchmarks and handwritten digits with errors of 2.53% and 0.35%, respectively. Ciregan, Meier, and Schmidhuber (2012) developed a deep CNN that firstly achieved near-human labeling performance. Wang et al. (2016) proposed a deep CNN–RNN on multi-labeling Microsoft COCO dataset.

Advances in high-performance computers and the application of GPUs significantly accelerate the training phase as well as the labeling phase in CNNs (Ovtcharov et al. 2015; Ho, Lam, and Leung 2008). With the improved computational abilities, the CNN has become one of the most popular picture labeling techniques. Few CNN studies, however, have been conducted to examine the massive and noisy pictures in tweets. While the unique homogeneous flooding texture in a flood picture can be easily recognized by CNNs, large commission errors may occur by mistakenly including pictures with flood-like texture (e.g. a dark-grey parking lot).

Texts tweeted along with the pictures provides useful information to enhance picture labeling. The common textual-based selection approach is to apply the keyword restriction. Sakaki, Okazaki, and Matsuo (2010), for instance, use keywords like 'earthquake' or 'shaking' to collect tweets that are related to an earthquake. Vieweg et al. (2010) applied a keyword matching approach on terms like 'fire' or 'grassfire' to sort out the tweets about grassfires in Oklahoma. Other existing applications of textual-based tweets selection can be found in studies of public health (Chew and Eysenbach 2010; Dredze 2012), extreme weather conditions (Lachlan et al. 2014; Martín, Li, and Cutter 2017; Caragea et al. 2014) and political science (Skoric et al. 2012; Small 2011). For a flood event, the most reliable keywords are those like 'flooding', 'flooded', or 'flood*' that includes the variations of the base word 'flood' (Li et al. 2018; Wang, Li, and Huang 2018). Keywords such as names of the geographic location and storm have also been used to achieve better identification of flood-related tweets (Vieweg et al. 2010). However, uncertainties of keyword selection are inevitable when only these predefined keywords are used (De Albuquerque et al. 2015). A large amount of flood-related tweets may be excluded if it lacks the predefined keywords. Thus an automatic method to comprehensively encompass the most potential keywords is needed to improve the flood-related tweets selection.

Taking the 2017 Houston Flood as a study case, this paper presents a visual–textual fused approach to labeling flood pictures during the event. Trained by a set of randomly selected flood pictures online, a CNN framework is firstly developed to maximally extract flood pictures in the Houston Flood. All tweeted words that have abnormal behaviors during the flood are extracted to build a flood-sensitive words pool that is later used to refine the flood picture selection. Finally, a duplication test is applied to identify pictures with identical visual content. Duplicate pictures posted later than its original in time line are removed to ensure the high spatial- and temporal relevance. This study significantly automates the selection of tweets with flood-related visual information and flood-sensitive textual information. The automatically derived geotagged flood pictures could seed a wide range of flood studies for improved situational awareness and rapid flood response.

2. Study case and datasets

2.1. Study case

This study selected the Houston Flood on 25th August –to 1st September 2017 as our study case (Figure 1). Affected by Hurricane Harvey, the City of Houston, TX, received at least 800 mm of precipitation within a few days (Blake and Zelinsky, n.d.). The record-breaking rainfall and crippled storm water infrastructures led to a devastating and widespread flood in this city, impacting all aspects of its society. The flood caused an estimate damage of $125 billion and was responsible for at least 64 deaths with over half (36) in the metropolitan area. The study area we chose fully covers the City of Houston.

2.2. Datasets

2.2.1. Training set of flood and non-flood pictures

The performance of CNNs largely depends on its training set. In this study we developed the comprehensive training sets of flood and non-flood pictures from a variety of sources. The training set of flood pictures comes from two major online sources: searching engine and social media. Via the Google searching engine, 1000 flood pictures are downloaded using the keyword 'flood' and manually verified. Via the Baidu searching engine, another set of 1000 flood pictures is extracted using 'Flooding' and 'Hongshui' (Chinese Pinyin of 'flooding'). For pictures in social media, we target on three popular, open source social media: Twitter & Instagram and Flickr. A predefined keyword, 'flood*', is used for Twitter & Instagram. A total of 29,578 tweets with pictures from 1st December 2015 to 1st December 2017 are downloaded, among which 4400 flood pictures are verified. For Flickr, a tag 'flooding' is used and 1000 verified flood pictures are extracted. Tweets in the period from 15th August to 15th September 2017 are not used to exclude pictures about the Houston

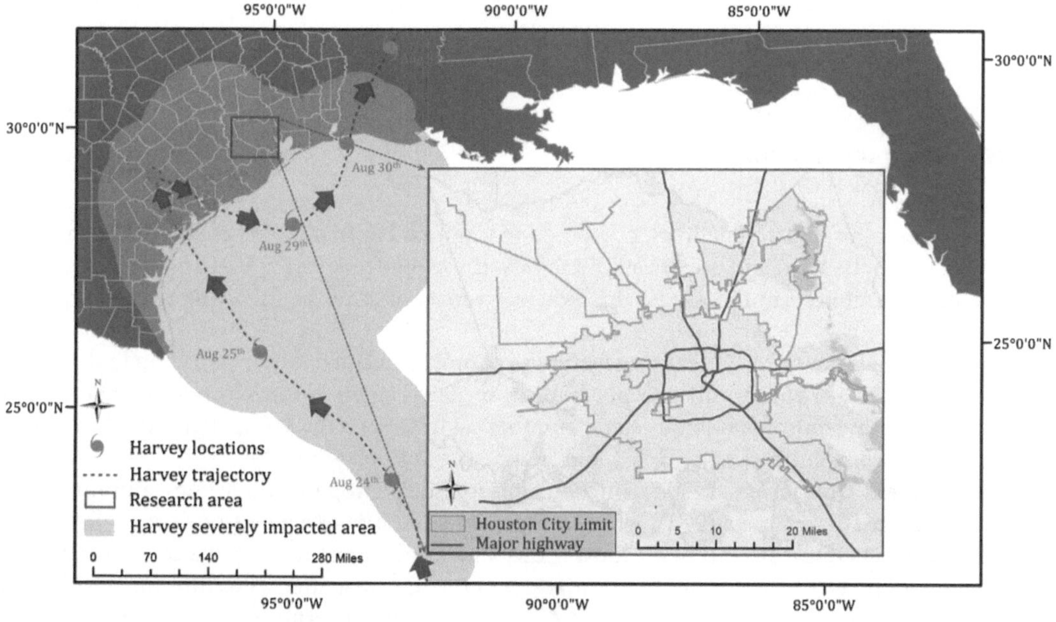

Figure 1. Hurricane Harvey and the study area of Houston, TX.

Flood from the training data selection. In total, 7400 verified flood pictures are collected and served as training inputs of the designed CNN for picture labeling. Note that retweeted pictures derived from social media are not included in this training set. Two-thirds of this dataset are used to train the CNN and one-third is used as the test set for validation analysis.

The non-flood training set contains pictures downloaded from Twitter & Instagram. To make the training sample size balanced, we randomly generate a same size set (7400) of non-flood pictures.

2.2.2. Tweets pool for word sensitivity test

To gain comprehensive understanding of the flood-sensitive words, a total of 1,048,576 tweets from 15th August to 15th September 2017 are collected across the state of Texas using Twitter Stream API (http://dev.twitter.com/streaming) and are stored in a Hadoop computer cluster. A state-wide geo-locational restriction provides more stable and reliable volunteered responses, which significantly reduces the impact of the intrinsic noises in Twitter.

2.2.3. Tweets pool for Houston Flood analysis

The geotagged tweets during the Houston Flood in the study area are extracted (retweeted posts are excluded). These tweets have the exact latitude/longitude coordinates (e.g. '29.76020, −95.39446') and therefore, provide high geocoding accuracy for better situational awareness at local level. A total of 38,992 picture-included tweets are extracted from a total of 133,762 streamed tweets during the Houston Flood period (25th August to 1st September). Both of their visual and textual information is analyzed for flood picture selection.

3. Methodology

In this study, the proposed methodology consists of three steps: (i) design a CNN and train it based on the flood and non-flood datasets derived from a variety of sources (CNN design and training). The trained CNN is then used to label the flood pictures during the Houston Flood; (ii) automatically extract the flood-sensitive keywords by applying a temporal analysis and use them to trim the CNN classified results (Word Sensitivity Test); (iii) identify the duplicated pictures and only keep the originally posted ones (Duplication Test). The final dataset is a refined collection of flood pictures with locational information during the Houston Flood. Detailed methodology overview is presented in Figure 2.

3.1. CNN design and training

3.1.1. CNN design

A CNN is a hierarchical neural network composed of input and output layers, as well as one or multiple hidden layers. Its form and utility vary upon how those hidden layers are set and realized. In the CNN architecture utilized in this study, the hidden layers consist of the following major types:

- Convolutional layers (CONV): learning features by applying a convolution operation to the input. Each convolutional neuron only accepts features in its receptive field. A convolutional layer is parameterized by four hyperparameters: number of filters (N_f), the spatial extent of those filters (F_{conv}), convoluting stride (S_{conv}) and number of zero padding (P).
- Rectified Linear Unit layers (RELU): introducing nonlinear property via a non-saturating activation function $f(x) = \max(0, x)$, where x represents the input to a neuron.
- Pooling layers (POOL): extracting the most robust and abstract feature within a convolution window in the prior layer and reducing the spatial size of the representation, thus limiting the amount of parameters used in computation and significantly controlling overfitting. It is parameterized by spatial extent (F_{pool}) and pooling stride (S_{pool}).

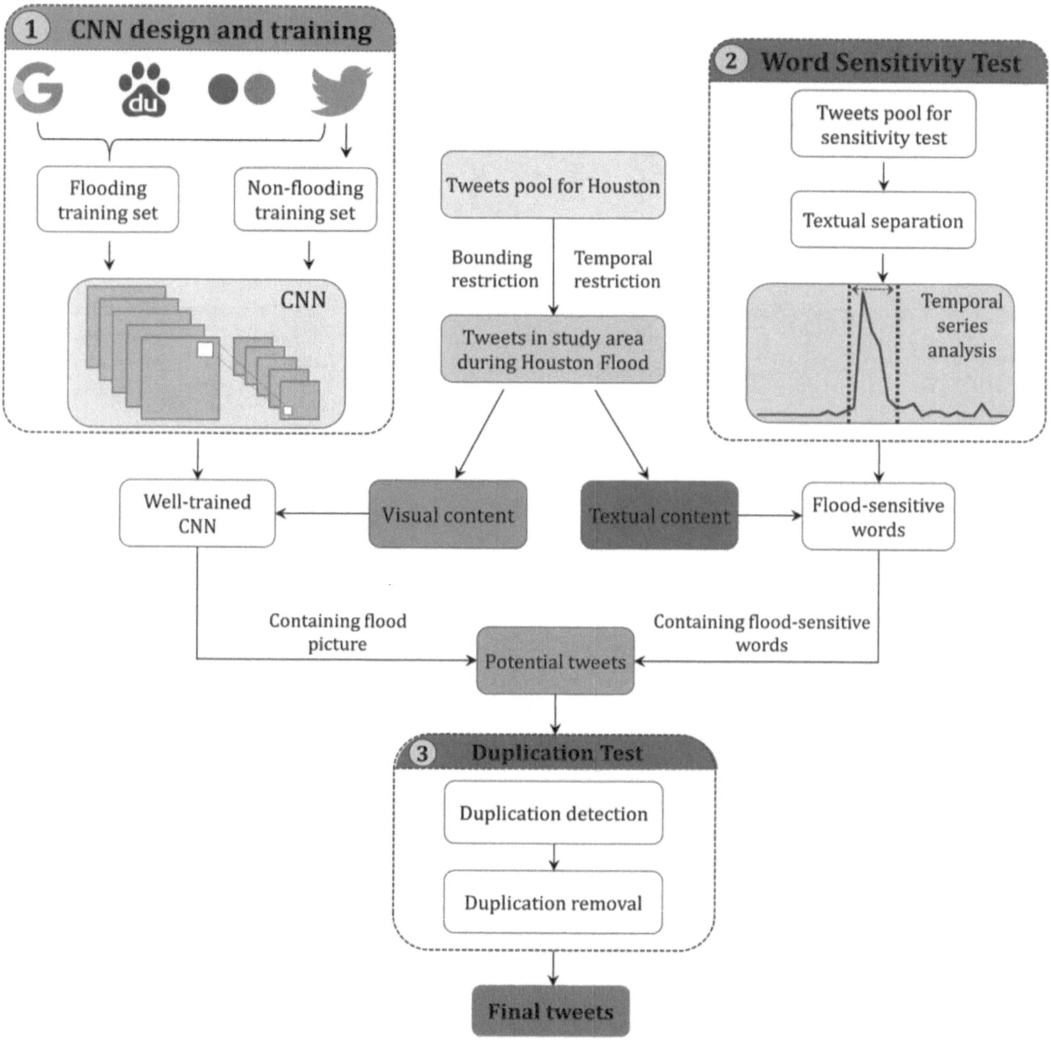

Figure 2. The workflow of the proposed approach. The methodology is composed of three steps: CNN design and training, word sensitivity test and duplication test.

- Fully connected layers (FC): flattening high dimensional features by connecting every neuron in the prior layer to next layer. A fully connected layer is parameterized by the number of its neurons (N_{fc}).

The CNN architecture designed in this study consists of eight layers (six CONVs and two FCs) (Figure 3). First, it resamples an input picture to a 3-D matrix at a 128*128*3 dimension (3 denotes that each picture has three channels, i.e. RGB), then passes it to numerous hidden layers and finally labels it as flood or non-flood. The pattern of hidden layers follows a general 'CONV–RELU–CONV–RELU–POOL–FC' scheme. The RELU layers are not presented in Figure 3 for simplicity and legibility of the figure. The double stacking of CONV aims to develop more complicated features before the pooling operation, which inevitably leads to information loss due to layer downsizing. Additionally, the progressive increase of CONV dimensions (64, 128, or 256) allows more possible representative features to be learned, greatly aiding in the richness of feature combinations. Such architectures have been commonly used in studies of machine learning for image processing (Simonyan and Zisserman 2014; Ronneberger et al. 2015).

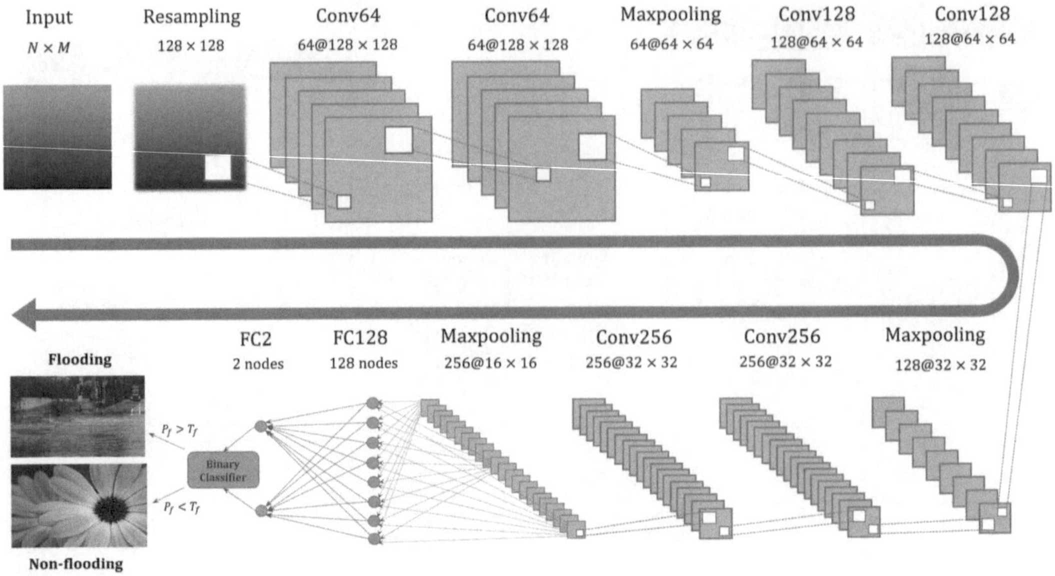

Figure 3. CNN architecture: the number before '@' denotes the depth of a layer.

The spatial pooling method in the proposed architecture is Maxpooling (Figure 3), a commonly used pooling method that finds the maximum value from each cluster of neurons at the prior layer (Nagi et al. 2011; Masci et al. 2013). It down-samples the prior layer, hence reducing the number of parameters by selecting a representative value in its matrix, in this case, the largest value. Due to the special characteristics of flood pictures, an additional Dropout layer (D-layer) is added after each CONV layer and the first FC layer (not showing in Figure 3) to prevent the overfitting problem.

The final outputs are two fully connected neurons that contain the probabilities of whether a picture is in the flood category (positive labeling) or non-flood category (negative labeling). P_f denotes the probability of positive labeling and T_f denotes the threshold for positive labeling. The picture is labeled as positive (flood) when $P_f > T_f$, and negative (non-flood) when $P_f < T_f$.

3.1.2. CNN training

To train a CNN, an organized training set is needed to feed the network. In this study, we applied a data augmentation approach to artificially enlarge the dataset, thus reducing the potential overfitting of the network. The data augmentation approach includes horizontal flipping, rotation and zooming. The augmented pictures are shuffled and resampled to a uniform size (128*128*3) before they are passed to the hidden layers. The network then learns different levels of the convoluted layers by constantly adjusting weights of its neurons via a backpropagation algorithm (Riedmiller and Braun 1993). A high-level neural networks API, Keras (Chollet 2015), is used as our coding platform and the parameter settings in our CNN are summarized in Table 1.

The total parameters in our designed CNN architecture are 9,538,506. The training of those parameters is accelerated by a NVIDIA GTX 1050 GPU with 8-GB RAM and the Computer Unified Device Architecture (CUDA). The whole training phrase finishes in 10 min.

The well-trained CNN is then applied to label tweet pictures from Houston Flood tweets pool. Tweet pictures in this pool are downloaded using Beautiful Soup, a data scratch library in Python, and they are fed directly to the network to generate their binary labels. Two categories of pictures are labeled: the flood pictures and the non-flood pictures tweeted during the Houston Flood.

Table 1. Parameter settings.

Notation	Parameter description	Value
Feeding phrase		
ϵ_z	Zooming factor	< 0.2
θ_r	Rotation factor	$< 40^\circ$
τ	Batch size	32
M	Input (resampled) image size ($M*M$)	128
N	Number of channels	3
Training phrase		
F_{conv}	Spatial extent of convolutional filters ($F_{\text{conv}} * F_{\text{conv}}$)	3
S_{conv}	Convoluting stride($S_{\text{conv}} * S_{\text{conv}}$)	2
F_{pool}	Spatial extent of pooling filters ($F_{\text{pool}} * F_{\text{pool}}$)	2
S_{pool}	Pooling stride ($S_{\text{pool}} * S_{\text{pool}}$)	2
D_r	Dropout rate of a D-layer	0.15
P	Zero padding	True[a]
e	Maximum epochs	200
l_r^i	Initial learning rate	0.00002
l_r^e	Progressive learning rate per epoch	l_r^i/epoch[b]
N_{fc}^1	Number of neurons in first FC	128
N_{fc}^2	Number of neurons in second FC	2

[a]True zero padding denotes the application of zero padding in CNN training phrase.
[b]l_r^e is a progressive variable that updates itself on each epoch.

3.2. *Word sensitivity test*

After the flood pictures are labeled by the CNN, we seek to improve the outputs by applying a restriction on their textual information. This process utilizes the textual messages of tweets, aiming to extract all the words with abnormal behaviors from a massive and noisy tweets pool during the Houston Flood. First, a textual separation is applied using white space and hashtag (#) as separators. Those segments after textual separation are organized in a database. However, using white space as separator inevitably allows other symbols (such as exclamation marks) in each segment. Here we remove all these tailing symbols while keeping the meaningful words. For example, 'flooding!!' is trimmed as 'flooding'. Moreover, we only use the segments occurring more than 100 times in the sensitivity analysis. The contribution of less frequently tweeted words is assumed trivial to flood awareness and thus not considered.

Words remained in each segment have different sensitivities regarding the Houston Flood event. Words with a remarkable increase in count exhibit a positive sensitivity to the Flood, while words mentioned remarkably less during the event exhibit a negative sensitivity. In this study, we summarize the word count in a daily level.

Due to the short acquisition period of the test dataset (15th August to 15th September) and the ubiquitous existence of extreme values, we apply a modified Z score to statistically evaluate the sensitiveness for each word. The modified Z score measures the outlier strength or how much a score differs from its median instead of mean, hence less influenced by extreme values. It is computed using the Median Absolute Deviation (MAD), modified from Iglewicz and Hoaglin (1993):

$$M_i = \frac{C_1(x_i - \tilde{x})}{\text{MAD}}, \tag{1}$$

where C_1 is a constant of 0.6745, x_i denotes the occurrence of a certain word on day i. $\tilde{x}$ denotes the sample median and M_i is the modified Z score on day i. MAD is calculated by taking the median of the absolute deviations from the median:

$$\text{MAD} = \text{median}\{|x_i - \tilde{x}|\}. \tag{2}$$

Specifically, when MAD = 0, the function is modified as

$$M_i = \frac{C_2(x_i - \tilde{x})}{\text{Mean AD}}, \tag{3}$$

$$\text{Mean AD} = \text{mean}\{|x_i - \tilde{x}|\}, \tag{4}$$

where C_2 is a constant of 0.7966. Similar as Equation (2), MeanAD is calculated by taking the mean of the absolute deviations from the median.

The modified Z scores provide the intensity of how daily count of a word deviates from the median in its time series from 15th August to 15th September. Given a single word, its overall flood sensitiveness to the Houston Flood, S_f, is the absolute value of its accumulated modified Z scores during the Flood period (25th August to 1st September):

$$S_f = \left|\sum_i M_i\right| \qquad (i \in [\text{25th August, } 1st \text{ September}]) \tag{5}$$

To bifurcate sensitive and non-sensitive words, we rank all the words based on accumulated modified Z score. Specifically, words with top 200 S_f are selected as flood-sensitive words.

For the CNN-extracted flood pictures during the Houston Flood, their textual information accompanying each tweet goes through the sensitivity test described above.

It is assumed that tweets of wrongly classified flood pictures are unlikely to contain flood-sensitive words. Therefore, only tweets that contain both labeled flood pictures and flood-sensitive words are selected into the refined tweets pool.

3.3. Duplication test

After the application of visual and textual restriction, tweets that contain positively labeled pictures and sensitive words were extracted. However, it is commonly observed that duplicate pictures are equally selected. For example, a flood picture may be shared multiple times by different users during the flood event. Unlike the original post (the first in timeline), the locational information of the tweets posting the duplicate pictures does not reflect the location and time when the picture is taken and therefore, does not help situational awareness. This process aims to identify and remove the duplicate pictures from the extracted flood pictures.

Given a picture pair, the duplicate identification algorithm computes the summation of the absolute difference in each channel:

$$S_{\text{score}} = \sum_{n}^{N} \sum_{i,j} |O_{i,j}^{n} - K_{i,j}^{n}| \tag{6}$$

where O and K represent the two pictures being compared. i and j are the row and column respectively. n denotes the channel number and N ($N = 3$) denotes the RGB channels of each picture. To make this process less computational expensive, all pictures are resampled to a size of 128*128*3 before feeding to the duplication test.

The S_{score} represents the similarity of the picture pair. A S_{score} with 0 means a perfect duplication. The higher the S_{score} is, the less similarity the picture pair has. In this study, only perfect duplications are considered. When a picture pair is identified as duplicate, the tweet with a later posting time is removed.

With a repetitive process, all tweets with duplicate pictures are removed. The final output of this study is a collection of flood-related tweets supported by their posted flood pictures.

3.4. Performance evaluation

A set of metrics, including the accuracy, loss, precision, recall and F_1 score, are used to evaluate the tweet labeling performance of using visual information only (CNN) and using both visual and textual information (CNN coupling with flood-sensitive words). For each evaluation, the classified results fall into four categories:

- True positive (TP): the number of tweets correctly identified as flood tweets
- False positive (FP): the number of tweets incorrectly identified as flood tweets
- True negative (TN): the number of tweets correctly identified as non-flood tweets
- False negative (FN): the number of tweets incorrectly identified as non-flood tweets.

The accuracy is frequently used to measure the ability of a model to differentiate binary classes, i.e. flood and non-flood in this study. It is calculated by combining TP, FP, TN, and FN:

$$\text{Accuracy} = \frac{\text{TP} + \text{TN}}{\text{TP} + \text{TN} + \text{FP} + \text{FN}} \tag{7}$$

The loss function is binary cross-entropy loss. It measures the performance of a binary labeling model as

$$\text{Loss} = -(y\log(p) + (1-y)\log(1-p)), \tag{8}$$

where y is a binary indicator (0 or 1). It equals 1 when a picture is labeled as flood and 0 when labeled as non-flood. p represents the predicted probability of a picture being labeled as flood.

Precision and recall are two indices commonly used in machine learning to quantify the performance of an algorithm. Precision measures the proportion of correctly identified flood pictures among all positive labels, or a measure of overestimation. Recall measures the proportion of correctly identified flood pictures among all flood pictures, i.e. a measure of omission:

$$\text{Precision} = \frac{\text{TP}}{\text{TP} + \text{FP}}, \tag{9}$$

$$\text{Recall} = \frac{\text{TP}}{\text{TP} + \text{FN}}. \tag{10}$$

Leveraging the precision and recall, the F_1 score is an index that measures the overall accuracy of a binary classifier. It is calculated as

$$F_1 = \frac{2}{\frac{1}{\text{precision}} + \frac{1}{\text{recall}}} \tag{11}$$

The F_1 score reaches its best value at 1 (perfect precision and recall) and the worst at 0.

4. Results and discussion

4.1. CNN performance

Accelerated by GPU and CUDA architecture, the designed CNN finished the training phrase in 10 min and 43 s through 200 epochs. The CNN reached an accuracy around 80% in both training and test sets (Figure 4a). Consecutive and stable training loss was observed throughout the training phrase (Figure 4b). The implementation of D-layer following each CONV layer has been proven efficient as it successfully prevents the model from being overfitted. Given the special characteristics in flood pictures and the ubiquitous existence of pictures with a similar texture (Feng and Sester

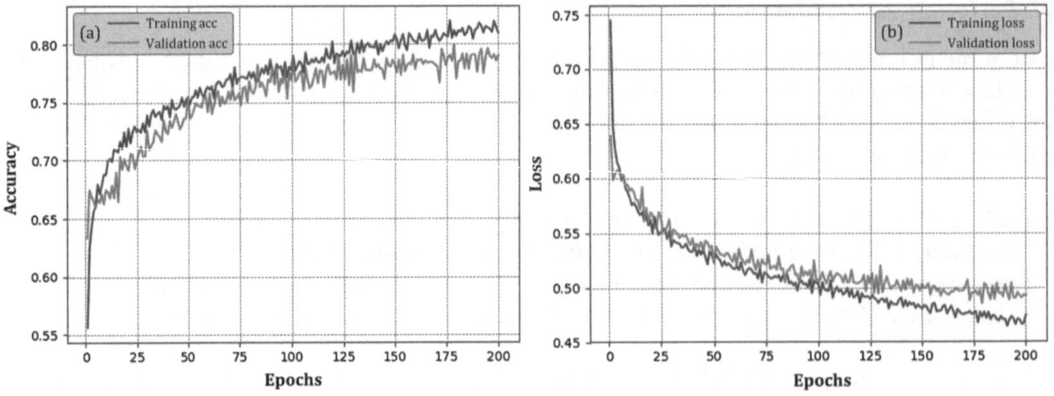

Figure 4. Training accuracy and loss of designed CNN architecture on flooding and non-flooding dataset supported by GTX 1050 GPU and CUDA.

2018), the accuracy around 80% using CNN alone is considered acceptable compared with other flood picture classification results (Avgerinakis et al. 2017; Bischke et al. 2017).

After the training, the CNN model with the optimal weight settings in its neurons was stored and then used to classify 38,992 tweet pictures derived from a total of 133,762 tweets during the Houston Flood (25th August to 1st September). To understand the performance of our model in this largely unbalanced Houston tweet dataset (the ratio between the number of flooding and non-flood pictures is much less than a balanced ratio 0.5), we gradually increased threshold T_f from 0.50 to 0.95 and recorded its performance with a varying T_f in positive labels (Figure 5). As expected, with the increasing of T_f, number of pictures with positive labels decrease dramatically (Figure 5a). Number of True positives (TPs) and False Positives (FPs) also decreases following the increasing T_f. For instance, when $T_f = 0.5$, 2545 pictures were labeled positive with 1301 TPs and 1244 FPs, while when $T_f = 0.7$, only 755 pictures were labeled positive with 663 TPs and 92 FPs. A larger T_f indicates a stricter criterion for positive labeling and results in a higher positive labeling precision at the expense of lower recall of positive labels (Figure 5b).

Figure 5(b) reveals strong negative correlation between precision and recall in positive labels, which demonstrates the dilemma of T_f setting if we purely rely on CNN for recognizing flood pictures. A lower T_f, for instance 0.5, leads to a high recall but low precision while a higher T_f, for instance 0.70, results in a low recall but high precision. In both cases, additional efforts are needed either to manually exclude the misclassified positive labels or include the misclassified negative labels. More details of negative labeling can be seen in the summary (Table 2).

In a flood event, a higher recall (less FNs) is preferred as it leads to a better coverage of true flood pictures, which is of great importance for agencies to gain a comprehensive situational awareness and for responders to evacuate as many people as possible. Given the importance of recall rate in flood events, a lower T_f value is preferred in this case. The low precision problem introduced by a low T_f value is compensated in next step.

4.2. Flood-sensitive words extraction

After the segmentation and tailing symbol removal of the tweeted messages across Texas during the Flood, 6415 words occurring over 100 times are remained in the selection. These segments contain not only regular English words but also commonly used modal particles such as 'OMG'. The summation of the modified Z score for each word S_f, represents the sensitivity of that word to the flood event. We manually categorize those words into four different sensitive levels based on their S_f value. Top 200 words with highest S_f value (ranging from 15 to 200) are then selected as flood-sensitive words.

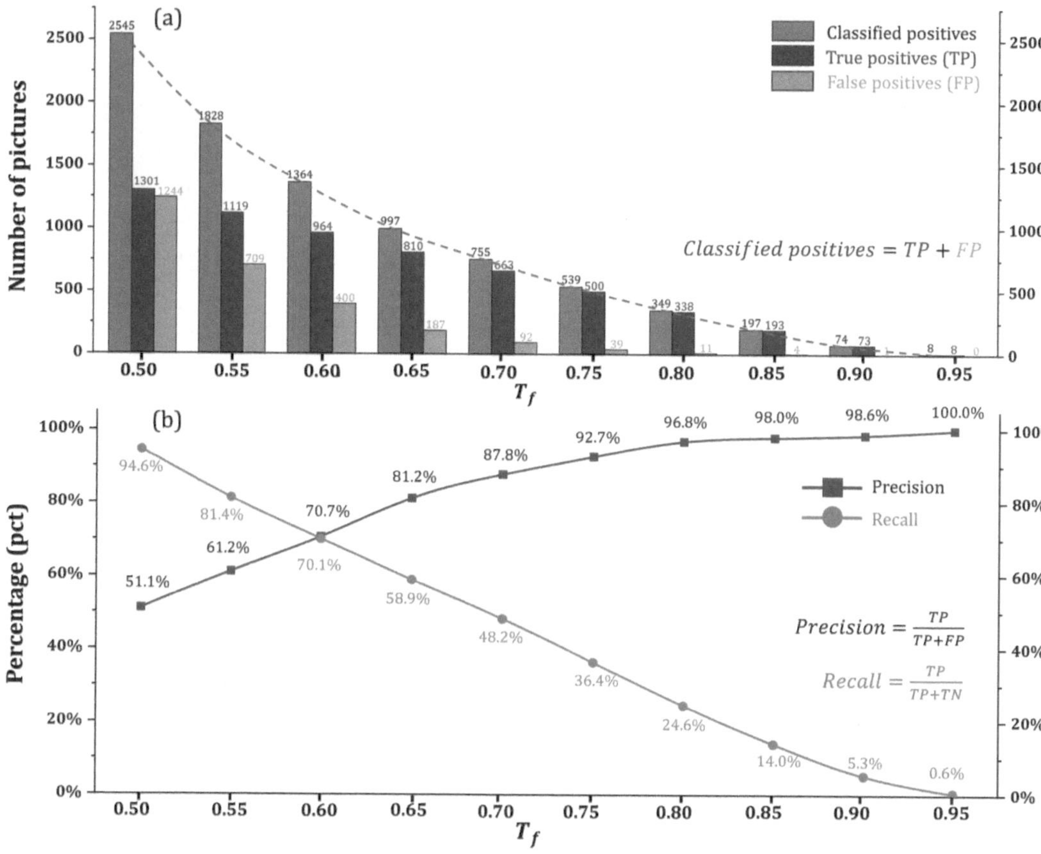

Figure 5. CNN performance with different T_f: (a) number of pictures classified as positive (flooding), verified as positive (TP) and verified as negative (FP) given different T_f and (b) trend of precision and recall given different T_f.

In Figure 6, the first tier of sensitive levels ($100 < S_f < 200$) includes words such as 'boat' ($S_f = 138.95$), 'safe' ($S_f = 126.57$) and 'raining' ($S_f = 199.77$). Words in this category were mentioned extremely more frequently during the Houston Flood. 'flood*', the most commonly used keyword to extract flood tweet in past studies (Li et al. 2018; Huang et al. 2018b), is in the second tier ($50 < S_f < 100$) (the S_f for 'flood*' is the average S_f for all the variations of 'flood'). Other words in this category include 'stuck', 'stay', 'shelters', 'storm', etc. It should be noted that tweets bearing sensitive words like 'stuck' or 'stay' may provide timely geolocational information about where potential flood victims are trapped. This information can greatly aid authorities and local responders for rapid

Table 2. CNN labeling result.

	Classified as positive			Classified as negative			Evaluation (positive)	
T_f	Total	TPs	FPs (Type I)	Total	TNs	FNs (Type II)	Precision (%)	Recall (%)
0.5	2545	1301	1244	36,447	36,373	74	51.1	94.6
0.55	1828	1119	709	37,164	36,908	256	61.2	81.4
0.6	1364	964	400	37,628	37,217	411	70.7	70.1
0.65	997	810	187	37,995	37,430	565	81.2	58.9
0.7	755	663	92	38,237	37,525	712	87.8	48.2
0.75	539	500	39	38,453	37,578	875	92.8	36.3
0.8	349	338	11	38,643	37,606	1037	96.8	24.6
0.85	197	193	4	38,795	37,613	1182	98.0	14.0
0.9	74	73	1	38,918	37,616	1302	98.6	05.3
0.95	8	8	0	38,984	37,617	1367	100.0	00.6

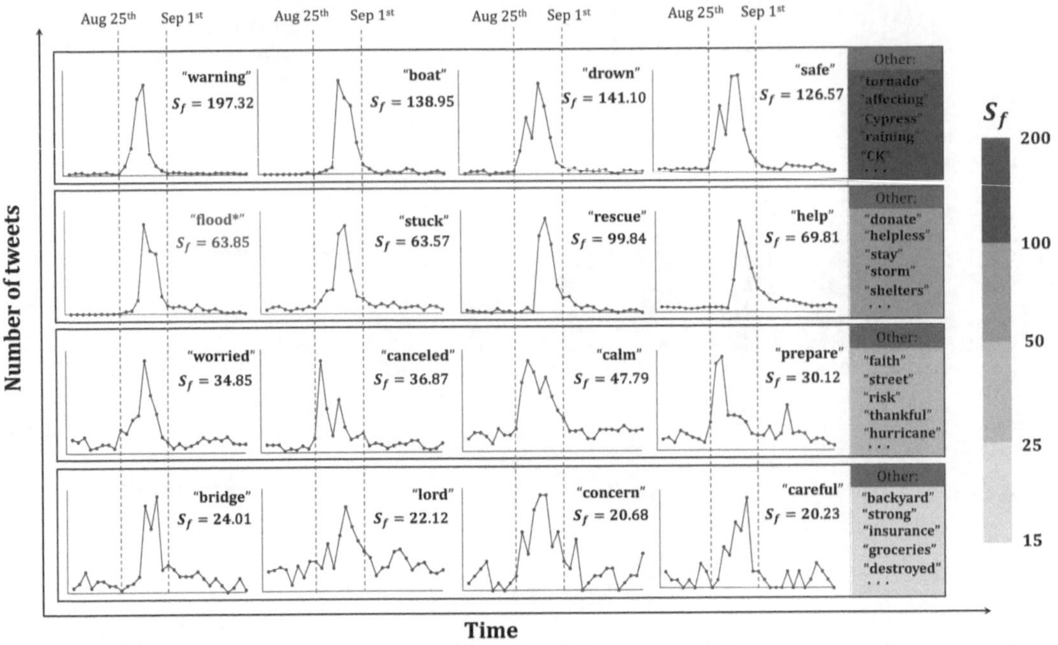

Figure 6. Temporal distribution of selected flood-sensitive words ($S_f > 15$). The *X*-axis and *Y*-axis in each subfigure denote time (from 15th August to 15th September in 2017) and number of tweets (count) respectively. Number of tweets for each word has been scaled for better visualization and comparison.

evacuation. Tweets containing words in the third ($25 < S_f < 50$) and fourth ($15 < S_f < 25$) tiers are also strongly flood related. However, they are largely ignored in traditional flood-related tweets selection processes, which only utilizes limited predefined keywords.

By quantifying the sensitivity of each word and creating an automatic, large pool of sensitive words for a flood event, this study greatly expands the scope of traditional flood selection process. As revealed in Figure 6, much more keywords are identified, which dramatically reduces the underestimation of flood tweets in previous studies. These extracted flood-sensitive words are used to refine the CNN-labeled flood pictures.

4.3. Refining CNN results based on sensitive words

After applying restrictions of sensitive words in textual tweets to address the low precision problem in CNN-labeled flood pictures, we observed a significant reduction in total positive labels after refining the CNN results with sensitive words (Table 3). After applying the textual restriction, the number

Table 3. Positive label (TPs and FPs) statistics in CNN only and CNN refined by sensitive words.

	CNN only			CNN refined by sensitive words		
T_f	Total extraction	TPs	FPs	Total extraction	TPs	FPs
0.5	**2545**	**1301**	**1244**	**1455**	**1272**	**183**
0.55	**1828**	**1119**	**709**	**1273**	**1141**	**132**
0.6	**1364**	**964**	**400**	**972**	**883**	**89**
0.65	997	810	187	812	739	73
0.7	755	663	92	658	608	50
0.75	539	500	39	513	472	41
0.8	349	338	11	321	294	27
0.85	**197**	**193**	**4**	**185**	**170**	**15**
0.9	**74**	**73**	**1**	**67**	**65**	**2**
0.95	**8**	**8**	**0**	**8**	**7**	**1**

of FPs dropped while the number of TPs remained relatively the same. The reduction is more dramatic at lower T_f thresholds. For instance, when $T_f = 0.5$, the number of FPs experienced a huge decrease from 1244 in the CNN labeling model itself to 183 after refining with sensitive words. Oppositely, the reduction of TPs is minimal (from 1301 to 1272). However, at a larger T_f value, the significant decrease in FPs over TPs becomes less obvious or even reversed. An extremely large T_f value, is not meaningful due to the limited number of TPs it can extract. The significant decrease in FPs over TPs when T_f is low indicates that coupling visual information with textual information can considerably remove non-flood tweets while keeping a majority of true flood tweets.

As shown in Table 4, the precision of the refined CNN model with sensitive words improves remarkably compared with the precision in using CNN only, especially when $T_f = 0.5$ where the precision increased by 36.3% (from 51.1% to 87.4%). The same pattern can be observed in all low T_f. Along with the significant boost in precision, the recall rate remains in a high level. An average that considers both precision and recall, F_1, also observes noticeable increased performance especially when T_f is low, which indicates more robustness in the model after refining the CNN results with sensitive words. To ensure the highest recall rate, we set parameter T_f as 0.5 for flood picture labeling in the Houston Flood.

In short, after applying the textual restriction on CNN classified results, we significantly increase the precision while keep the recall in a high level in low T_f. We successfully keep most TPs in the selection and remove majority of FPs which were wrongly classified by CNN.

4.4. Duplication removal and the flood tweets geotagging

After applying the visual-textual restriction, the figure-attached tweets that satisfy both requirements were selected and stored in a potential tweets pool. With the algorithm described in Section 3.3, we identified a total of 214 duplicate pictures in the potential tweets pool of the Houston Flood. Figure 7 shows several examples of the detected originals and their duplications in the timeline (the posting sequence). The originals, posted first in the timeline, are represented in red dots, while the duplications in blue dot. We found that pictures with the most duplications were those people used to share their concerns, for example the picture of 'PRAY FOR HOUSTON' in Figure 7, which has been repeatedly posted 34 times during the flood period. Exaggerated flood pictures that do not belong to the study area or modified flood pictures for amusement were also found heavily duplicated.

As expected, flood pictures in the original post often well indicated the flooded situation in the posted locations. The duplicate pictures fail to provide timely and accurate locational information for flood awareness. By only keeping the original post, this study significantly reduced the number of these duplicate pictures. The resulted tweets pool guarantees a high spatio-temporal relevance, which renders more confidence when assessing flood extent, evaluating flood damage and gauging evacuation.

Table 4. Precision, recall and F_1 score in CNN only and CNN refined by sensitive words.

	CNN only			CNN refined by sensitive words		
T_f	Precision	Recall	F_1	Precision	Recall	F_1
0.5	**51.1%**	**94.6%**	**0.664**	**87.4%**	**92.5%**	**0.899**
0.55	**61.2%**	**81.4%**	**0.699**	**89.6%**	**83.0%**	**0.862**
0.6	**70.7%**	**70.1%**	**0.704**	**90.8%**	**64.2%**	**0.752**
0.65	81.2%	58.9%	0.683	91.0%	53.7%	0.676
0.7	87.8%	48.2%	0.622	92.4%	44.2%	0.598
0.75	92.8%	36.4%	0.522	92.0%	34.3%	0.500
0.8	96.8%	24.6%	0.392	91.6%	21.4%	0.347
0.85	98.0%	14.0%	0.246	91.9%	12.4%	0.218
0.9	98.6%	5.3%	0.101	97.0%	4.7%	0.902
0.95	100.0%	0.6%	0.012	87.5%	0.5%	0.101

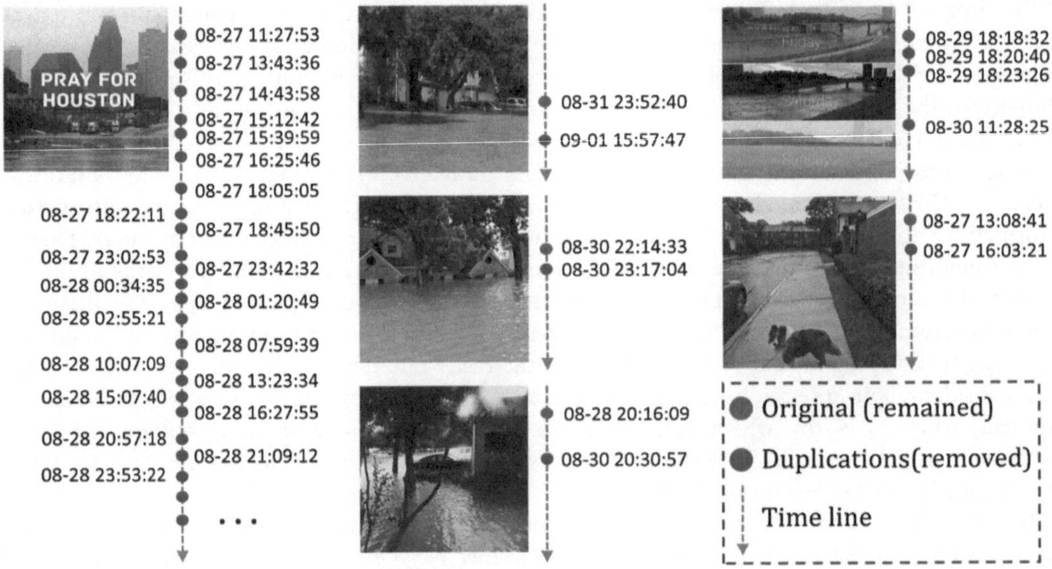

Figure 7. Examples of selected originals and their duplications presented in their timeline. Only the originals (the first in timeline) were kept in the final selection.

The final pool contains the duplication-removed tweets (a total of 1241) in the Houston Flood that meet both visual and textual restrictions. A point map of these tweets is displayed in Figure 8. Each tweet indicates the high possibility of being flooded at its location. In general, the distribution of final selection correlates well with population density as densely populated area tends to have more Twitter users, and therefore, generates more tweets.

Some geotagged tweets contain useful water height information that can be drawn from either visual content, textual content or both. If validated, these water heights could serve as hundreds of additional 'rivers gauges' that are widely distributed across the study area, providing valuable flooding awareness in such a huge area in a timely manner. The water height information derived from a huge volume of automatically generated flood tweets could largely compensate the limitations of river gauges due to its spatial isolated characteristic and water overflow induced data missing problem. Besides, given their real-time merit, the extracted flood tweets provide timely locational flood awareness, significantly assisting rapid response by local authorities and first responders to provide evacuations to those who need. Moreover, supported with local Digital Elevation Model (DEM), the geolocational information rendered by extracted flood tweets can serve as an important input to generate timely flood extent map with a large spatial coverage, significantly aiding the understanding of flood situation in a holistic way.

5. Conclusion

Nowadays social media such as Twitter has become a popular data source for rapid flood evaluation. Serving as an *in situ* source, the highly up-to-date responses and large coverage from those citizen sensors greatly benefit a variety of real-time flood analysis. Relying on both visual and textual information a tweet provides, this study presents an automated approach to tagging flood tweets during a flood event. It could be easily generalized to other flood cases with a larger Twitter sample. We collected a comprehensive training dataset of flood pictures from various online searching engines and social media and used it to train our designed CNN model. The CNN labeling results in the Houston Flood were later refined by flood-sensitive words derived from a word sensitivity test. The remaining tweets that satisfied both visual and textual requirements were further trimmed by a duplication test

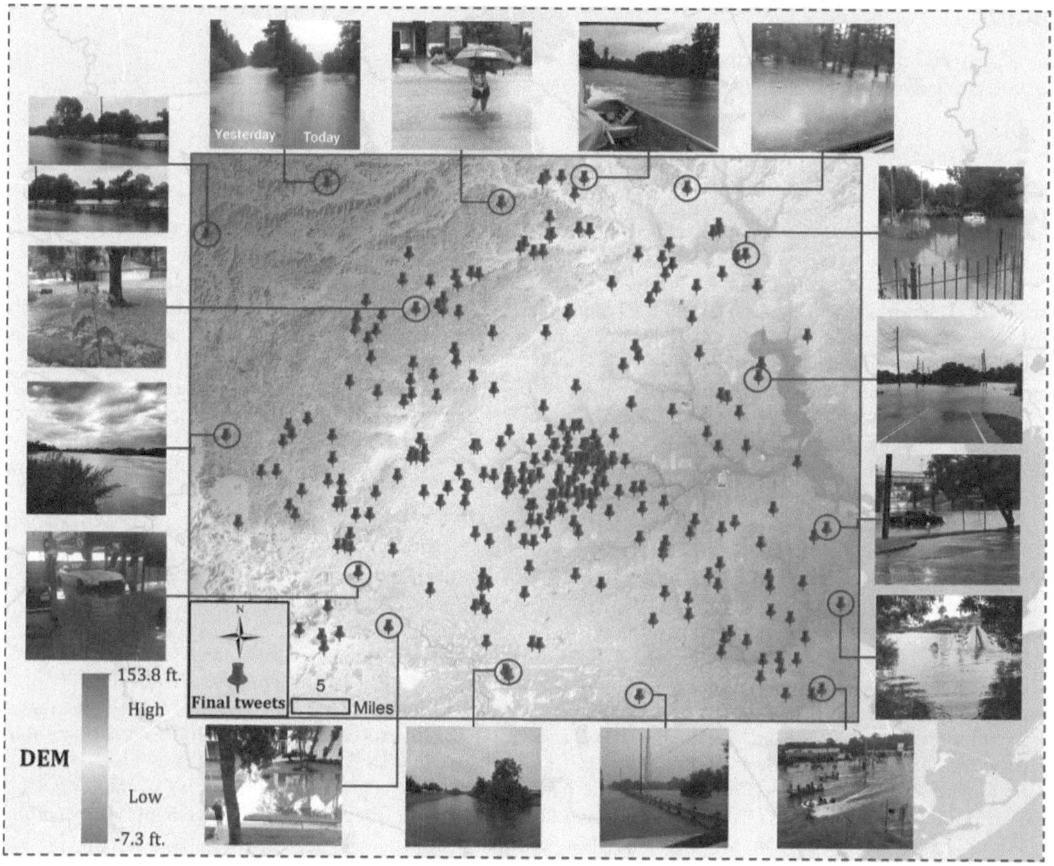

Figure 8. Selected examples of final tweets geotagging.

model. Only the originally posted tweets were remained, which composed the final tweets pool of the metropolitan Houston during the flood.

The primary findings of this study include: (i) While the commonly used keywords can effectively extract flood tweets, automatic selection of sensitive words is found practical to ensure a better flood tweets pool; (ii) The temporal distribution patterns vary in different words and the summation of the modified Z score during the flood period can effectively quantify the sensitivity of a certain word; (iii) The tradeoff between precision and recall in a CNN picture classifier can be compensated by coupling CNN classified results with flood-sensitive words, leading to a significant precision improvement while keeping the recall in a high level; and (iv) The removal of duplicated pictures guarantees the inclusion of more spatio-temporal flood relevant tweets.

The automation of geotagging flood tweets can greatly improve traditional selection process which is labor- and time-consuming. The usage of statistically extracted flood-sensitive words instead of a few predefined keywords significantly expands the searching scope and ensures a better coverage of selected flood tweets from millions of tweets in the pool. The proposed method could be generalized to other events and seed a wide range of social media based disaster studies.

Disclosure statement

No potential conflict of interest was reported by the authors.

ORCID

Xiao Huang http://orcid.org/0000-0002-4323-382X
Cuizhen Wang http://orcid.org/0000-0002-0306-9535

References

Avgerinakis, K., A. Moumtzidou, S. Andreadis, E. Michail, I. Gialampoukidis, S. Vrochidis, and I. Kompatsiaris. 2017. "Visual and Textual Analysis of Social Media and Satellite Images for Flood Detection@ Multimedia Satellite Task MediaEval 2017." Proceedings of the working notes proceeding MediaEval workshop, Dublin, Ireland (pp. 13–15).

Bischke, B., P. Bhardwaj, A. Gautam, P. Helber, D. Borth, and A. Dengel. 2017. "Detection of Flooding Events in Social Multimedia and Satellite Imagery Using Deep Neural Networks." Proceedings of the working notes proceeding MediaEval workshop, Dublin, Ireland (pp. 13–15).

Blake, E., and D. Zelinsky. n.d. "National Hurricane Center Tropical Cyclone Report: Hurricane Harvey (Rep.)." Accessed May 11, 2018 from National Hurricane Center Website: https://www.nhc.noaa.gov/data/tcr/AL092017_Harvey.pdf.

Boiman, O., E. Shechtman, and M. Irani. 2008. "Defense of Nearest-Neighbor Based Image Classification." IEEE conference on Computer Vision and Pattern Recognition (CVPR'08) (pp. 1–8). IEEE.

Caragea, C., A. C. Squicciarini, S. Stehle, K. Neppalli, and A. H. Tapia. 2014. "Mapping Moods: Geo-Mapped Sentiment Analysis During Hurricane Sandy." ISCRAM.

Chew, C., and G. Eysenbach. 2010. "Pandemics in the age of Twitter: Content Analysis of Tweets During the 2009 H1N1 Outbreak." *PloS One* 5 (11): e14118.

Chollet, F. (2015). Keras. https://keras.io/.

Ciregan, D., U. Meier, and J. Schmidhuber. 2012. "Multi-Column Deep Neural Networks for Image Classification." IEEE conference on Computer Vision and Pattern Recognition (CVPR'12), June (pp. 3642–3649). IEEE.

Ciresan, D. C., U. Meier, J. Masci, L. Maria Gambardella, and J. Schmidhuber. 2011. "Flexible, High Performance Convolutional Neural Networks for Image Classification." IJCAI'11 Proceedings of the 22nd international joint conference on Artificial Intelligence, Barcelona, Catalonia, Spain, July 16–22 (pp. 1237–1242).

De Albuquerque, J. P., B. Herfort, A. Brenning, and A. Zipf. 2015. "A Geographic Approach for Combining Social Media and Authoritative Data Towards Identifying Useful Information for Disaster Management." *International Journal of Geographical Information Science* 29 (4): 667–689.

Dredze, M. 2012. "How Social Media Will Change Public Health." *IEEE Intelligent Systems* 27 (4): 81–84.

Fakhari, A., and A. M. E. Moghadam. 2013. "Combination of Classification and Regression in Decision Tree for Multi-Labeling Image Annotation and Retrieval." *Applied Soft Computing* 13 (2): 1292–1302.

Feick, R., and S. Roche. 2013. "Understanding the Value of VGI." In *Crowdsourcing Geographic Knowledge*, 15–29. Dordrecht: Springer.

Feng, Y., and M. Sester. 2018. "Extraction of Pluvial Flood Relevant Volunteered Geographic Information (VGI) by Deep Learning from User Generated Texts and Photos." *ISPRS International Journal of Geo-Information* 7 (2): 39.

Fohringer, J., D. Dransch, H. Kreibich, and K. Schröter. 2015. "Social Media as an Information Source for Rapid Flood Inundation Mapping." *Natural Hazards and Earth System Sciences* 15 (12): 2725–2738.

Goodchild, M. F. 2007. "Citizens as Sensors: The World of Volunteered Geography." *GeoJournal* 69 (4): 211–221.

Herfort, B., J. P. de Albuquerque, S. J. Schelhorn, and A. Zipf. 2014. "Exploring the Geographical Relations Between Social Media and Flood Phenomena to Improve Situational Awareness." Connecting a Digital Europe Through Location and Place (pp. 55–71). Cham: Springer.

Ho, T. Y., P. M. Lam, and C. S. Leung. 2008. "Parallelization of Cellular Neural Networks on GPU." *Pattern Recognition* 41 (8): 2684–2692.

Huang, X., C. Wang, and Z. Li. 2018a. "A Near Real-Time Flood Mapping Approach by Integrating Social Media and Post-Event Satellite Imagery." *Annals of GIS, 24(2)*, 113–123.

Huang, X., C. Wang, and Z. Li. 2018b. "Reconstructing Flood Inundation Probability by Enhancing Near Real-Time Imagery with Real-Time Gauges and Tweets." *IEEE Transactions on Geoscience and Remote Sensing*, *56*(8), 4691–4701.

Iglewicz, B., and D. C. Hoaglin. 1993. *How to Detect and Handle Outliers*. Vol. 16. Milwaukee, WI: ASQC Quality Press.

Jancsary, J., S. Nowozin, T. Sharp, and C. Rother. 2012. "Regression Tree Fields – An Efficient, Non-Parametric Approach to Image Labeling Problems." 2012 IEEE conference on Computer Vision and Pattern Recognition (CVPR) (pp. 2376–2383). IEEE.

Krizhevsky, A., I. Sutskever, and G. E. Hinton. 2012. "ImageNet Classification with Deep Convolutional Neural Networks." Proceedings of the 25th International Conference on Neural Information Processing Systems, Lake Tahoe, NV, December 3–6 (pp. 1097–1105).

Lachlan, K. A., P. R. Spence, X. Lin, and M. Del Greco. 2014. "Screaming Into the Wind: Examining the Volume and Content of Tweets Associated with Hurricane Sandy." *Communication Studies* 65 (5): 500–518.

Li, Z., C. Wang, C. T. Emrich, and D. Guo. 2018. "A Novel Approach to Leveraging Social Media for Rapid Flood Mapping: A Case Study of the 2015 South Carolina Floods." *Cartography and Geographic Information Science* 45 (2): 97–110.

Longueville, B. D., G. Luraschi, P. Smits, S. Peedell, and T. D. Groeve. 2010. "Citizens as Sensors for Natural Hazards: A VGI Integration Workflow." *Geomatica* 64 (1): 41–59.

Martín, Y., Z. Li, and S. L. Cutter. 2017. "Leveraging Twitter to Gauge Evacuation Compliance: Spatiotemporal Analysis of Hurricane Matthew." *PLoS One* 12 (7): e0181701.

Masci, J., A. Giusti, D. Ciresan, G. Fricout, and J. Schmidhuber. 2013. "A Fast Learning Algorithm for Image Segmentation with Max-Pooling Convolutional Networks." 20th IEEE International Conference on Image Processing (ICIP'13) (pp. 2713–2717). IEEE.

McCann, S., and D. G. Lowe. 2012. "Local Naive Bayes Nearest Neighbor for Image Classification." 2012 IEEE conference on Computer Vision and Pattern Recognition (CVPR), June (pp. 3650–3656). IEEE.

McDougall, K. 2011. "Using Volunteered Information to Map the Queensland Floods." Proceedings of the 2011 Surveying and Spatial Sciences Conference: Innovation in Action: Working Smarter (SSSC'11) (pp. 13–23). Surveying and Spatial Sciences Institute.

Nagi, J., F. Ducatelle, G. A. Di Caro, D. Cireşan, U. Meier, A. Giusti, and L. M. Gambardella. 2011. "Max-Pooling Convolutional Neural Networks for Vision-Based Hand Gesture Recognition." 2011 IEEE International Conference on Signal and Image Processing Applications (ICSIPA), November (pp. 342–347). IEEE.

Ovtcharov, K., O. Ruwase, J. Y. Kim, J. Fowers, K. Strauss, and E. S. Chung. 2015. "Accelerating Deep Convolutional Neural Networks Using Specialized Hardware." *Microsoft Research Whitepaper* 2 (11): 1–4.

Pinheiro, P., and R. Collobert. 2014. "Recurrent Convolutional Neural Networks for Scene Labeling." *International Conference on Machine Learning* 32: 82–90.

Riedmiller, M., and H. Braun. 1993. "A Direct Adaptive Method for Faster Backpropagation Learning: The RPROP Algorithm." IEEE international conference on Neural Networks (pp. 586–591). IEEE.

Ronneberger, O., P. Fischer, and T. Brox. 2015. "U-NET: Convolutional Networks for Biomedical Image Segmentation." International conference on Medical Image Computing and Computer-assisted Intervention, October (pp. 234–241). Cham: Springer.

Sakaki, T., M. Okazaki, and Y. Matsuo. 2010. "Earthquake Shakes Twitter Users: Real-Time Event Detection by Social Sensors." Proceedings of the 19th International Conference on World Wide Web, April (pp. 851–860). ACM.

Schnebele, E., G. Cervone, S. Kumar, and N. Waters. 2014. "Real Time Estimation of the Calgary Floods Using Limited Remote Sensing Data." *Water* 6 (2): 381–398.

Schnebele, E., and N. Waters. 2014. "Road Assessment After Flood Events Using Non-Authoritative Data." *Natural Hazards and Earth System Sciences* 14 (4): 1007.

Simard, P. Y., D. Steinkraus, and J. C. Platt. 2003, August. "Best Practices for Convolutional Neural Networks Applied to Visual Document Analysis." *ICDAR* 3: 958–962.

Simonyan, K., and A. Zisserman. 2014. "Very Deep Convolutional Networks for Large-Scale Image Recognition." arXiv preprint arXiv:1409.1556.

Skoric, M., N. Poor, P. Achananuparp, E. P. Lim, and J. Jiang. 2012. "Tweets and Votes: A Study of the 2011 Singapore General Election." 45th Hawaii International Conference on System Science (HICSS'12), January (pp. 2583–2591). IEEE.

Small, T. A. 2011. "What the Hashtag? A Content Analysis of Canadian Politics on Twitter. Information." *Communication & Society* 14 (6): 872–895.

Suykens, J. A., and J. Vandewalle. 1999. "Least Squares Support Vector Machine Classifiers." *Neural Processing Letters* 9 (3): 293–300.

Triglav-Čekada, M., and D. Radovan. 2013. "Using Volunteered Geographical Information to Map the November 2012 Floods in Slovenia." *Natural Hazards and Earth System Sciences* 13 (11): 2753.

Vieweg, S., A. L. Hughes, K. Starbird, and L. Palen. 2010. "Microblogging During Two Natural Hazards Events: What Twitter May Contribute to Situational Awareness." Proceedings of the SIGCHI conference on Human Factors in Computing Systems, April (pp. 1079–1088). ACM.

Wang, C., Z. Li, and X. Huang. 2018. "Geospatial Assessment of Wetness Dynamics in the October 2015 SC Flood with Remote Sensing and Social Media." *Southeastern Geographer* 58 (2): 164–180.

Wang, J., Y. Yang, J. Mao, Z. Huang, C. Huang, and W. Xu. 2016. "CNN-RNN: A Unified Framework for Multi-Label Image Classification." 2016 IEEE conference on Computer Vision and Pattern Recognition (CVPR) (pp. 2285–2294). IEEE.

Wei, Z., J. Chen, W. Gao, B. Li, L. Zhou, Y. He, and K. F. Wong. 2013. "An Empirical Study on Uncertainty Identification in Social Media Context." Proceedings of the 51st Annual Meeting of the Association for Computational Linguistics (Volume 2: Short Papers) (pp. 58–62).

Wei, Y., W. Xia, M. Lin, J. Huang, B. Ni, J. Dong, and S. Yan. 2016. "HCP: A Flexible CNN Framework for Multi-Label Image Classification." *IEEE Transactions on Pattern Analysis and Machine Intelligence* 38 (9): 1901–1907.

Wu, K., and K. H. Yap. 2006. "Fuzzy SVM for Content-Based Image Retrieval: A Pseudo-Label Support Vector Machine Framework." *IEEE Computational Intelligence Magazine* 1 (2): 10–16.

Rapid estimation of an earthquake impact area using a spatial logistic growth model based on social media data

Yandong Wang, Shisi Ruan, Teng Wang and Mengling Qiao

ABSTRACT
Rapid estimates of impact areas following large earthquakes constitute the cornerstone of emergency response scenarios. However, collecting information through traditional practices usually requires a large amount of manpower and material resources, slowing the response time. Social media has emerged as a source of real-time 'citizen-sensor data' for disasters and can thus contribute to the rapid acquisition of disaster information. This paper proposes an approach to quickly estimate the impact area following a large earthquake via social media. Specifically, a spatial logistic growth model (SLGM) is proposed to describe the spatial growth of citizen-sensor data influenced by the earthquake impact strength after an earthquake; a framework is then developed to estimate the earthquake impact area by combining social media data and other auxiliary data based on the SLGM. The reliability of our approach is demonstrated in two earthquake cases by comparing the detected areas with official intensity maps, and the time sensitivity of the social media data in the SLGM is discussed. The results illustrate that our approach can effectively estimate the earthquake impact area. We verify the external validity of our model across other earthquake events and provide further insights into extracting more valuable earthquake information using social media.

1. Introduction

Large earthquakes often occur suddenly with little warning; consequently, disaster management can be a difficult task. A quick and efficient estimation of the impact area in which an earthquake will generate casualties and property damage is essential for emergency response protocols and loss estimation (Badal, Vázquez-prada, and González 2005). However, obtaining such information quickly is often challenging, and substantial amounts of manpower and material resources are needed to collect real-time information through traditional data acquisition practices such as spot investigations, telephone reporting or remote sensing (Panagiota, Jocelyn, and Erwan 2011; Li et al. 2018).

Because an increasing number of people use the Internet to publish and share information following natural disasters, it seems intuitive to employ the social web ('Web 2.0') as a new information source (Goodchild 2007; Goodchild and Glennon 2010). Unlike traditional data, these online user-generated data (e.g. Twitter messages and Flickr photos) enable low-cost data acquisition on an unprecedented temporal scale during disaster emergencies; thus, these data cannot be ignored in disaster-related decision-making endeavors (Enrico, Porto, and Alexander 2015; Huang and Xiao 2015; Kryvasheyeu et al. 2016).

Of particular interest here is that humans can be regarded as citizen sensors when they share their observations and views using mobile devices and social media platforms. The online messages generated by citizen sensors pertaining to real-world events are considered 'citizen-sensor data', which can be used to monitor user activities and ongoing event indicators (Goodchild 2007; Nagarajan et al. 2009; Sheth 2009). Particularly, during disaster emergencies, real-time online sensor data can promote the identification of disaster signals and enhance situational awareness (Earle, Bowden, and Guy 2011; Amit et al. 2014).

In earthquake management, the feedback from citizen sensors after an earthquake can effectively and qualitatively describe the ground motion effects and provide a possible way to identify the area throughout which an earthquake will generate casualties and property damage (Atkinson and Wald 2007). A number of recent research efforts have shown that online citizen-sensor data such as social media messages and other crowdsourcing data can be employed to explore the impact area after an earthquake (Crooks et al. 2013; Resch, Usländer, and Havas 2018). The abovementioned studies performed in this field used spatial clustering techniques and visualization approaches in spatial analysis to intuitively detect the impact areas; however, they generally focused on data hotspots while ignoring the spatial growth processes of citizen-sensor data following an earthquake, and they failed to reveal the relationship between the spatial distribution of the data and the impact area.

In this article, we argue that the spatial growth trend of post-earthquake citizen-sensor data influenced by the earthquake impact strength is similar to the logistic growth trend of a biological population, which can be quantitatively described. Based on this finding, we present an approach to quickly detect the impact area after a large earthquake via social media. Specifically, a theoretical model known as the spatial logistic growth model (SLGM) based on a logistic function is proposed to describe the spatial growth processes of citizen-sensor data following an earthquake, and the connection between the spatial growth trend of citizen-sensor data and the earthquake impact area is quantified. Furthermore, a framework is provided based on the SLGM to rapidly estimate an earthquake's impact area by combining social media data with other auxiliary data.

To demonstrate the reliability of our approach, we implemented the proposed framework for two large earthquake cases and compared the detection results with official intensity maps from the United States Geological Survey (USGS). Moreover, we discussed the temporal sensitivity of the SLGM to social media data. The experimental results illustrate that the proposed SLGM is suitable for describing the spatial growth processes of citizen-sensor data and for effectively estimating the impact area in a short time. Our approach could represent an alternative option for authorities to better cope with earthquake emergencies in a timely manner. The external validity of the SLGM was verified across 18 other earthquake events, the limitations of the model were evaluated, and an additional text analysis was performed to obtain further insights into the quick extraction of more valuable information regarding earthquake emergencies.

This paper is structured as follows. Section 2 provides background information for the current work related to the use of social media for detecting earthquake information and the logistic function. Section 3 illustrates the construction of our model, while section 4 describes the proposed framework for quickly estimating the earthquake impact area via social media based on the proposed model. Section 5 presents two case studies to prove the feasibility of our approach. Section 6 discusses the performance and limitations of our model, and section 7 concludes with possible perspectives and extensions of our research.

2. Background

2.1. Social media analysis for detecting earthquake emergency information

The global diffusion of the Internet and citizen crowdsourcing has allowed social media to become a hotspot for detecting earthquake disaster information at a low cost (Steiger, Westerholt, and Zipf 2016). Given the tremendous volume of social media data, such data are limited in the information provided is not very precise or important (Ryoo and Moon 2014). However, this deficiency does not diminish the enthusiasm of scholars who study the detection of earthquake information based on social media data (Goolsby 2010). Over the past few years, numerous studies have examined the use of social media data for detecting earthquake emergency information. The main goals of these studies can be grouped into two types: real-time earthquake event detection and damage assessment.

For detection purposes, time series analysis and burst detectors based on a binomial model have been adopted to identify real-time seismic events (Sakaki, Okazaki, and Matsuo 2010; Sakaki, Toriumi, and Matsuo 2011; Robinson, Power, and Cameron 2013). Furthermore, Earle et al. (2010) assessed the speed with which tweeters responded to the 30 March 2009 Morgan Hill earthquake and found that monitoring the earthquake by Twitter was faster than traditional monitoring. Similarly, Guy et al. (2010) explored Twitter to detect seismic events and developed a software application called the Twitter Earthquake Detector (TED) for earthquake emergencies, and Fontugne et al. (2011) detected earthquake occurrences from the behaviors of social media users during the outbreak of natural disasters.

In studies on damage assessment techniques, methods such as text mining, natural language processing techniques (e.g. support vector machines (SVMs), naive Bayesian classifications, and decision trees), spatial clustering analysis, network analysis, and visualization approaches have been applied (Sakaki, Okazaki, and Matsuo 2010; Judith and Nikolai 2011; Murakami and Nasukawa 2012). For example, Kim, Lee, and Zettsu (2011) designed an mTrend system, which constructs and visualizes spatiotemporal trends in topics, based on geotagged tweets after the Tokyo earthquake in 2011 to explore the locations of and changes in disaster hotspots. Moreover, Fontugne et al. (2011) described the nature of these events from tags posted on the Flickr website and evaluated the aftermaths of such disasters.

In the particular field of earthquake impact area detection, many researchers have attempted to analyze the areal extent of earthquake damage from social media. Some scholars utilized text analysis to determine the intensity level of each social media text and then used spatial interpolation to construct an earthquake intensity map (Cresci, Cimino, et al. 2015; Yang et al. 2016). However, despite the volume of data, only a few messages contained information regarding the damage. Crooks et al. (2013) analyzed Twitter messages during the 2011 earthquake in Mineral, Virginia, to discern activity patterns, and they found that tweets within 10 min following an earthquake provided a good approximation of the impact area; however, their study lacked a quantitative basis, as they used only visualization methods to analyze the impact area. Burks, Miller, and Zadeh (2014) used historical official intensity observations and earthquake-related tweets to construct regression models and then used those models to construct an earthquake shaking intensity map. Kropivnitskaya et al. (2017) produced an accurate intensity map in real time by analyzing the relationship between the tweet rates and the earthquake intensity observations within 10 min after the earthquake. However, due to technological limitations, historical seismic data, ground motion recordings and corresponding social media data are not easily accessible. In addition, these methods exhibit some restrictions under different geographical conditions.

It should be noted that while these previous studies demonstrated great potential for using social media to identify earthquake-impacted areas, they nevertheless ignored compelling theoretical evidence on the correlation between the spatial distribution of data and the impact area.

2.2. Logistic function

The logistic function, which commonly exhibits an S-shaped (i.e. sigmoid-shaped) curve, boasts a lengthy history in statistics and modeling. The logistic function was proposed by the Belgian mathematician Verhulst, who studied it in relation to population growth (Tsoularis and Wallace 2002). Figure 1 displays the logistic function curve, the equation for which is as follows:

$$F(x) = \frac{K}{1 + e^{-l(x-x_0)}} \tag{1}$$

where e is the natural logarithm base, x_0 represents the x-value of the sigmoid midpoint, l is the steepness of the curve, and K is the carrying capacity. The growth rate first increases and then begins to decelerate when x_0 is reached, but the F value continues to increase until it reaches an asymptote, thereby forming an S-shaped curve.

Researchers in a variety of disciplines have found that not only biological processes but also phenomena in other fields exhibit this S-shaped growth pattern (Bhowmick and Bhattacharya 2014). The simple structure of the classic logistic growth function has attracted the attention of many scientists who are eager to model data with S-shaped curves in their own work. Thus, the main emphasis of the logistic growth model is no longer the evolution of biological populations but rather the application of a model to real datasets from different disciplines, including demography (Dawed, Koya, and Goshu 2014), economic statistics (Tourres 2003), bacterial growth (Baranyi and Pin 2001), and behavioral ecology (Symonds and Moussalli 2011).

The logistic function has also appeared in earthquake management research related to social media (Zhu et al. 2011; Xu et al. 2012). In this field, the logistic function is used in regression models for message classification and landslide hazard mapping purposes. For instance, some studies used logistic regression to determine whether a set of tweets can describe target earthquake events (Carlos, Marcelo, and Barbara 2013; Sakaki, Okazaki, and Matsuo 2013). Additionally, others have used logistic regression to build a model of the factors affecting earthquake bursts based on tweets (Mizunuma et al. 2014). Furthermore, the logistic function has been used in neural network models for landslide susceptibility mapping to delineate landslide hazard areas following an earthquake occurrence (Xu et al. 2013).

These researchers applied the logistic function during several phases of analysis of social media data after an earthquake, and they especially focused on its regression and prediction abilities. In contrast, few studies have applied the logistic function to analyze the spatial growth of the distribution of social media data. In our work, we focus on the growth trend of the logistic function and use it to model the spatial growth of the social media citizen-sensor data distribution after an earthquake.

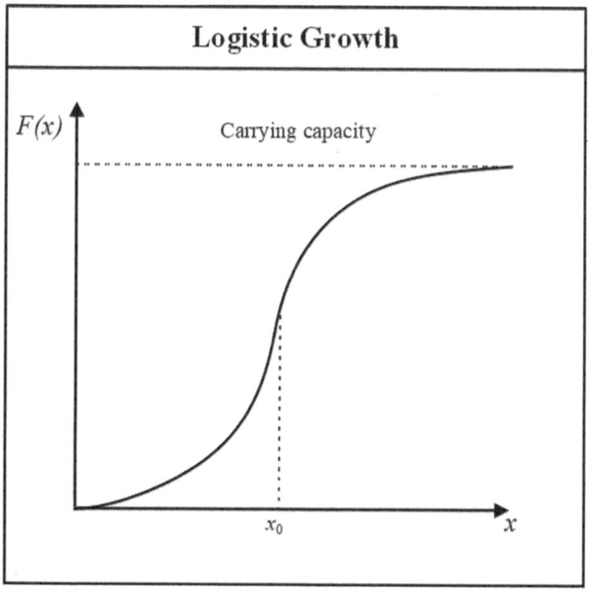

Figure 1. The logistic function curve.

3. Spatial logistic growth model

We consider a generic model to reflect the spatial growth process of citizen-sensor data following an earthquake based on our online citizen responses assumptions, in which the connection between the data spatial growth trend and the earthquake impact area can be quantitatively described. Specifically, we find that the spatial growth trend of citizen-sensor data is similar to the logistic growth of a biological population. In virtue of the population logistic growth, we develop the proposed SLGM to describe the spatial growth process of citizen-sensor data. To better understand the proposed SLGM, we first introduce the model variables and analyze the data spatial growth trend, after which we present the proposed model.

3.1. Model variables

Our SLGM is a theoretical model that is used to delineate the spatial growth of the distribution of citizen-sensor data after an earthquake. The proposed model is presented as the mathematical evolution of the normalized citizen-sensor (by population) accumulation as a function of the distance from the earthquake epicenter. The variables included in our model are illustrated in Figure 2.

In Figure 2a, the red point (c) represents the epicenter, and the blue points depict the citizen-sensor data. The interval between any two adjacent circles is constant. Each differently colored annulus represents a single region corresponding to a distance interval from the epicenter, where r represents the distance from the epicenter. The amount of citizen-sensor data in the region of an annulus (c; r_i, r_{i+1}) is represented by $N_{ri\text{-}ri+1}$. The population size in the region of an annulus (c; r_i, r_{i+1}) is represented by $P_{ri\text{-}ri+1}$, while $NP_{ri\text{-}ri+1}$ is the amount of data normalized by the population (i.e. the higher the population is, the higher the probability of citizen-sensor data in the region). The amount of normalized data within a radius r is represented by MP, which is actually the accumulation of NP. Figure 2b displays a table of the mathematical expressions of these variables.

3.2. Citizen-sensor data spatial growth trend analysis

Some researchers have discovered a certain connection between the willingness of people to publish relevant online messages after an earthquake and their distance from the epicenter (Sakaki, Okazaki, and Matsuo 2010; Sakaki, Toriumi, and Matsuo 2011). This connection ensures that citizen-sensor data is regularly spatially distributed and can thus be quantitatively described. Sakaki, Toriumi, and Matsuo (2011) analyzed how people used Twitter after earthquakes in Japan. They found that in heavily damaged areas close to the epicenter, people were trapped in situations that were too difficult to post sensor readings or use the Internet; regarding people in damaged areas, those who were

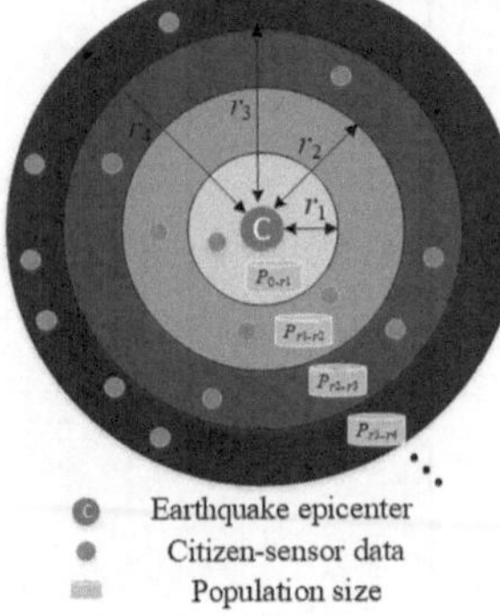

(a)

Distance from the epicenter (r)	Range	Sensor reading count (N)	Population count (P)	Normalized data count (NP)	Normalized data accumulation (MP)
r_1	$(0, r_1)$	$N_{0\text{-}r1} = 1$	$P_{0\text{-}r1}$	$NP_{0\text{-}r1} = N_{0\text{-}r1} / P_{0\text{-}r1}$	$MP_{r1} = NP_{0\text{-}r1}$
r_2	(r_1, r_2)	$N_{r1\text{-}r2} = 3$	$P_{r1\text{-}r2}$	$NP_{r1\text{-}r2} = N_{r1\text{-}r2} / P_{r1\text{-}r2}$	$MP_{r2} = MP_{r1} + NP_{r1\text{-}r2}$
r_3	(r_2, r_3)	$N_{r2\text{-}r3} = 5$	$P_{r2\text{-}r3}$	$NP_{r2\text{-}r3} = N_{r2\text{-}r3} / P_{r2\text{-}r3}$	$MP_{r3} = MP_{r2} + NP_{r2\text{-}r3}$
r_4	(r_3, r_4)	$N_{r3\text{-}r4} = 6$	$P_{r3\text{-}r4}$	$NP_{r3\text{-}r4} = N_{r3\text{-}r4} / P_{r3\text{-}r4}$	$MP_{r4} = MP_{r3} + NP_{r3\text{-}r4}$
⋮	⋮	⋮	⋮	⋮	⋮

(b)

Figure 2. Schematic diagram of each variable: (a) data distribution simulation; and (b) corresponding statistics.

located far from the earthquake epicenter were more inclined to post relevant tweets than people who were closer to the epicenter; finally, people in safe areas were less inclined to post geolocated tweets than people in damaged areas during the initial stages of earthquake propagation.

Based on their findings, we conclude the following: in the earthquake impact area, with an increase in the distance from the earthquake epicenter (r), the amount of normalized citizen-sensor data (NP) and the data spatial growth rate (dMP/dr) will increase, as people farther away from the epicenter are more likely to post sensor data; when r reaches a critical distance, NP and the data spatial growth rate begin to decrease until the growth rate reaches 0, as people are less inclined to post relevant geolocated data in the area where the earthquake may not cause casualties and damage. The growth rate first increases and then begins to decrease when a certain distance is reached; this characteristic gives the relationship between the distance from the earthquake epicenter (r) and the normalized data accumulation (MP) an 'S-shape'.

This S-shaped growth curve, which represents the abovementioned mathematical logistic curve, is often used to describe the logistic growth of a biological population: the population growth is rapid at the beginning, after which the growth begins to slow down when a certain indicator is reached, and then the growth stops completely (Tsoularis and Wallace 2002). Our assumption regarding the spatial growth of citizen-sensor data is similar to the logistic growth of biological populations over time. As Re (2000) noted, 'a statement need not match reality in all its aspects, relations, and/or details, but has to match it only as far as it goes'. Time and space are two basic forms of physical movement; therefore, applying a time-based theoretical approach to spatial analysis can enrich spatial data mining endeavors (Wang et al. 2007). Thus, we construct an analogy between the change in the biological population size over time and the change in the citizen-sensor data size over a given distance, thereby establishing a spatial growth model for citizen-sensor data after an earthquake.

3.3. Construction of the spatial logistic growth model

The three key features of logistic population growth are the following: a population must have a numerical upper bound on the growth size called the carrying capacity K; the growth rate (the increase in the population size per unit time) first increases and then decreases with increasing population size; and the population at the inflection point (where the growth rate is a maximum) is exactly half the carrying capacity. Similar to these, our model is developed under the following assumptions:

(1) There is an upper limit to the number of the normalized citizen-sensor data the spatial environment can support; this limit is defined as the carrying capacity K of the spatial environment.
(2) The data spatial growth rate (dMP/dr) first increases and then decreases with an increase in the accumulation of normalized sensor data MP.
(3) The data accumulation at the inflection point (where the spatial growth rate is a maximum, at which r reaches the impact area critical distance) is exactly half the carrying capacity $K/2$.

Figure 3 shows the spatial growth pattern of citizen-sensor data. We use a simple structure of the classic logistic growth function to model our S-shaped citizen-sensor data growth curve. The SLGM is defined as follows:

$$MP(r) = \frac{K}{1 + e^{-l_0(r - r_m)}} \tag{2}$$

All of the symbols used to describe the model formulation are listed in Table 1.

r_m is the epicentral distance indicator of the impact area. In Figure 3, we mark r_m with a red dashed line for easy reference. To the left of the red dashed line, the spatial growth rate increases, whereas to the right of the red dashed line, the spatial growth rate begins to decelerate; the data accumulation MP keeps growing until it reaches K. The safe area where the earthquake will not cause casualties and property damage resides at distances r that exceed r_m; meanwhile, distances r

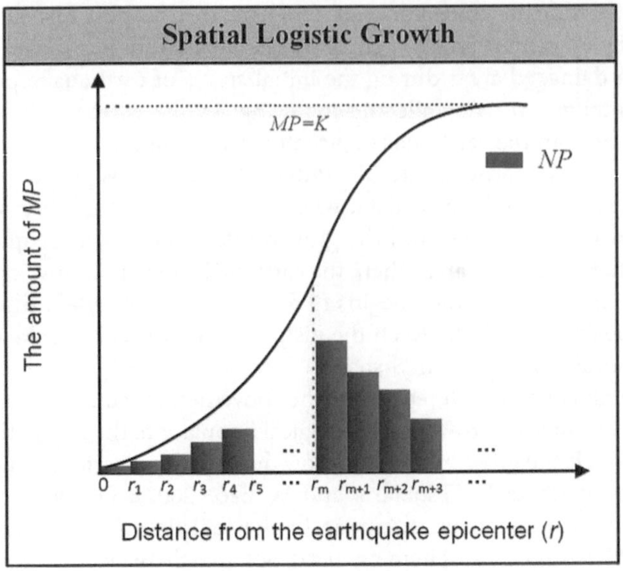

Figure 3. Spatial logistic growth of citizen-sensor data: the growth patterns of *NP* (the amount of normalized data) and *MP* (the accumulation of normalized data) are functions of the distance from the earthquake epicenter (*r*).

that are less than r_m belong to the impact area where the earthquake will cause casualties and property damage.

The reliability of this proposed theoretical model will be validated in section 5 and section 6.

4. Framework implementation

This section explains the framework for rapidly estimating the earthquake impact area via social media data based on our SLGM. Figure 4 schematically depicts our workflow, which is divided into three main components:

(1) Data collection and processing: gathering social media data (keyword filtration) and population data, social media data preprocessing (tokenization, removing URLs, numbers, stop words, etc.), and sensor data extraction (text classification and time threshold filtering);
(2) Building the SLGM for the social media data: determining the spatial statistics (choosing the model distance interval, creating the annular buffers, data normalization, recording the *MP* value as a function of *r*), and building the SLGM (estimating the model parameters by a maximum likelihood estimation (MLE) method);
(3) Validating the estimated impact area: mapping the earthquake impact area based on the parameter r_m and validating the detection result by using an official intensity map.

Table 1. The symbols used in the construction of the SLGM.

Symbols	Description
MP	Accumulation of normalized data
R	Distance from the earthquake epicenter
r_m	Impact area epicentral distance indicator
K	Total amount of normalized data (carrying capacity)
l_0	Maximum spatial growth rate
E	Natural logarithm base (also known as Euler's number)

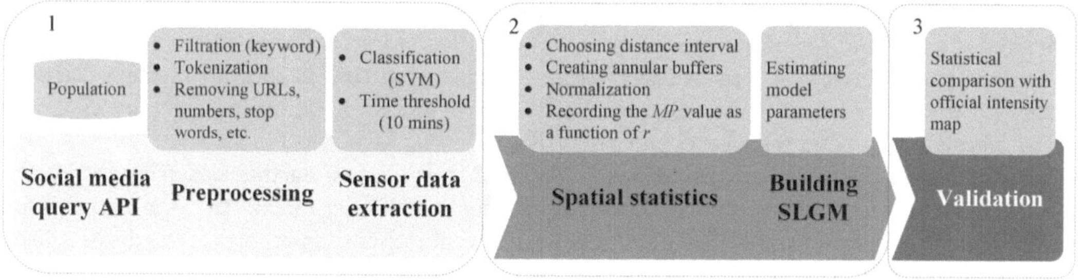

Figure 4. Impact area estimation workflow.

4.1. Data collection and processing

4.1.1. Social media data and population data

The social media data we employed in this study are derived from two services: Twitter and Sina-Weibo (a Twitter-like microblogging site widely used in China). The geo-tweets can be collected via the Twitter search application program interface (API) within a given radius of a given latitude/longitude. The georeferenced Weibo messages can be collected by the 'nearby_timeline' interface on the Sina-Weibo API. We transformed the collected messages into a geo-tweet-type internal data structure that contains only a user identification, a time-stamp, a geo-location, and a text message. It is worth mentioning that social media data with exact latitude and longitude coordinates are posted by the individuals using a mobile device application opt-in to geolocate their message. These social media data with exact coordinates do not include the reposted messages.

The population distribution data we used was provided by the LandScan Global Population Database (http://web.ornl.gov/sci/landscan/). LandScan represents the ambient population at a 1∗1 km grid resolution, which is the finest available resolution of global population distribution data (Bright, Rose, and Urban 2013).

4.1.2. Social media data preprocessing

Social media data containing the keywords 'earthquake' (地震) will be retained. However, this keyword selection method does not necessarily guarantee that all messages collected are actually relevant to a target event. Some users might send messages such as 'Earthquake drill dawn' or 'Earthquake Conference'. This would influence the precision of the emergency information extraction; therefore, these messages have to be filtered out, and a semantic analysis of the text is needed.

Before the actual semantic text analysis of the social media data, the original social media texts need to be preprocessed to reduce statistical noise. The tweets and Weibo text should be split into individual words, and URLs, numbers and stop words (commonly used words that do not carry a distinct semantical meaning) should be removed before performing the actual semantic analysis. We removed stop words using a standard predefined list provided by the Natural Language Toolkit (NLTK; http://www.nltk.org/).

4.1.3. Earthquake-related social media sensor data extraction

To extract earthquake-related messages from our dataset, we built a text classifier model. An SVM was used as the machine-learning algorithm to build the classifier (Suykens and Vandewalle 1999). We trained our classifier with a free and reliably labeled dataset (http://socialsensing.eu/datasets) and used the classifier to predict our collected social media data. This free dataset contains tweets about two earthquakes (the L'Aquila earthquake and the Emilia earthquake), and it is manually annotated in the following way: (i) messages related to the disaster carrying damage information (damage class); (ii) messages without damage information but are relevant to the disaster (no damage class); and (iii) messages that are not relevant (not relevant class) (Cresci, Tesconi, et al. 2015).

Our classifier aims to distinguish between relevant messages and nonrelevant messages. Therefore, we merged the first and second classes of the annotated dataset into a relevant class. Consequently, the relevant and nonrelevant free datasets became the learning corpus to train the text classifier model, and the trained model was used to classify our tweets and Weibo messages by assigning each of them a given class: relevant or not relevant.

Moreover, even if a social media message is related to the target earthquake, it might not be appropriate as an event sensor report. Unlike real sensor data, social media data originating from unaffected areas are responses to media coverage of the event rather than reports of the disaster itself (Sakaki, Okazaki, and Matsuo 2010; Crooks et al. 2013; Ryoo and Moon 2014). Such responses had a negative effect on the attempts to extract sensor data in our analysis. Crooks et al. (2013) defined a signal-to-noise ratio (SNR_T) for filtering social media sensor observations; they found that messages within 10 min following an earthquake had a higher SNR_T. Therefore, we chose the relevant messages that occurred in the 10 min after an earthquake as the social media sensor data for our study. The time sensitivity of our method will be further discussed in section 6.

4.2. Building the SLGM for social media data

The component is composed of two parts made up of spatial statistics determination and model parameters estimation. The first part (i.e. the spatial statistics determination) consists of choosing the distance interval (based on the 1-km population resolution, we chose 1 km as the distance interval in our model; that is, $r_{i+1} - r_i = 1$ km). Then, the number of social media messages and the populations at different distances in the annular buffer from the earthquake epicenter with the distance interval are counted. Subsequently, the amount of normalized sensor data in each annular buffer area is calculated, and thus, we can record the MP value as a function of r.

The second part consists of estimating the model parameters. The MLE method is used to estimate the parameters of our SLGM. The MLE technique first proceeds to maximize a likelihood function, which in turn maximizes the agreement between the model and the data. This method intuitively selects the parameter values that maximize the probability of the data. After computing the maximum likelihood estimate of the unknown parameter r_m, we mapped the estimated earthquake impact area.

4.3. Validating the estimated impact area

For validation purposes, the impact area detected by our model was compared with the official ShakeMap published by the USGS, which can be obtained from their website (http://earthquake.usgs.gov/shakemap). ShakeMap provides a map of the shaking intensity following significant earthquakes using the modified Mercalli intensity (MMI) scale to evaluate the effects of earthquakes. The MMI scale is composed of 12 levels of intensity on a scale from I (not felt) to XII (total destruction), and it is based on observable earthquake damage.

Of the 12 levels on the MMI scale, grade V represents the critical level. Walls will buckle and rupture in an area where the intensity is below grade V, but generally, the earthquake will not cause casualties and property damage. In contrast, in an area where the intensity is grade V or higher, the amount of damage increases considerably; some injuries will occur, dishes and windows will be broken, pendulum clocks may stop, and heavy furniture might begin to move (Wood and Neumann 1931). As our model detects the impact area where the earthquake will cause casualties and damage, we compare our detection result with the region whose intensity grade is greater than or equal to V in the ShakeMap. The following indexes are computed for the accuracy evaluation:

X, the ratio of the correctly detected area among the detected result:

$$X = \frac{\text{the area of the intersection of detected result and ShakeMap(intensity} \geq \text{V)}}{\text{the area of the detected result}}$$

Y, the ratio of the correctly detected area based on the ShakeMap, where the earthquake intensity ≥ V:

$$Y = \frac{\text{the area of the intersection of detected result and ShakeMap(intensity} \geq \text{V)}}{\text{the area of the ShakeMap where the earthquake intensity} \geq \text{V}}$$

Z-measure, the degree of agreement between the detected result and the ShakeMap:

$$\text{Z - } measure = \frac{X + Y}{2}$$

The more accurate the detection result, the closer the *Z-measure* value is to unity.

In this manner, our approach seeks to quickly estimate the impact area following a large earthquake to enhance situational awareness and shorten response times. This approach is applied to two large earthquake cases, as described in the next section.

5. Experimental results

To validate our approach, we applied the SLGM to estimate the impact areas of two large earthquakes: the 2015 Nepal earthquake and the 2017 Jiuzhaigou earthquake. The 2015 Nepal earthquake occurred with a magnitude of 8.1 at 11:56:25 local time on 25 April 2015, and it was the largest earthquake to strike Nepal since 1944. The 2017 Jiuzhaigou earthquake struck with a magnitude of 7.0 at 21:19:46 local time on 8 August 2017 in Jiuzhaigou County, China. The dominant social media platforms in these two regions are Twitter and Sina-Weibo (https://weibo.com, the most popular microblogging service in China). Using these two earthquake cases, we show the applicability of our approach.

5.1. Data

We collected 264763 geotagged tweets posted in a radius of 1609 kilometers (1000 miles) centered on the Nepal earthquake epicenter (84.731°E, 28.230°N) from 11:56 on 25 April 2015 to 23:59 on 28 April 2015; among those tweets, 8988 included the keyword 'earthquake'. For the Jiuzhaigou earthquake, we collected 19779 geotagged Weibo messages posted in a radius of 1000 kilometers centered on the Jiuzhaigou earthquake epicenter (103.855°E, 33.193°N) from 21:19 on 8 August 2017 to 23: 59 on 8 August 2017; 5852 messages contained the keyword '地震'(earthquake). In Figure 5a, we show the study area and the distributions of the tweets and population for the Nepal earthquake. Few tweets were collected to the north of Nepal, as Twitter is blocked in China. Figure 5b provides an overview of the study area for the Jiuzhaigou earthquake with the locations of geolocated Weibo messages and the population distribution.

Table 2 shows the performance of our text classifier. The free dataset containing both earthquakes (the L'Aquila earthquake and the Emilia earthquake) used in this study was split into groups: 80% was used for training, and 20% was used for testing. The classifier accurately detected relevant messages (the F-scores reached 0.902 and 0.908 for the free test dataset). We used this trained model to classify our preprocessed tweets and Weibo messages. As shown in Table 3, each social media message for the two earthquakes was assigned to a given class: relevant or not relevant. We chose the relevant messages that occurred within 10 min after the target earthquakes as the social media sensor data for our study.

5.2. Detection results

We chose 1 km as the distance interval, recorded the *MP* value as a function of *r*, and built the SLGM separately for these two earthquakes. To measure the goodness-of-fit of the SLGM, we used the ***R***-

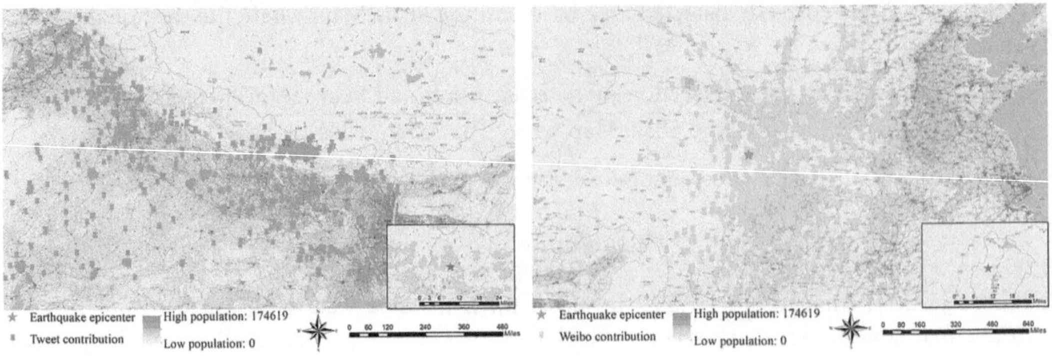

(a) 2015 Nepal earthquake (b) 2017 Jiuzhaigou earthquake

Figure 5. Overview of the study areas and data distributions.

Table 2. Classification experiments performed using the free dataset as a training set. (Accuracy: a global score referring to the percentage of data correctly classified. Precision: the ratio of the number of correctly classified data over the total number of data classified as belonging to a particular class. Recall: the ratio of the number of correctly classified tweets over the total number of tweets belonging to a particular class in the test sets. *F*-score: the harmonic mean of precision and recall).

		Relevant			Not relevant		
Dataset	Accuracy	Precision	Recall	F-score	Precision	Recall	F-score
L'Aquila	0.845	0.831	0.987	0.902	0.933	0.475	0.629
Emilia	0.838	0.862	0.958	0.908	0.511	0.211	0.309

squared value, which takes a value between zero and one. The closer the data points lie to the curve of our model, the closer the ***R**-squared* value is to one. The modeling results were as follows.

5.2.1. 2015 Nepal earthquake

Figure 6a shows the logistic spatial growth curve between *MP* and *r*. To facilitate the calculation, we calculated the percentages of *MP* rather than the actual values. As shown in Figure 6a, the ***R**-squared*

Table 3. Results of the classification of our tweets and Weibo messages.

2015 Nepal earthquake		2017 Jiuzhaigou earthquake	
Tweets	Type	Weibo messages	Type
Felt my floor and chair shaking. Earthquake in Delhi?	relevant	感觉刚才地震了呀 http://t.cn/Rt5qDX3 (I felt the earthquake just now.)	relevant
just felt earthquake for almost 5–10 s …	relevant	朋友圈地震了。我在这里很安全哈哈哈 http://t.cn/R2W6G4R (Know that an earthquake is happening from WeChat Moments. I'm safe here, hahaha)	not relevant
Only the gift by god is earthquake. #	relevant	抄经为地震地区祈福 (Pray for the people in the earthquake-stricken areas.)	relevant
Oh god EARTHQUAKE. More than 30sec.	relevant	是在地震了吗？怎么这么摇，我还以为是我头晕了。(Earthquake? I was dizzy from the shake.)	relevant
Still remember conversation with colleagues from @bbcnepali. Who said Kathmandu valley was prone to earthquake and destruction	not relevant	妈的，躺在床上地震了，吓死我了，一个鲤鱼打挺，但是又没有了，不敢睡了 (Damn, I was lying in bed when the earthquake happened just now, I jumped out of bed and was scared to death. Dare not sleep)	relevant
My goodness earthquake, every thing shaked for a bit … Have u realized that …	relevant	我滴妈呀！吓死了！吓死了！感觉超级强烈的……地震 (My god! Scared to death! Scared to death! Feel super strong … earthquake)	relevant
…	…	…	…

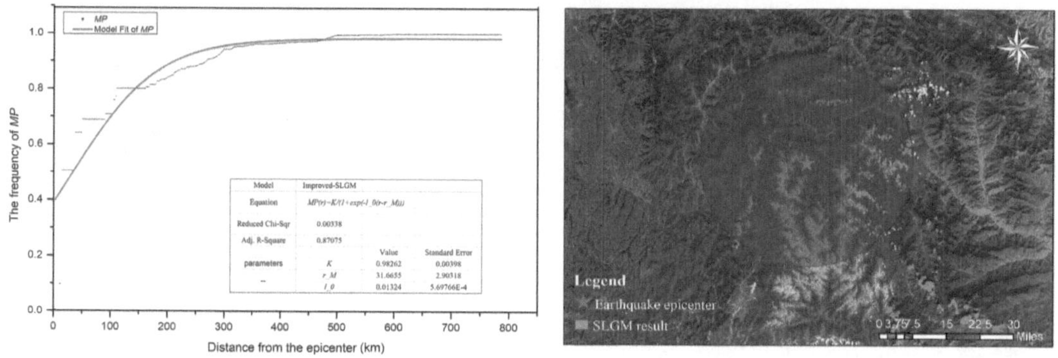

(a) (b)

Figure 6. Earthquake impact area assessment for the 2015 Nepal earthquake as derived from the SLGM analysis: (a) the model fitting result; (b) the estimated impact area.

value reached 0.959, which means that our model fits the social media data for this earthquake, while the MLE method yielded a value of $r_m = 293$ km. We show the impact area detected by the SLGM in Figure 6b.

5.2.2. 2017 Jiuzhaigou earthquake

A similar experiment was performed for the Jiuzhaigou earthquake. Figure 7a displays the *MP* versus *r* curve for the logistic spatial growth model. The SLGM also shows a good performance (0.871) in fitting the *MP*-*r* relationship of this earthquake, while r_m was estimated to be 32 km. The detection result of this earthquake is depicted in Figure 7b.

To demonstrate the practicality of our approach, the areas estimated by our model are compared with the official intensity maps in the next section.

5.3. Verification of results

In Figures 8 and 9, we project the detected areas on top of the USGS ShakeMaps for both earthquakes; the red circles represent the boundaries of the detected areas based on the SLGM.

As shown in Figure 8, for the Nepal earthquake, the region whose earthquake intensity is greater than or equal to V in the ShakeMap corresponds well to the area detected by the SLGM (with a radius of 293 km). More specifically, 52.37% of the detected area corresponds to the area where the

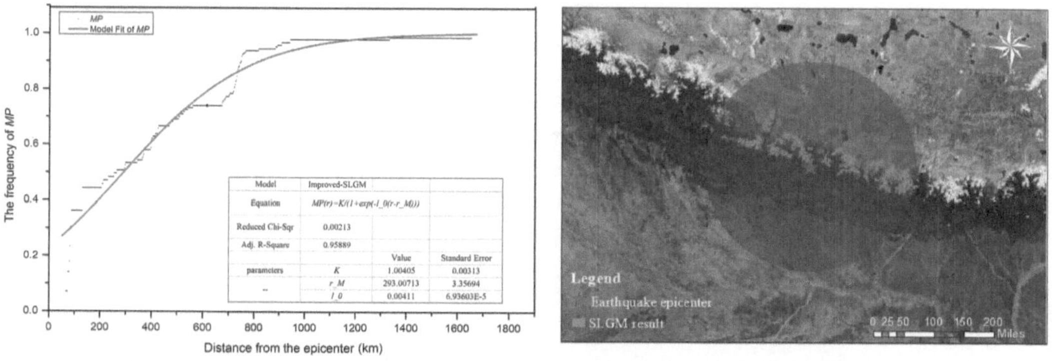

(a) (b)

Figure 7. Earthquake impact area assessment for the 2017 Jiuzhaigou earthquake as derived from the SLGM analysis: (a) the model fitting result; (b) the estimated impact area.

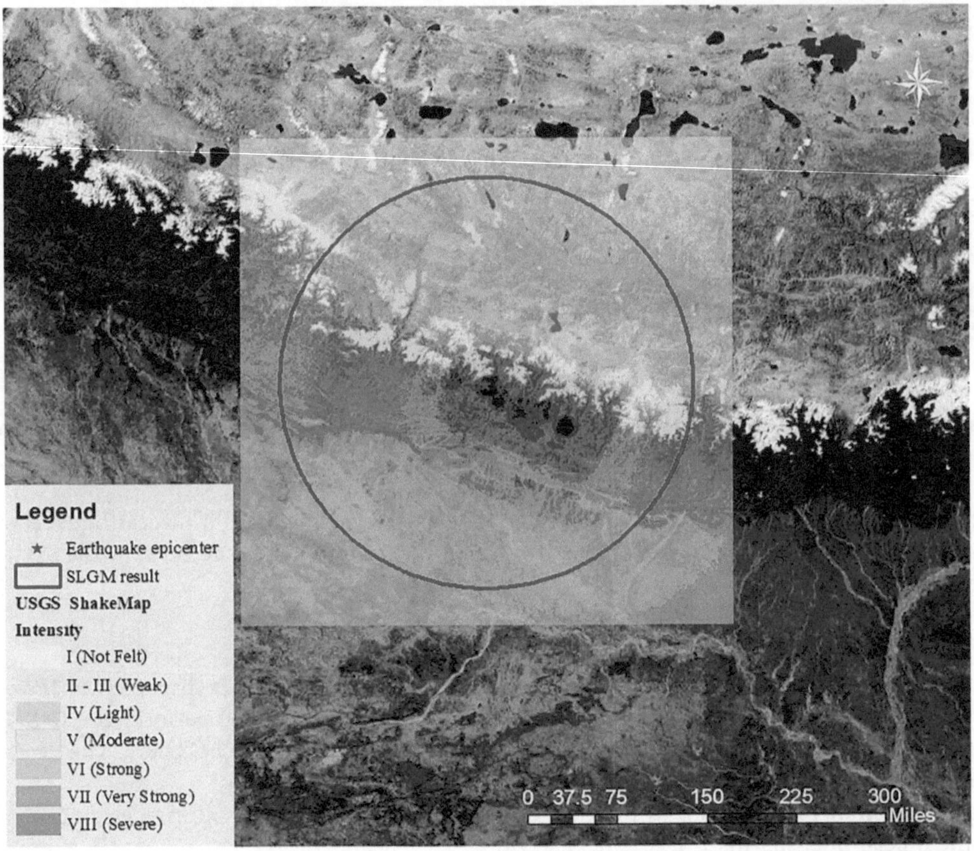

Figure 8. Comparison of the official intensity map produced by the USGS with the detection result of the impact area derived from the SLGM analysis for the 2015 Nepal earthquake.

earthquake intensity grade ≥ V in the ShakeMap ($X = 52.37\%$), while 91.11% of the ShakeMap (where the earthquake intensity grade ≥ V) corresponds to the detected area ($Y = 91.11\%$); the *Z-measure* was 71.74%. For the Jiuzhaigou earthquake, as shown in Figure 9, we can observe that the model is also efficient in estimating the impact area (with a radius of 32 km), as demonstrated by the encouraging values ($X = 98.81\%$, $Y = 87.36\%$, *Z-measure* = 93.11%), similar to Figure 8. These evaluation results show that our proposed SLGM effectively and quantitatively expresses the spatial growth process of social media citizen sensor data after an earthquake and that it can rapidly estimate the impact area with a reasonable accuracy.

The results show that we produced a more accurate estimate in the impact area of the Jiuzhaigou earthquake (93.11%) than in that of the Nepal earthquake (71.74%). Since the China firewall blocks Twitter, there were few tweets from the region to the north of Nepal. This might have affected the accuracy of the detected results for the Nepal earthquake. Moreover, the population around Jiuzhaigou is slightly more evenly distributed, and people in this study area use Sina-Weibo without limits. Thus, our result is reasonable: the more uniform the distribution of sensors is, the more precise the estimation.

6. Discussion

6.1. Evaluation of the SLGM time sensitivity to social media data

To test the time sensitivity of our proposed method, we also built the SLGM for the social media disaster-related data posted at different times after the target earthquakes. Comparisons of the

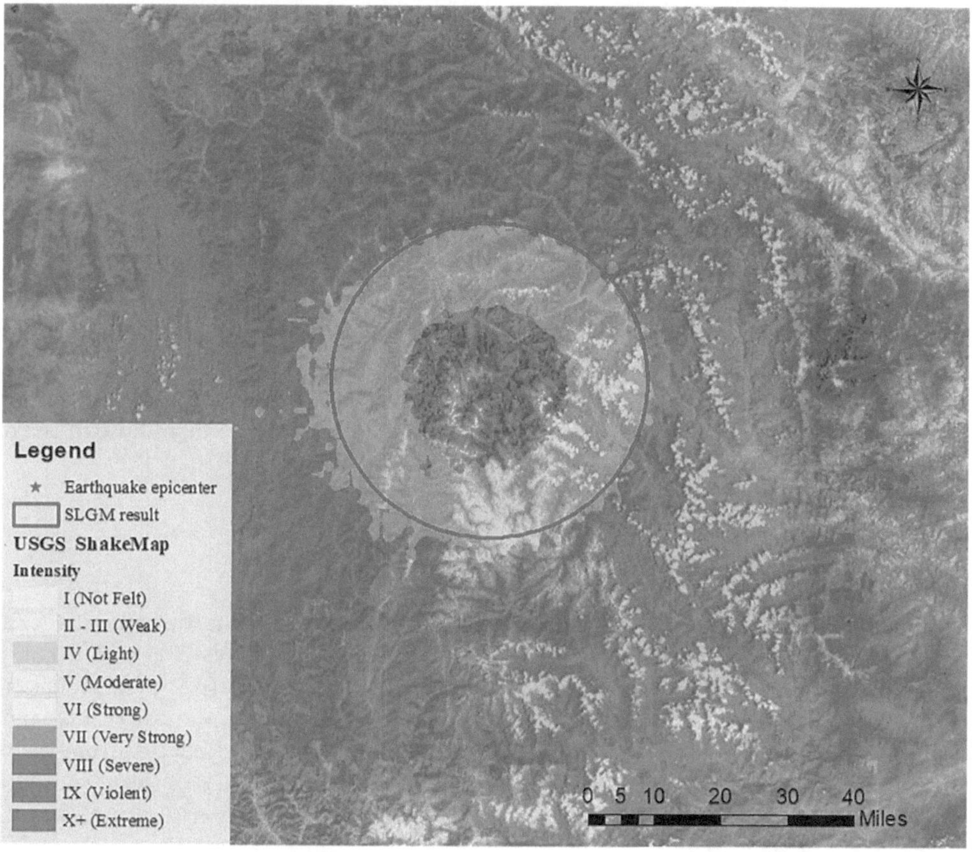

Figure 9. Comparison of the official intensity map produced by the USGS with the detection result of the impact area derived from the SLGM analysis for the Jiuzhaigou earthquake.

fitting accuracies and their effects on the impact area detection results between different time periods are summarized in Table 4.

The detection accuracies both steadily increase in the first few minutes of the two earthquakes (for the Nepal earthquake, $R^2 = 0.422$ to 0.959, *Z-measure* = 50.49% to 71.74%; for the Jiuzhaigou earthquake, $R^2 = 0.919$ to 0.976, *Z-measure* = 51.95% to 93.41%). After the first 10 min, the

Table 4. The effect of time threshold filters on the impact area detection result produced by the SLGM.

	2015 Nepal earthquake			2017 Jiuzhaigou earthquake		
Time period (min.)	R^2	r_m (km)	*Z-measure*	R^2	r_m (km)	*Z-measure*
0–4	0.422	22	50.49%	0.919	172	51.95%
0–5	0.907	152	61.88%	0.964	149	52.61%
0–6	0.883	154	61.91%	0.960	166	52.10%
0–7	0.925	222	68.08%	0.987	182	51.75%
0–8	0.951	263	71.32%	0.976	113	54.54%
0–9	0.932	286	71.20%	0.854	17	62.46%
0–10	0.959	293	71.74%	0.871	32	93.41%
0–11	0.941	318	70.31%	0.876	29	86.27%
0–12	0.963	−1960	–	0.885	37	90.57%
0–13	0.962	−2049	–	0.832	−19	–
0–14	0.959	−2112	–	0.686	−33	–
0–16	0.956	−2130	–	0.724	−58	–
0–20	−0.24	–	–	0.774	−55	–

detection accuracies of the two events both fluctuate around a moderate level. However, in the following 12–20 min, the accuracies decrease rapidly, and our model is no longer suitable for detecting the impact areas for the two earthquake events. That is, despite having the highest total count of messages, the data posted in the first 20 min after the earthquakes are unsuitable for estimating the impact areas.

The model accuracy first increases a few minutes after the earthquake and then decreases. Arguably, this trend could be explained by the distribution of social media data, which is fueled not only by the impact of the earthquake but also by the widespread coverage of the earthquake on social media. Earthquake-related data posted a few minutes after an earthquake can be regarded as sensor values. As the amount of sensor data increases, the detected information becomes more reliable. After a certain period, user-posted data might represent not only the reaction to the earthquake but also the responses to the keyword 'earthquake', which cannot be regarded as sensor readings. This will result in a reduced credibility of the detected results. In the period we studied, the *Z-measure* values of the two earthquakes in the first 10 minutes were higher than those in the other periods.

6.2. Model external validity and limitations analysis

To determine the external validity and limitations of the SLGM, we tested it with other large earthquake events. We analyzed major U.S. earthquakes that occurred during 2000–2017 with magnitudes greater than 5 and depths of less than 60 km (the characteristics of destructive earthquakes), as declared by the USGS. Due to the lack of social media data for these historical events, another type of online citizen-sensor data was used in this paper. These data were collected from the dedicated crowdsourcing 'Did You Feel It?' (DYFI) program tapped by the USGS (http://earthquake.usgs.gov/dyfi/), where citizens use the Internet website to report their observations of any earthquakes (regardless of whether they felt them) with geolocation information. The number of DYFI citizen-sensors for these events followed a long-tailed distribution. We analyzed only the events where the number of citizen-sensor data points was greater than the average number of 8512. In total, 18 separate earthquake events were noted. We followed the same method as described above to estimate the impact areas from the DYFI data. Table 5 shows the SLGM detection results for these 18 events

Table 5. SLGM detection results for additional earthquake events. (Event ID: each reported earthquake event is given a unique ID by the USGS).

Event ID	Magnitude	Depth (km)	Responses	R^2	r_m (km)	*Z-measure*
uw10530748	6.8	51.8	26568	0.946	125	0.872
ci14151344	5.2	13.1	29319	0.979	32	0.835
ld2002042000	5.3	4.8	9950	0.988	53	0.833
nc40204628	5.5	9.7	64070	0.989	55	0.757
nm606657	5.2	14.3	42194	0.995	117	0.751
nc71863625	5.3	8.6	10114	0.996	23	0.734
se609212	5.8	6	144178	0.987	174	0.711
usp000jadn	5.7	5.2	65329	0.994	121	0.710
us10009757	5.8	12.2	15718	0.988	74	0.707
nc21323712	6.5	8.4	21984	0.906	37	0.666
nc72282711	6	11.1	41364	0.986	50	0.663
nc71996906	5.7	8	11254	0.97	44	0.656
nc30228270	6	8.1	14406	0.867	17	0.628
ci10736069	5.4	12.3	24940	0.951	54	0.621
us10006jxs	5.8	5.6	60407	0.998	157	0.611
ci14745580	5.7	8.8	38686	0.99	110	0.560
ci14383980	5.4	15.5	41176	0.996	103	0.549
ci15481673	5.1	5.1	16155	0.993	73	0.524

All of the events exhibited strong logistic spatial growth processes in the citizen-sensor data; high R^2 values were obtained for our SLGM fitting results (as shown in Table 5). We found that for the events whose *Z-measures* were less than 0.6 (events ci14745580, ci14383980, and ci15481673), their epicenters were all around the West Coast of the United States. These resulted in lower sensor values around the epicenters; thus, our model results were not significant. For all earthquake events whose epicenters were far away from the ocean (events uw10530748, ld2002042000, nm606657, se609212, usp000jadn, and us10009757), their *Z-measure* values all reached 0.7. This shows that one may expect even more accruable results with a relatively uniform distribution of sensor values around the earthquake epicenter.

In general, our SLGM is reliable, especially for earthquake cases where the earthquake epicenter is located on a continental landmass and has a relatively uniform distribution of sensor values around it. Since the messages that people post on social media a few minutes after an earthquake can be regarded as sensor observations, social media data based on our model can enable a rapid assessment of the earthquake impact area (e.g. the DYFI online citizen-sensor data acquisition took several hours or even days.).

6.3. Social media data word cloud analysis

In addition to analyzing the SLGM time sensitivity to social media data and its limitations, we performed an additional social media text analysis to explore more useful information based on our detected results for the Nepal earthquake and Jiuzhaigou earthquake. We analyzed the contents of all collected messages and the messages in the corresponding detection areas. The word cloud shown in Figure 10 visualizes the frequency of individual words in our data corpus, where larger words are mentioned more frequently. In Figure 10a, we show the word cloud for the Nepal earthquake; the left image was created with 264763 Twitter messages posted in a radius of 1609 km (1000 miles) centered on the epicenter, and the right image was created with 10249 tweets posted in the detected impact area (293 km). Figure 10b shows the equivalent word cloud for the Jiuzhaigou

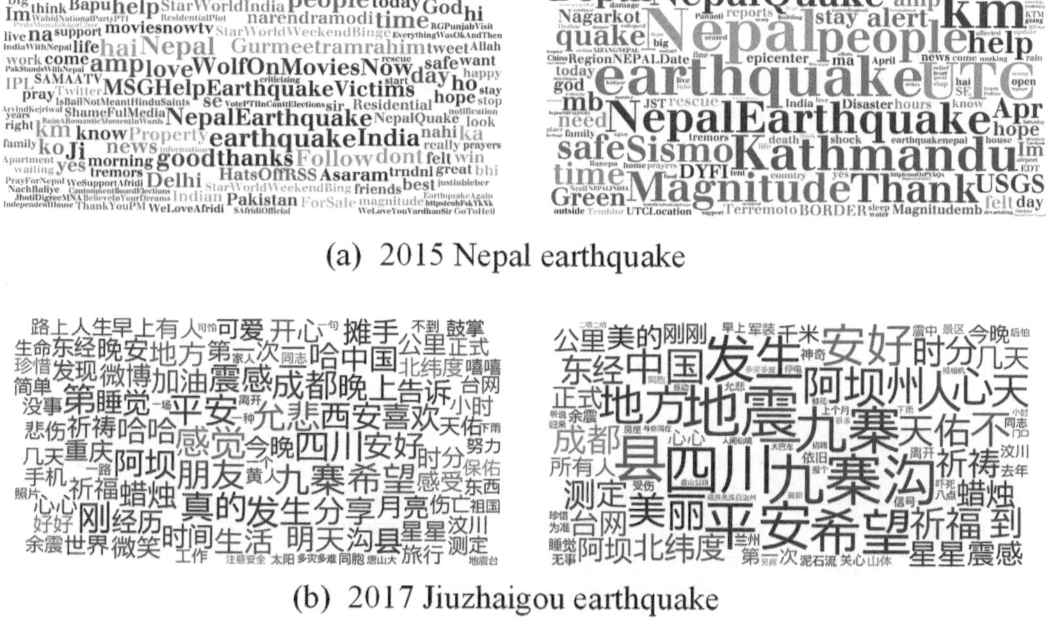

(a) 2015 Nepal earthquake

(b) 2017 Jiuzhaigou earthquake

Figure 10. Word cloud of the most frequent terms during the two earthquakes.

earthquake; the left image was created with 19779 Weibo messages posted in a radius of 1000 km centered on the epicenter, and the right image was created with 245 Weibo messages posted in the detected impact area (32 km). It is easy to see that the social media data in the detected area are more relevant to the disaster.

Thus, our model can not only quickly estimate the impact area after a large earthquake but also provide a reference for future work with a text-based or an image-based analysis of social media data. One could obtain a higher probability of identifying useful information in the detected impact area. Messages can be weighted based on the detected range prior to data processing. Hence, the efficiency of social media data analysis can be improved, and we can perform a more precise, quantitative assessment of the social media messages posted in the detected area.

7. Conclusions

Social media data that carry geographic references are easy to obtain, and they might represent real-time citizen-sensor data for a disaster; consequently, they could be used to enhance situational awareness. In this article, we present an approach that quickly estimates the impact area following a large earthquake via social media. Specifically, we apply a time-based theory to spatial analysis. We construct an analogy between changes in the biological population size over time and changes in citizen-sensor data size over distance, thereby proposing a novel model (i.e. the SLGM) to describe the spatial growth of citizen-sensor data after an earthquake. Then, we develop a framework to quickly estimate the earthquake impact area via social media based on the proposed SLGM. In this approach, we reveal the connection between the citizen-sensor data spatial growth trend and the earthquake impact area and apply this connection to model the social media citizen-sensor data, thereby estimating the impact area through spatial statistical analysis.

To demonstrate how our approach leverages citizen-sensor data from social media to quickly locate earthquake impact areas, we used representative case studies, namely, the 2015 Nepal earthquake and the 2017 Jiuzhaigou earthquake. The reliability and restriction of our SLGM was further discussed across 18 other earthquake events. The experimental results illustrate that the SLGM can effectively express the spatial growth of citizen-sensor data following a large earthquake, and the proposed model is suitable for producing a quick estimation of the earthquake impact area via social media. This approach allows us to quickly extract valuable earthquake emergency information following a large earthquake without requiring prior knowledge or feasible cases.

The citizen-sensor observations (derived from social media) in our cases were obtained after keyword-based filtering. However, it is possible that some messages were relevant to the earthquake but did not include the selected keyword and were thus not collected. This is a limitation of our study that results from using a keyword selection strategy to collect social media data. Additional key words, such as 'shaking' and 'disaster', will be included in our future work. Moreover, we used an external dataset as a training set to train the classification model for the classification of our messages, and we did not perform a manual verification of the resulting dataset to identify any misclassified relevant messages. This is an avenue for future research.

Additionally, we illustrated that the results obtained from social media were nevertheless useful for detecting the earthquake impact areas for the Nepal and Jiuzhaigou earthquakes. In future work, additional earthquake events will be included to further verify the effectiveness of our method. In addition, we will explore how different earthquake characteristics can influence the credibility and generalizability of the results obtained from social media data based on the SLGM.

Moreover, social media sensors are different from real sensors, and their distribution will be affected by demographic, economic, and community disparities among many other factors. Hence, further studies are needed to explore the possibility of incorporating more factors into our model to make it more reasonable and thus provide even more meaningful information to better support disaster management endeavors.

Acknowledgements

This product was made utilizing the LandScan 2012™ High Resolution global Population Data Set copyrighted by UT-Battelle, LLC, operator of Oak Ridge National Laboratory under Contract No. DE-AC05-00OR22725 with the United States Department of Energy. The United States Government has certain rights in this Data Set. Neither UT-BATTELLE, LLC NOR THE UNITED STATES DEPARTMENT OF ENERGY, NOR ANY OF THEIR EMPLOYEES, MAKES ANY WARRANTY, EXPRESS OR IMPLIED, OR ASSUMES ANY LEGAL LIABILITY OR RESPONSIBILITY FOR THE ACCURACY, COMPLETENESS, OR USEFULNESS OF THE DATA SET.

Disclosure statement

No potential conflict of interest was reported by the authors.

Funding

This work was supported by National Natural Science Foundation of China [grant number 41271399].

References

Amit, Sheth, Purohit Hemant, Smith Gary Alan, Brunn Jeremy, Jadhav Ashutosh, Kapanipathi Pavan, Lu Chen, and Wang Wenbo. 2014. "Twitris: A System for Collective Social Intelligence." In *Encyclopedia of Social Network Analysis and Mining*, edited by Alhajj Reda, and Rokne Jon, 2240–2253. New York: Springer-Verlag New York.

Atkinson, Gail M., and David J. Wald. 2007. ""Did You Feel It?" Intensity Data: A Surprisingly Good Measure of Earthquake Ground Motion." *Seismological Research Letters* 78 (3): 362–368. doi:10.1785/gssrl.78.3.362.

Badal, José, Miguel Vázquez-prada, and Álvaro González. 2005. "Preliminary Quantitative Assessment of Earthquake Casualties and Damages." *Natural Hazards* 34 (3): 353–374. doi:10.1007/s11069-004-3656-6.

Baranyi, József, and Carmen Pin. 2001. "A Parallel Study on Bacterial Growth and Inactivation." *Journal of Theoretical Biology* 210 (3): 327–336. doi:10.1006/jtbi.2001.2312.

Bhowmick, Amiya Ranjan, and Sabyasachi Bhattacharya. 2014. "A new Growth Curve Model for Biological Growth: Some Inferential Studies on the Growth of Cirrhinus Mrigala." *Mathematical Biosciences* 254: 28–41. doi:10.1016/j.mbs.2014.06.004.

Bright, Eddie A., Amy N. Rose, and Marie L. Urban. 2013. "LandScan 2012." In. Oak Ridge, TN: Oak Ridge National Laboratory. http://www.ornl.gov/landscan/.

Burks, L., M. Miller, and R. Zadeh. 2014. "Rapid Estimate of Ground Shaking Intensity by Combining Simple Earthquake Characteristics with Tweets." Paper presented at the 10th US Nat. conf. earthquake Eng., front. earthquake Eng., Anchorage, AK, USA, Jul. 21Y25.

Carlos, Castillo, Mendoza Marcelo, and Poblete Barbara. 2013. "Predicting Information Credibility in Time-Sensitive Social Media." *Internet Research: Electronic Networking Applications and Policy* 23 (5): 560–588. doi:10.1108/IntR-05-2012-0095.

Cresci, Stefano, Andrea Cimino, Felice Dell'Orletta, and Maurizio Tesconi. 2015. "Crisis Mapping During Natural Disasters via Text Analysis of Social Media Messages." In *Web Information Systems Engineering – WISE 2015: 16th International Conference, Miami, FL, USA, November 1-3, 2015, Proceedings, Part II*, edited by Jianyong Wang, Wojciech Cellary, Dingding Wang, Hua Wang, Shu-Ching Chen, Tao Li and Yanchun Zhang, 250–258. Cham: Springer International Publishing. doi:10.1007/978-3-319-26187-4_21.

Cresci, Stefano, Maurizio Tesconi, Andrea Cimino, and Felice Dell'Orletta. 2015. "A Linguistically-driven Approach to Cross-Event Damage Assessment of Natural Disasters From Social Media Messages." In *Proceedings of the 24th International Conference on World Wide Web*, ACM, New York, NY, USA, 1195–1200. Florence: ACM. doi:10.1145/2740908.2741722.

Crooks, Andrew, Arie Croitoru, Anthony Stefanidis, and Jacek Radzikowski. 2013. "#Earthquake: Twitter as a Distributed Sensor System." *Transactions in GIS* 17 (1): 124–147. doi:10.1111/j.1467-9671.2012.01359.x.

Dawed, Mohammed Yiha, Purnachandra Rao Koya, and Ayele Taye Goshu. 2014. "Mathematical Modelling of Population Growth: The Case of Logistic and Von Bertalanffy Models." *Open Journal of Modelling and Simulation* 2 (04): 113–126. doi:10.4236/ojmsi.2014.24013.

Earle, Paul S., Daniel C. Bowden, and Michelle R. Guy. 2011. "Twitter Earthquake Detection: Earthquake Monitoring in a Social World." *Annals of Geophysics* 54 (6): 708–715. doi:10.4401/ag-5364.

Earle, Paul, Michelle Guy, Richard Buckmaster, Chris Ostrum, Scott Horvath, and Amy Vaughan. 2010. "OMG Earthquake! Can Twitter Improve Earthquake Response?" *Seismological Research Letters* 81 (2): 246–251. doi:10.1785/gssrl.81.2.246.

Enrico, Steiger, Albuquerque João Porto, and Zipf Alexander. 2015. "An Advanced Systematic Literature Review on Spatiotemporal Analyses of Twitter Data." *Transactions in GIS* 19 (6): 809–834. doi:10.1111/tgis.12132.

Fontugne, Romain, Kenjiro Cho, Youngjoon Won, and Kensuke Fukuda. 2011. "Disasters Seen Through Flickr Cameras." In *Proceedings of the Special Workshop on Internet and Disasters*, ACM, New York, NY, USA, 1–10. Tokyo: ACM. doi:10.1145/2079360.2079365.

Goodchild, Michael F. 2007. "Citizens as Sensors: the World of Volunteered Geography." *GeoJournal* 69 (4): 211–221. doi:10.1007/s10708-007-9111-y.

Goodchild, Michael F., and J. Alan Glennon. 2010. "Crowdsourcing Geographic Information for Disaster Response: A Research Frontier." *International Journal of Digital Earth* 3 (3): 231–241. doi:10.1080/17538941003759255.

Goolsby, Rebecca. 2010. "Social Media as Crisis Platform: The Future of Community Maps/Crisis Maps." *ACM Transactions on Intelligent Systems and Technology* 1 (1): 1–11. doi:10.1145/1858948.1858955.

Guy, Michelle, Paul Earle, Chris Ostrum, Kenny Gruchalla, and Scott Horvath. 2010. "Integration and Dissemination of Citizen Reported and Seismically Derived Earthquake Information via Social Network Technologies." *International Conference on Advances in Intelligent Data Analysis* 2010: 42–53. doi:10.1007/978-3-642-13062-5_6.

Huang, Qunying, and Yu Xiao. 2015. "Geographic Situational Awareness: Mining Tweets for Disaster Preparedness, Emergency Response, Impact, and Recovery." *ISPRS International Journal of Geo-Information* 4 (3): 1549–1568. doi:10.3390/ijgi4031549.

Judith, Gelernter, and Mushegian Nikolai. 2011. "Geo-Parsing Messages From Microtext." *Transactions in GIS* 15 (6): 753–773. doi:10.1111/j.1467-9671.2011.01294.x.

Kim, Kyoung-Sook, Ryong Lee, and Koji Zettsu. 2011. "mTrend:Discovery of Topic Movements on geo-Microblogging Messages." *ACM Sigspatial International Conference on Advances in Geographic Information Systems* 2011: 529–532. doi:10.1145/2093973.2094066.

Kropivnitskaya, Yelena, Kristy F. Tiampo, Jinhui Qin, and Michael A. Bauer. 2017. "The Predictive Relationship Between Earthquake Intensity and Tweets Rate for Real-Time Ground-Motion Estimation." *Seismological Research Letters* 88 (3): 840–850. doi:10.1785/0220160215.

Kryvasheyeu, Yury, Haohui Chen, Nick Obradovich, Esteban Moro, Pascal Van Hentenryck, James Fowler, and Manuel Cebrian. 2016. "Rapid Assessment of Disaster Damage Using Social Media Activity." *Science Advances* 2 (3), doi:10.1126/sciadv.1500779.

Li, Zhenlong, Cuizhen Wang, Christopher T. Emrich, and Diansheng Guo. 2018. "A Novel Approach to Leveraging Social Media for Rapid Flood Mapping: a Case Study of the 2015 South Carolina Floods." *Cartography and Geographic Information Science* 45 (2): 97–110. doi:10.1080/15230406.2016.1271356.

Mizunuma, Yuhiro, Shuhei Yamamoto, Yutaro Yamaguchi, Atsushi Ikeuchi, Tetsuji Satoh, and Satoshi Shimada. 2014. "Twitter bursts: Analysis of their occurrences and classifications." Paper presented at the ICDS. http://www.thinkmind.org/download.php?articleid=icds_2014_7_40_10139.

Murakami, Akiko, and Tetsuya Nasukawa. 2012. "Tweeting About the Tsunami?: Mining Twitter for Information on the Tohoku Earthquake and Tsunami." *International World Wide Web Conferences* 2012: 709–710. doi:10.1145/2187980.2188187.

Nagarajan, Meenakshi, Karthik Gomadam, Amit P. Sheth, Ajith Ranabahu, Raghava Mutharaju, and Ashutosh Jadhav. 2009. "Spatio-Temporal-Thematic Analysis of Citizen Sensor Data: Challenges and Experiences." *Lecture Notes in Computer Science* 5802: 539–553. doi:10.1007/978-3-642-04409-0_52.

Panagiota, Matsuka, Chanussot Jocelyn, and Pathier Erwan. 2011. "State of the art on remote sensing for vulnerability and damage assessment on urban context." *URBASIS Consort: Grenoble, France.*

Re, Giuseppe Del. 2000. "Models and Analogies in Science." *Hyle International Journal for Philosophy of Chemistry* 6 (1): 5–15. http://www.hyle.org/journal/issues/6/delre.pdf.

Resch, Bernd, Florian Usländer, and Clemens Havas. 2018. "Combining Machine-Learning Topic Models and Spatiotemporal Analysis of Social Media Data for Disaster Footprint and Damage Assessment." *Cartography and Geographic Information Science* 45 (4): 362–376. doi:10.1080/15230406.2017.1356242.

Robinson, Bella, Robert Power, and Mark Cameron. 2013. "A Sensitive Twitter Earthquake Detector." In *Proceedings of the 22nd International Conference on World Wide Web*, ACM, New York, NY, USA, 999–1002. Rio de Janeiro: ACM. doi:10.1145/2487788.2488101.

Ryoo, KyoungMin, and Sue Moon. 2014. "Inferring Twitter User Locations with 10 km Accuracy." In *Proceedings of the 23rd International Conference on World Wide Web*, ACM, New York, NY, USA, 643–648. Seoul: ACM. doi:10.1145/2567948.2579236.

Sakaki, Takeshi, Makoto Okazaki, and Yutaka Matsuo. 2010. "Earthquake Shakes Twitter Users: Real-Time Event Detection by Social Sensors." In *Proceedings of the 19th International Conference on World Wide Web*, ACM, New York, NY, USA, 851–860. Raleigh, NC: ACM. doi:10.1145/1772690.1772777.

Sakaki, T., M. Okazaki, and Y. Matsuo. 2013. "Tweet Analysis for Real-Time Event Detection and Earthquake Reporting System Development." *IEEE Transactions on Knowledge and Data Engineering* 25 (4): 919–931. doi:10.1109/TKDE.2012.29.

Sakaki, Takeshi, Fujio Toriumi, and Yutaka Matsuo. 2011. "Tweet Trend Analysis in an Emergency Situation." In *Proceedings of the Special Workshop on Internet and Disasters*, ACM, New York, NY, USA, 1–8. Tokyo: ACM. doi:10.1145/2079360.2079363.

Sheth, Amit. 2009. "Citizen Sensing, Social Signals, and Enriching Human Experience." *IEEE Internet Computing* 13 (4): 87–92. doi:10.1109/mic.2009.77.

Steiger, Enrico, Rene Westerholt, and Alexander Zipf. 2016. "Research on Social Media Feeds – A GIScience Perspective." In *European Handbook of Crowdsourced Geographic Information*, edited by C. Capineri, M. Haklay, H. Huang, V. Antoniou, J. Kettunen, F. Ostermann, and R. Purves, 237–254. London: Ubiquity Press.

Suykens, J. A. K., and J. Vandewalle. 1999. "Least Squares Support Vector Machine Classifiers." *Neural Processing Letters* 9 (3): 293–300. doi:10.1023/a:1018628609742.

Symonds, Matthew R. E., and Adnan Moussalli. 2011. "A Brief Guide to Model Selection, Multimodel Inference and Model Averaging in Behavioural Ecology Using Akaike's Information Criterion." *Behavioral Ecology and Sociobiology* 65 (1): 13–21. doi:10.1007/s00265-010-1037-6.

Tourres, Marie-Aimee. 2003. "Technological Revolutions and Financial Capital: The Dynamics of Bubbles and Golden Ages." *Journal of Economic Issues* 37 (3): 833–835. doi:10.1080/00213624.2003.11506628.

Tsoularis, A., and J. Wallace. 2002. "Analysis of Logistic Growth Models." *Mathematical Biosciences* 179 (1): 21–55. doi:10.1016/S0025-5564(02)00096-2.

Wang, Yuan Zhi, Jia Jie Sun, Lian Xing Liu, and Fei Sheng. 2007. "Thinking About Introducing Space Sequence Concept." [In Chinese.] *Journal of Shenyang Institute of Engineering* 3 (2): 179–180. doi:10.13888/j.cnki.jsie(ns).2007.02.026.

Wood, H. O., and F., Neumann. 1931. "Modified Mercalli Intensity Scale of 1931." *Bulletin of the Seismological Society of America* 21 (4): 277–283. http://www.bssaonline.org/content/21/4/277.accessible-long.

Xu, Chong, Xiwei Xu, Fuchu Dai, Zhide Wu, Honglin He, Feng Shi, Xiyan Wu, and Suning Xu. 2013. "Application of an Incomplete Landslide Inventory, Logistic Regression Model and its Validation for Landslide Susceptibility Mapping Related to the May 12, 2008 Wenchuan Earthquake of China." *Natural Hazards* 68 (2): 883-900. doi:10.1007/s11069-013-0661-7.

Xu, Zhiheng, Yang Zhang, Yao Wu, and Qing Yang. 2012. "Modeling User Posting Behavior on Social Media." In *Proceedings of the 35th International ACM SIGIR Conference on Research and Development in Information Retrieval*, ACM, New York, NY, USA, 545–554. Portland, OR: ACM. doi:10.1145/2348283.2348358.

Yang, Tian Qing, X. I. Nan, Yi Zhang, and L. I. Jie-Fei. 2016. "Fast Determination Method of Seismic Intensity Distribution Based on Discrete Disaster Information." [In Chinese.] *Earthquake* 36 (2): 48–59. doi:10.3969/j.issn.1000-3274.2016.02.005.

Zhu, J., F. Xiong, D. Piao, Y. Liu, and Y. Zhang. 2011. "*Statistically Modeling the Effectiveness of Disaster Information in Social Media.*" Paper presented at the 2011 IEEE global humanitarian technology conference, Oct. 30 2011–Nov. 1 2011.

Mapping near-real-time power outages from social media

Huina Mao, Gautam Thakur, Kevin Sparks, Jibonananda Sanyal
and Budhendra Bhaduri

ABSTRACT

Social media, including Twitter, has become an important source for disaster response. Yet most studies focus on a very limited amount of geotagged data (approximately 1% of all tweets) while discarding a rich body of data that contains location expressions in text. Location information is crucial to understanding the impact of disasters, including where damage has occurred and where the people who need help are situated. In this paper, we propose a novel two-stage machine learning- and deep learning-based framework for power outage detection from Twitter. First, we apply a probabilistic classification model using bag-of-ngrams features to find true power outage tweets. Second, we implement a new deep learning method–bidirectional long short-term memory networks–to extract outage locations from text. Results show a promising classification accuracy (86%) in identifying true power outage tweets, and approximately 20 times more usable tweets can be located compared with simply relying on geotagged tweets. The method of identifying location names used in this paper does not require language- or domain-specific external resources such as gazetteers or handcrafted features, so it can be extended to other situational awareness analyzes and new applications.

1. Introduction

Power outages cause tremendous economic loss and threaten people's health during disasters. In 2012 Hurricane Sandy caused 8.5 million power outages, leading to $35 billion in damages; many victims waited more than 2 weeks for power to be restored (http://www.pennlive.com/midstate/index.ssf/2012/11/restoring_power_to_hurricane_s.html). Restoration was slow not only due to the large-scale and severe damages caused by the hurricane, but also because utility companies experienced significant delays in gathering efficient power outage information and allocating their limited power resources (Lee, Shin, and Zadeh 2013). Timely power outage detection is critical for utility companies and emergency responders to ensure efficient power restoration. Companies rely almost exclusively on customers to report electricity failures, and restoration is verified by checking back in with reporting customers or dispatching a field crew to the site. This traditional method is ineffective, especially during times of massive outages. In contrast to telephone calls, advanced metering infrastructure (AMI) technology enables utilities to monitor power outages more precisely and restore power more quickly. Existing AMI methodologies involve detecting power outages based on data generated by smart meters, which measure and record customers' real-time electricity usage (Chen et al. 2015; Liu and Schulz 2002). For example, an outage would be detected when a smart meter stopped recording usage, and the utility company would be automatically alerted. Chai et al. (2016) developed the frequency monitoring network, GridEye, based on frequency disturbance recorder sensors that monitor the stability of grid networks. Despite the accuracy of electronic sensors, large-scale AMI deployments are expensive and lengthy. According to the IEI report (Electric Company Smart Meter Deployments: Foundation for A Smart Grid, October 2016), only 30 utility companies in the United States (out of thousands) have fully deployed smart meters.

Unlike expensive and proprietary physical sensors, *social sensors* (i.e. social media) are free and in the public domain. Sakaki, Okazaki, and Matsuo (2010) compare Twitter users to *social sensors* and tweets to *sensory information*. Takeichi et al. (2014) further state that Twitter often behaves like a *social sensor* in which users actively sense real-world events and spontaneously mention these events in cyberspace. The data generated by millions of social sensors are user-created content or volunteered geographic information (VGI) (Goodchild 2007). Social media, as an important channel for VGI information, has emerged as an efficient two-way communication medium through which utility companies can share information related to outages with their customers and customers can report outages. During a disaster and its aftermath, social media are widely used to spread important information. Twitter operates as a real-time, public microblogging network with more than 300 million active monthly users. Hence, it has become an ideal data source for monitoring and accessing the extent of disasters (Huang and Xiao 2015). For example, Twitter has been used to detect events such as earthquakes (Sakaki, Okazaki, and Matsuo 2010), wildfires (Boulton, Shotton, and Williams 2016), and power outages (Boulton, Shotton, and Williams 2016; Sun et al. 2016). However, most existing work focuses on geotagged data (i.e. known latitude/longitude coordinates); unfortunately, this body of data represents only about 1% of all tweets (Middleton, Middleton, and Modafferi 2014). Even though less than 0.5% of Twitter users turn on the geotagging feature in tweets due to privacy concerns (Li, Wang, and Chang 2012), they may still mention location information in their tweet content. Location information could be harvested from millions of additional tweets using location detection technologies.

Location detection can be considered as a subtask of named entity recognition (NER). One of the most fundamental tasks in information extraction, NER aims to classify words within text into different categories (e.g. person, organization, and location). The conditional random field (CRF) has served as the dominant machine learning method for NER, with well-known Stanford NER model developed at Stanford University (Finkel, Grenager, and Manning 2005). The original Stanford NER tool is trained on formal texts such as news data, which are remarkably different from short, informal, and noisy Twitter data. Hence, applying an NER model that focuses on complex data can generate poor performance in analyzing tweets. Lingad, Karimi, and Yin (2013) showed that NER tools retrained on Twitter data can significantly boost location detection from tweets. In contrast to existing methods, deep learning methods–specifically, recurrent neural networks (RNNs) and convolutional neural networks–are usually trained in an end-to-end fashion, significantly reducing dependency on hand-engineered features. In this paper, different NER models are applied to tweets regarding power outages and compared for location detection. Models include the original Stanford NER, retrained Stanford NER with tweets, and a deep learning RNN (i.e. bidirectional long short-term memory (LSTM) networks.

This paper most notably presents a machine learning- and deep learning-based framework for mapping power outages using Twitter data that has not been geotagged. A deep learning method is applied to enhance location detection from tweets. Because geotagged tweets are scarce, we use public address data to augment the limited training data set, a method that improves location detection. Our method significantly increases the availability of location-based tweets, enabling power outage monitoring on a much larger scale. The advanced natural language processing (NLP) techniques and deep learning methods used in this study may also improve future location-based social media studies of many kinds.

2. Literature review

The method of harvesting location information from tweets is related to work on geoparsing (Jones and Purves 2009). *Geoparsing* allows a researcher to convert place names and qualitative details in text into specific spatial references such as latitude and longitude coordinates. Given the massive amount of data available, it is important to develop efficient and accurate methods to automatically identify various types of geographic information. Place names include geopolitical entities and administrative divisions (e.g. country, city names), street names, buildings, and points of interest (POI) (e.g. hotels, restaurants, parks); and colloquial place names. Identifying vernacular descriptions of locations becomes an especially important task considering the amount and value of VGI information (Hu, Mao, and McKenzie 2018). Geoparsing methods can be generally classified into three categories: (1) gazetteer matching (i.e. a dictionary of place names is used to match location names in text), (2) linguistic pattern and semantic rules matching, and (3) machine learning methods. Also, hybrid methods are often used (Gelernter and Shilpa 2013). Detailed discussions of geoparsing methods and challenges are provided in Leidner and Lieberman (2011) and Monteiro, Davis Jr, and Fonseca (2016).

This paper focuses on location extraction from short text (i.e. *tweets*). Due to the brevity, lack of structure, and commonly use of abbreviations, and prevalence of misspellings, machine-based recognition of geographic information in tweets is a difficult task. Kinsella, Murdock, and O'Hare (2011) used coordinates extracted from geotagged tweets to create a language model that infers the location of the tweet based on the term distribution within geotagged tweets. Gelernter and Mushegian (2011) applied Stanford NER software, based on a CRF model to recognize place names in tweets. However, because the standard NER model uses formal text such as news data, it cannot recognize locations that are noncapitalized, abbreviated, misspelled, or colloquial (Gelernter and Mushegian 2011). To improve the out-of-the-box Stanford NER, Lingad, Karimi, and Yin (2013) retrained the NER model using labeled tweets, where location names are manually annotated. Results have shown a significant improvement in detecting location information in tweets, including the names of POIs such as hotels, restaurants, and malls. Malmasi and Dras (2015) reported the best result in the ALTA 2014 Twitter Location Detection competition task (Molla and Karimi 2014), but their method relied heavily on a mix of NLP tools (i.e. noun phrase parser), gazetteers, and manually defined rules, which cannot be easily scaled to other data sets.

In recent years, deep learning methods have been used successfully in computer vision applications including image recognition and object detection (LeCun, Bengio, and Hinton 2015). Deep learning methods have significantly reduced dependency on hand-engineered features for linguistic pattern and semantic analysis. Collobert et al. (2011) introduced the first deep learning method for NER; it is a simple but effective feed-forward neutral network containing only one convolution layer. This model, however, is restricted to a fixed window size (e.g. three to five words in close proximity) but disregards 'long-distance' relationships between words in a sentence. In contrast, RNNs, in particular, LSTM networks (Hochreiter and Schmidhuber 1997), can take into account an infinite amount of context for a word to predict the current output through purpose-built memory cells. While the networks consider word- and sentence-level information to predict the named entities independently, they discard dependencies across sentence-level tags. To further improve NER performance, a CRF layer, which can model a sequence of labels by considering neighbors' labels, is usually added to further process network outputs to find the best sequence of labels (Huang, Xu, and Yu 2015; Lample et al. 2016). In this paper, we adopt the bidirectional LSTM model with a CRF layer to identify location names from tweets.

3. Methodology and results

Figure 1 shows the framework of our methodology. First, we classified tweets into true and false power outage tweets. Second, tweets with a high probability were considered true power outage

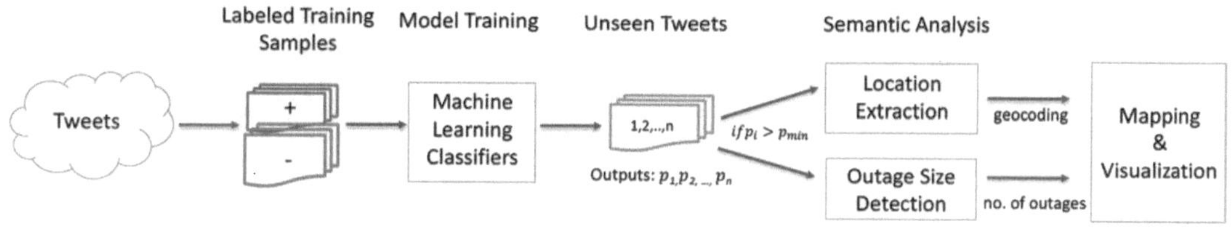

Figure 1. The proposed framework of power outage classification and location detection.

reports for the purposes of this research and passed to the second step for location detection. If the number of power outages for the event was mentioned in tweets, we extracted that information based on regular expression patterns.

3.1. Data collection and preprocessing

We used Twitter Streaming API to collect tweets containing at least one of the following keywords: 'power outage', 'power outages','power cut', 'power blackout', and 'blackout' over the period from 3 September to 9 October 2016. We also set a geographic boundary to collect tweets posted from the United States. In total, 612,178 tweets were obtained, with non-English tweets removed. A special word tokenizer was written to split each tweet into tokens. The tokens included URLs; hashtags (i.e. a string preceded by #); user name mentions (i.e. a string preceded by @); date and time strings (e.g. 9/1/2016, 10:30 AM); numbers (e.g. 13.99, 2000); and emoticons. To reduce feature space, we standardized each tweet by transforming text into lowercase, replacing the URL with special character '<URL>', and replacing mentioned usernames with '@username'. Emoticons were converted to Unicode strings (e.g. the term 'smiley' to replace smiley faces, the term 'sad' to replace frowny faces). With this conversion, emotions are treated the same as other normal words. Because very short tweets rarely contain useful information, we removed tweets containing less than three words.

3.2. Machine learning classification of power outage tweets

To train the classifiers, we manually labeled positive and negative tweets, and duplicate tweets were removed. In order to avoid overlap with testing data, training data is randomly selected from a different time period, i.e. from November 2016 to February 2017. We defined positive tweets as reporting an actual power outage event and negative as not about a real power outage or otherwise being irrelevant (see examples in Table 1). Positive tweets clearly stated a power outage event and usually contained location, number of outages, and expressions of sad emotion. Conversely, even though all the negative tweets included power outage-related keywords, they were not actually discussing a real power outage. For example, alternate meanings for the term 'power' sometimes caused a false positive tweet to enter the data pool. See the sixth example given in Table 1: '@realDonaldTrump I would like to c u cut government in half and give states the power back', in which 'power' refers political power rather than electricity.

Table 1. Positive and negative power outage tweet examples.

1	*1am update to power outages: 13,240 in Brevard County. Numberkeeps climbing. #HurricaneMatthew*	positive
2	*@ohioedison power outage in Jackson Twp. 44718. Not on map. Any news?*	positive
3	*I've never lived in an area with so many power outages. I'm sopissed off right now!*	positive
4	*This power outage isn't cool at all*	positive
5	*I had a dream Texas state had a huge power outage and my testwas canceled...*	negative
6	*@realDonaldTrump I would like to c u cut government in half andgive states the power back*	negative
7	*Did you lose power yesterday? We were lucky and never lost ours*	negative
8	*Are you prepared in the event of a power outage? https://t.co/GK8TyNdxu4*	negative

Three annotators manually labeled a random sample of 2355 positive and 2355 negative tweets. Classification results were compared based on k-fold cross-validation; here, we set $k = 5$. We randomly split the data set into five equally sized samples (i.e. 20% each) and left one sample as test data and the other $k - 1$ folds of datasets for training. k iterations need to be conducted, and then the performance is the average of the values computed in the loop.

Table 2 compares classification results across multiple machine learning models, including logistic regression, multinomial and Bernoulli naive bayes, and linear support vector classification (SVC) (i.e. a linear support vector machine for classification). These classifiers are implemented by Python scikit-learn libraries, where the default parameter settings are adopted (parameter updates do not show a significant change in classification performance). For each model, we compared different feature sets including (1) features generated by the bag-of-words model (i.e. individual word of the data set) weighted by the term frequency and term-frequency inverse-document-frequency (TFIDF) (Sparck Jones 1972), and (2) features generated by the bag-of-ngrams model weighted by TFIDF. TFIDF measures the importance of a term (t) to a document (d) in a collection of documents (D), which is calculated by $tfidf(t, d, D) = tf(t, d) \times idf(t, D)$, where $tf(t, d)$ is the occurrence frequency of t in d and $idf(t, D)$ is the logarithm of the inverse of the division between the number of documents containing t and the total number of documents in D. TFIDF of ngrams is generated by the bag-of-ngrams model, where ngrams is a contiguous sequence of n words. Specifically, we used a combinations of unigrams (i.e. a single word) and bigrams (i.e. two consequent words). We also added trigrams (i.e. three words) as features, but performance did not improve. Results were evaluated in terms of four metrics: precision, recall, F-score, and accuracy, calculated as shown below.

$$
\begin{aligned}
precision &= \frac{tp}{tp + fp}, \\
recall &= \frac{tp}{tp + fn}, \\
F &= \frac{2 \cdot precision \cdot recall}{precision + recall}, \\
accuracy &= \frac{tp + tn}{tp + tn + fp + fn},
\end{aligned} \tag{1}
$$

where tp represents true positive, tn represents true negative, fp represents false positive, and fn represents false negative. The set of tweets labeled as true power outages and also are correctly detected by our method constitutes true positive data. The set of true power outage tweets that were misclassified as negative constitutes true negative data. The set of true power outage tweets missed by our method constitutes false negative. The set of false power outage data misclassified as true power

Table 2. Comparison of machine learning classifiers on power outage tweet classification.

Model	Precision	Recall	F-score	Accuracy
logistic_tf	0.8335	0.8713	0.8485	0.8486
logistic_tfidf	0.8281	0.8760	0.8470	0.8471
logistic_tfidf_ngram	0.8193	0.9146	**0.8560**	**0.8565**
mult_nb_tf	0.8148	0.8900	0.8434	0.8437
mult_nb_tfidf	0.8132	0.8790	0.8381	0.8384
mult_nb_tfidf_ngram	0.8224	0.9172	**0.8590**	**0.8594**
bern_nb_tf	0.8107	0.9134	0.8493	0.8499
bern_nb_tfidf	0.8107	0.9134	0.8493	0.8499
bern_nb_tfidf_ngram	0.8121	0.9427	**0.8613**	**0.8622**
svc_tf	0.8231	0.8433	0.8310	0.8310
svc_tfidf_tf	0.8288	0.8815	0.8495	0.8497
svc_tfidf_ngram	0.8253	0.9236	**0.8636**	**0.8641**

outage tweets (i.e. false alarms) constitutes false positive data. According to the definition in Equation 1, *precision* measures the fraction of relevant tweets from all the tweets identified as relevant, while *recall* is the fraction of relevant tweets within all the actual relevant documents. The precision decreases or increases as the recall increases or decreases. *F-score* is a weighted average of the precision and recall, which we use to compare the performance of various machine learning classifiers.

As shown in Table 2, all the models tested generate promising results, with a *F*-score of higher than 0.8. Among all the classifiers (i.e. logistic regression, naive bayes, and SVC), using tfidf-ngram features give better results than tf- and tfidf-unigram (i.e. bag-of-words) features. It can be because adding bigrams is useful in identifying more informative phrases (such as 'outage in', 'outage at', 'internet outage') than single word for classification. With the same tfidf-ngram features, classifiers perform almost equally well in terms of *F*-score and accuracy. In this study, we adopted the *SVC-tfidf-ngram* model for the classification of large-scale power outage tweets, which uses a linear L2-squared hinge loss function and a penalty parameter of $C = 1.0$.

Figure 2 plots the top most informative positive and negative features. Location related features such as 'outage in', 'outage at', 'at', 'in', 'area', and other words such as 'due to', 'please', and 'closed' are important positive features, while 'blackout' (because it often co-occurs with the word 'drunk'), 'obama', 'internet outage' are negative features.

In practice, not only do we want to know if a tweet is actually about a power outage, but we also want to know the confidence score on the prediction. To obtain the confidence score, we performed probability calibration for the SVM outputs based on Platt's sigmoid model (Platt 1999), as shown in Equation (2).

$$Pr(y = 1|x) = \frac{1}{1 + exp(Af(x) + B)}, \tag{2}$$

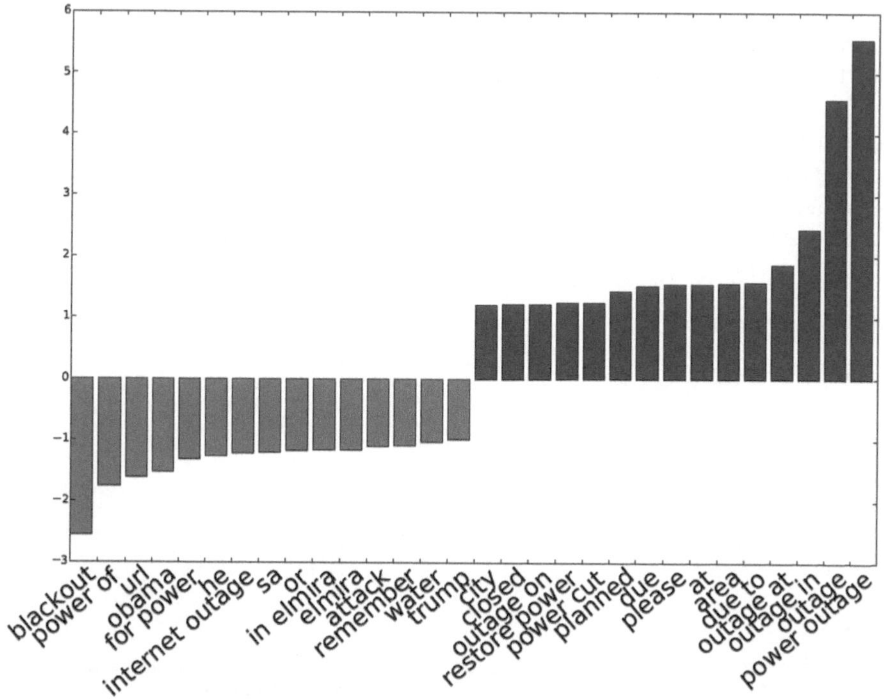

Figure 2. Top positive and negative features for power outage classification using the SVC and bag-of-ngram model.

where x is the input and $f(x)$ is the output of the classifier. A and B are parameters learned by the algorithm. Briefly speaking, Platt scaling fits a logistic regression model to classifier outputs and produces probability estimates (from 0 to 1). Classification results have shown that, among the set of 621,544 tweets containing power outage-related keywords, 90,060 (14%) tweets had a probability $p \geq .5$, which we consider positive tweets. Among the set of positive tweets, the percentage of tweets with $p \geq .6$ is 85% (76,588); $p \geq .7$, 73% (65,667); $p \geq .8$, 55,234 (61%); $p \geq .9$, 40,316 (45%). Figure 3 shows the number of power outage tweets with a probability of $p \geq .9$ over the study period. Spikes of outage tweets can clearly be seen across two hurricanes–Hermine (28 August to 8 September 2016) and Matthew (28 September to 10 October 2016)–and on September 13, when there were major power outages in north New Jersey (including the Newark Airport), as well as on 22 September 2016, when a fire at a Puerto Rico power plant caused utility failure.

3.3. Location extraction

After identifying the set of true power outage tweets, we identified and mapped outage locations. As stated earlier, location detection can be considered an NER subtask. We compared several NER models for their usefulness in performing location detection in tweets: the original Stanford NER, Stanford NER retrained with tweets, and a deep learning neural network model.

3.3.1. Stanford NER: conditional random field classifier

Stanford's NER (https://nlp.stanford.edu/software/CRF-NER.shtml), a cutting-edge NER tool, uses CRF and distributional similarity-based features to improve NER (Finkel, Grenager, and Manning 2005). CRF is a conditional sequence model: Given a sequence of observations (e.g. words), it aims to find the highest possible sequence of states (e.g. named entity tags). The entity tag for each word in a sentence is predicted, not independently, but by considering the tags of neighboring words. For this research, we examined two versions of CRF models: the original Stanford NER that is pretrained based on news and Wikipedia data, and the retrained model based on the annotated Twitter data. The features that we adopt to train CRF models include current word, previous word, next word, part-of-speech (POS) tag of current word, surrounding-sequence POS tag, word shape features, character n-gram, and left and right words within the window size of 6 words.

3.3.2. Deep learning method: bidirectional LSTM

LSTM can be used to model sequential data. Given an input vector $\mathbf{X} = (x_1, x_2, \ldots, x_n)$, it returns another sequence vector $\mathbf{h} = (h_1, h_2, \ldots, h_n)$, where h_t is a state corresponding to x_t at time step t. The key component of LSTMs is the memory cell, which contains several 'forget gates' that can

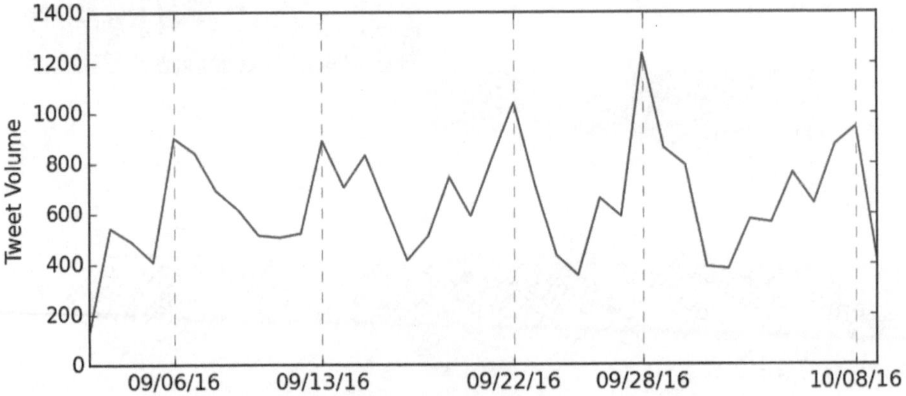

Figure 3. Daily power outage tweet volume (6 September to 8 October 2016).

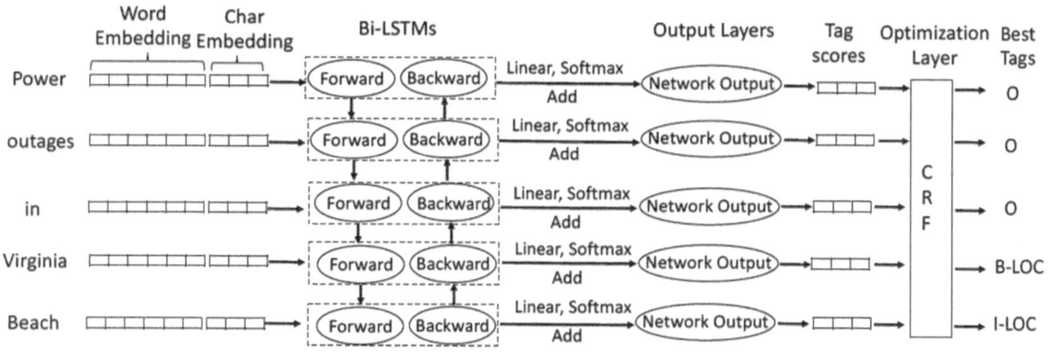

Figure 4. The network architecture of bi-LSTM-CRF for location detection.

control the amount of information flow into or out of their memory (Hochreiter and Schmidhuber 1997). This design allows LSTM to account for long-distance dependencies, while the traditional RNNs tend to be biased toward the most recent inputs in the sequence (Bengio, Simard, and Frasconi 1994). To understand a unit in a sequence, considering both the left- and right-hand sides are useful. To model the left (or right) context of a word x_t in a sentence, a forward (or backward) LSTM model can be used. The bidirectional LSTM model proposed in (Graves and Schmidhuber 2005) combines the forward and backward LSTMs, offering advantages compared to the single-direction methodologies. In the NER problem, bidirectional LSTM networks allow the model to take into account a long-range context on both left- and right-hand sides of a word. Figure 4 shows the network architecture of this deep learning method, which includes a word and character embedding layer, bi-LSTM (followed by a linear and softmax layer), and a CRF layer that optimizes sequence labeling at the end.

First, the embedding layer transforms discrete features into a continuous vector representation that combines both word- and character-based embeddings. Here, the discrete features are word and character indices: Word index is the word ID in the vocabulary, and character index is the ID in a unique character set, including *abcdefghijklmnopqrstuvwxyz0123456789-* ,;!?:′ ″ /|_#$%&* ~′ + =<> ()[]{}. Real-valued word embeddings learned by neural language models have been shown to successfully capture syntactic and semantic information about words, which has significantly improved the performance of a variety of NLP tasks such as syntactic parsing, NER, and sentiment analysis (Collobert et al. 2011; Lample et al. 2016). The word embeddings adopted in this paper are pretrained using an unsupervised learning algorithm, GloVe, which aims to learn vector representations for words based on aggregated global word-word co-occurrence statistics from a corpus (see (Pennington, Socher, and Manning 2014) for details on the model and training process). The word-embedding dimension we use is 100; it can be downloaded from https://nlp.stanford.edu/projects/glove/. While word embedding represents what a term means, character-level embedding captures information about word morphology and shape (i.e. what the word looks like) (Santos and Zadrozny 2014). Character embeddings along 25 dimensions were randomly initialized in this task. Then, each word, w_n, in a sentence was converted into a vector $v_n = [v^{word}, v^{char}]$ to concatenate the word– and character-level embeddings.

Second, a sequence of word vector representations obtained from the embedding layers was given to a bidirectional LSTM. The output of each network at each time step was decoded by a linear layer and a log-softmax layer into log probabilities for each tag category. These two vectors were then added together to produce the final output of the network: a sequence of vectors containing the probability of each named entity label for each corresponding token. Third, the sequence of probability vectors from the previous output layer was given to the final sequence optimization layer, CRF, to identify the best sequence of predicted named entity labels for the whole sequence.

From the first to the final layers, word features were transformed into named entity tag scores. The parameters of all layers were learned during the training process. Our model is trained based on a back-propagation algorithm using stochastic gradient descent with a learning rate of 0.005. The dropout rate was 0.5 (i.e. the smaller dropout rate led to longer training time and higher rates reduced performance). Our implementation of this bidirectional LSTM and CRF architecture was based on a cutting-edge deep learning NER system, NeuroNER (Dernoncourt, Lee, and Szolovits 2017) developed by MIT. Unlike the traditional machine learning classifiers, deep learning method can effectively learn features directly from training data instead of relying on handcrafted features.

3.3.3. NER model comparisons

The location-annotated Twitter data we used for our research was provided by the ALTA 2014 Twitter Location Detection competition (Molla and Karimi 2014). The original training and testing sets included 2000 and 1000 tweet IDs, respectively. Some tweets were deleted or became invalid, making our data set 1851 for training and 930 tweets for testing. The training data was randomly split into two subsets: 1600 for training and 251 for validation. The training set is used to train the neural network parameters, and the validation set is used to determine when to stop training. Annotated locations included place names and POIs, in either main text strings or hashtags.

The named entity tags were encoded with the standard Beginning/Inside/Outside (BIO) format, where *B-* represents the beginning of a named entity, *I-* indicates the token is inside an entity but is not the first word, and *O-* is used for tokens outside of an entity. Table 3 provides an example of a BIO-encoded sentence.

Table 4 shows the NER test results based on four models: (1) the original Stanford NER, (2) the Stanford NER model retrained with location-annotated Twitter data, (3) the winner of the ALTA 2014 task, and (4) NeuroNER. The original Stanford NER model generated high precision (0.9451) but very low recall (0.3207). It can identify formal location names (e.g. city and county names) but fails to recognize street names, zip code, city or place name abbreviations, and location hashtags. Retraining Stanford NER with location-annotated tweets can significantly boost its recall and improve the *F*-score to 0.7728. The method used by Malmasi and Dras (2015) for the ALTA 2014 competition task relied heavily on a mix of NLP tools (i.e. noun phrase extraction), gazetteers, and manually defined rules, which cannot be easily scaled to other datasets. In contrast, the deep learning bidirectional LSTM-CRF model outperforms all other methods in terms of *F*-score (0.8094) without using any manually defined features, rules, or gazetteers. This method generalizes to different languages and new data and domains. Therefore, in the following session, we utilized the deep learning model to detect location names from tweets.

3.3.4. Applying NER model for location detection in power outage tweets

Our starting data set of 1600 tweets included a very limited amount of labeled data. To increase the number of training samples, we leveraged the open data source OpenAddresses (OA) (https://openaddresses.io/), a VGI project integrating address data contributed by worldwide volunteers.

Table 3. Example of sentence encoded with BIO tagging scheme.

A	fire	destroyed	cabins	In	Great	Smoky	Mountains	area	this	morning	.
O	O	O	O	O	B-LOC	I-LOC	I-LOC	O	O	O	O

Table 4. Model comparisons: location detection results on the ALTA 2014 datasets.

Model	Precision	Recall	*F*-Score
Stanford NER	0.9451	0.3207	0.4788
Stanford NER retrained	0.8668	0.6972	0.7728
ALTA2014 Winner (Malmasi and Dras 2015)	–	–	0.7920
Deep learning method (bi-LSTM-CRF)	0.8481	0.7741	**0.8094**

Table 5. Datasets used for training, validating, and testing the location detection model.

Data	Train	Validation	Test
Tweets (ALTA 2014)	1851	251	930
Address names (OpenAddresses)	8107	1673	772

We considered the street names to be true samples of location names. We collected 8,107 unique street addresses in New York City as training data, 1673 addresses in Rochester, New York, as validation data, and 772 addresses in Cambridge, MA, as testing data. We took steps to ensure minimal overlap between the training and test data; only one address appears both in training and test data sets. Table 5 lists the data sets used to train, validate, and test NeuroNER for location detection. To test whether the addition of street address as training data can improve location recognition, we compared two models: the first with ALTA Twitter data as training data only, and the second with both ALTA and street address data as training data. Results in Table 6 show that adding street addresses for training improved the *F*-score from 0.81 to 0.85. Therefore, in the following, we used the deep learning model trained using a combination of ALTA data and street addresses to detect power outage locations in tweets.

As discussed in Section 3.2, 40,316 power outage tweets in the data set had a classification probability of $p \geq .9$. We then applied our location detection method on these strongly positive power outage tweets. Here, we must clarify two terms: 'geotagged' and 'geodetected'. Tweets with known coordinates are *geotagged* data, and tweets for which a location could be detected from text by our method are *geodetected* data. Table 7 shows that, of all the positive power outage tweets, only 1386 (3.44%) were geotagged tweets, while 28,825 (71.5%) were geodetected, after applying our location detection method. In other words, nearly 20 times more geodetected tweets are available compared to geotagged data. Our location detection method will enrich the pool of geographic data available for mapping large-scale power outages. Table 8 shows examples of location strings detected power outage tweets. As we can see the location strings identified are different types including state/county/city names, streets, place/area names, and zipcode, expressed either in text strings or hashtags.

After locations were detected, we geoparsed unstructured location information into geographical coordinates. Bing Maps Location API was used via Python's geocoding library (i.e. geopy), which accepts a location string as the input and generates address name and latitude/longitude coordinates. For example, Battleship Road extracted from a tweet is converted to (Battleship Rd, Camden, SC 29020, United States, [34.26021, – 80.62693]). We acknowledge the limitation of Bing Maps API for geocoding ambiguous place names (e.g. location names that exist in multiple regions). So, when multiple results are returned, the most likely one is selected in our results. With the coordinates obtained, we then mapped power outage locations.

As shown in Section 3.2, results include 40,316 positive power outage tweets with a classification probability of $p \geq .9$, and among which, 71.5% of the data is geodetected. Because there are too many points to visualize in one graph, we only selected a small fraction of data for a better visualization.

Table 6. Location detection using the bidirectional LSTM-CRF model with different training datasets.

Training Data	Precision	Recall	*F*-Score
Only Tweets (ALTA 2014)	0.8971	0.7438	0.8133
Tweets + OpenAddresses	0.8936	0.8172	**0.8537**

Table 7. Number of geotagged and geodetected tweets among 40,316 power outage tweets with $p \geq .9$.

Data	No. Tweets	Percentage
Geotagged	1386	3.44%
Geodetected	**28,825**	**71.50%**

Table 8. Examples of locations detected from tweets by using bidirectional LSTM-CRF.

1	*Hamilton County still experiencing power outages.*
2	*Another power outage just popped up. This time between Scotts Hill and Hampstead. Over 100 affected.*
3	*Hey @username or @username any news on a power outage near wynwood?*
4	*UPDATE — Power restored at #SCState. Downed power line caused by a fallen tree resulted in outage. @username #scnews*
5	*@username trying to call you and not working. Power outage in Sun Valley Ca.. Valerio St / Cartwright Ave.*
6	*It appears we've had a power outage here at Smith Stadium. There were some problems with the fire alarm earlier. Mounties on move #arpreps*
7	*McHenry Outdoor Theater cancels double feature due to power outage - Northwest Herald*
8	*@username There's been a power outage & now I have no internet. I'm in the red hills road area. When will service be restored?*
9	*@username power outage in the 90504 area. Any updates?*

Here, we set the classification probability filter as: $p \geq .998$, which gives 1790 tweets, i.e. 4% of all the 40,316 positive power outage tweets. Figure 5 shows the selected small subset of all the geodetected power outage tweets obtained in Section 3.3.4. Each pinpoint represents a power outage tweet at the geodetected location, and a pop-up window shows its key attributes (e.g. tweet, time posted, user who posted the tweet, and location names detected by our method). These pieces of information enable emergency responders to better understand power outage situations.

3.4. Outage size detection

Understanding the damage level incurred with power outage events is also important for emergency responders and utility companies. Some tweets mention how many power outages have occurred as of the time of the tweet (e.g. '9,200 customers have power outages, and the number is climbing'; 'At least more than one hundred people have power outages in our neighborhood; 1000 outages in King County') To capture such patterns, we defined linguistic rules, which can be written in regular

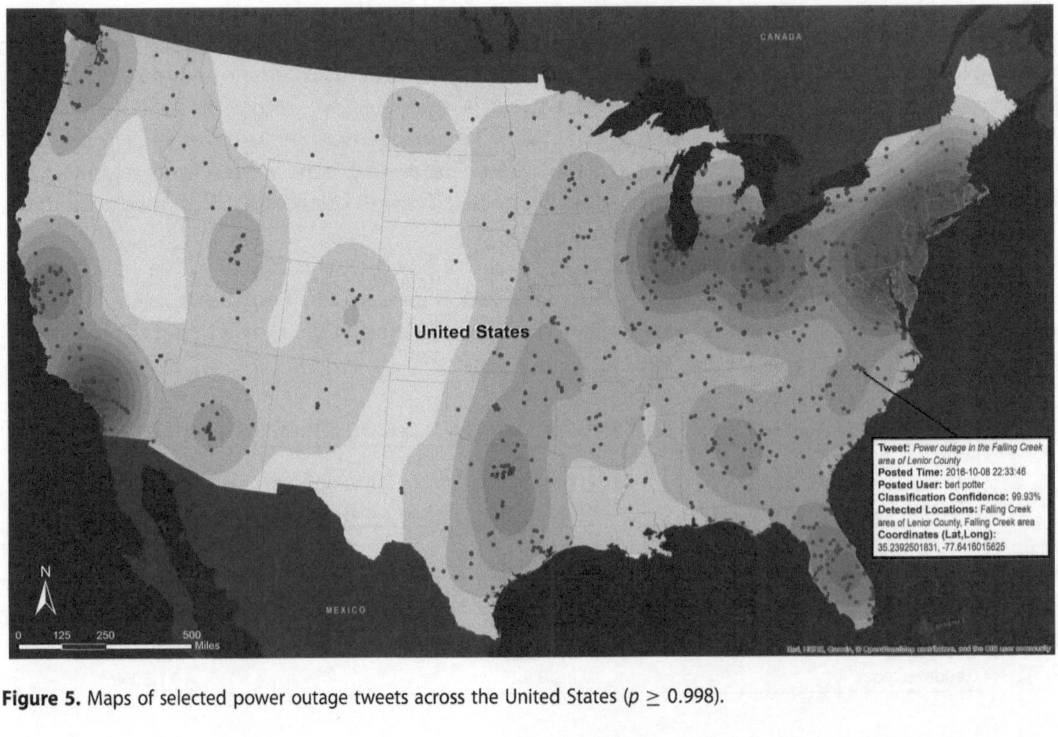

Figure 5. Maps of selected power outage tweets across the United States ($p \geq 0.998$).

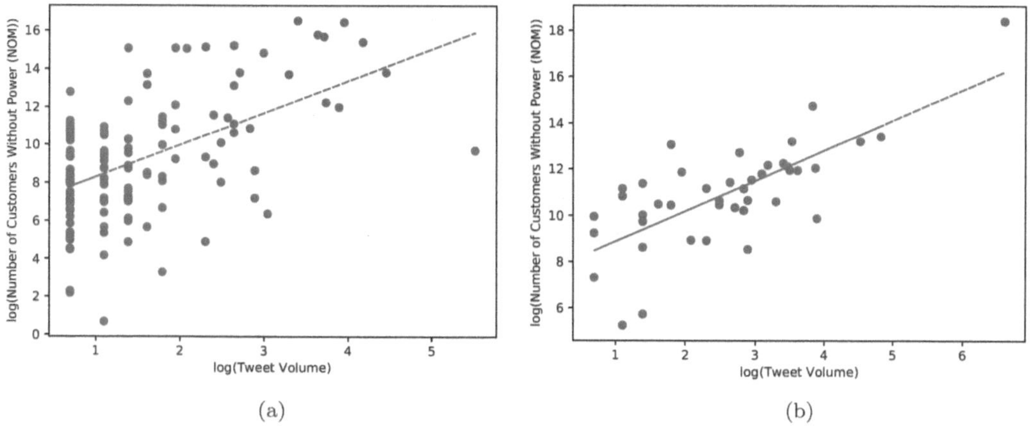

Figure 6. Scatter plot of power outage tweets versus number of customers without power, as reported by National Outage Map, for the period during which Matthew Hurricane affected Florida (6–7 October 2016) at both the US county and state levels. (a) County-level. (b) State-level.

expression patterns (e.g. numbers or digits followed by keywords such as 'customers', 'people', 'members', 'outages', 'power cuts') within a small window size (i.e. there words). Phrases matching such patterns were extracted and reported as the size of outages.

3.5. Validation

To validate our results, we compared the number of power outage tweets with the official number of customers without power, as reported in the National Outage Map (NOM) developed by Oak Ridge National Laboratory (ORNL) for the US Department of Energy. NOM data is available at a county level, but due to the data scarcity, we also conduct the correlation analysis based on the aggregated data at the state-level.

Figure 6 shows a strong positive linear correlation between the number of geodetected power outage tweets and the number of customers in the counties and states recorded by NOM during Matthew Hurricane. At the county-level, it shows a statistically significant correlation with a Pearson correlation coefficient $\gamma = 0.53$ ($p \ll 0.01$, $n = 131$). At the state level, the linear correlation is stronger, with a Pearson correlation coefficient $\gamma = 0.73$ ($p \ll 0.01$, $n = 42$). The positive correlations show the effectiveness of our method for power outage detection using social media. The number of power outages reported by NOM is collected from hundreds of utility companies' websites. However, any change to a website's structure can reduce how effective crawling it is; this can be avoided only with consistent manual updates – a difficult and time-consuming endeavor. It should be noted that not every utility company has or actively maintains their websites, and website updates often come after a time lag. Therefore, Twitter – real-time, public, and widely used by millions – can provide a valuable data source for power outage monitoring and can complement information gathered from utility companies' websites.

4. Conclusion

This paper provides a novel machine learning- and deep learning-based framework for locating and classifying near-real-time power outages from social media. This research is different from existing work that focuses on geotagged tweets (i.e. those with known latitude/longitude coordinates), and our work promises to enrich the data available for power outage research by enlarging the pool of available data. We are the first to use a deep learning model to identify locations associated with power outage tweets. Results have shown an improvement compared to other methods: (1) the original Stanford NER model trained on newswire text performed poorly on tweets that are full of informal languages, misspellings, abbreviations, hashtags, and unreliable capitalization and punctuation; (2) traditional rule- and machine learning-based methods rely on handcrafted features which are time-consuming to develop and may not generalize well to new corpus. In contrast, the great advantage of deep learning methods can be attributed to their effectiveness to learn useful features automatically from the training dataset. The deep learning model used in this paper, bidirectional LSTM model coupled with a CRF top layer, does not require any prior knowledge, nor a gazetteer, and it does not depend on other NLP tools for preprocessing (e.g. part-of-speech, chunking). Therefore, it can be easily adapted to new data and domains.

Detecting locations within social media text automatically and accurately help solve the problem of geotagged data scarcity and significantly adds a rich amount of geographic data that emergency responders can leverage to find people who need help. Our results show a classification accuracy of 86% and significantly increase the amount of location data from 1386 geotagged tweets to 28,825 geodetected tweets, providing about 20 times more data than geotagged only. This work highlights the value of using free and public social sensor data (i.e. Twitter) to locate areas experiencing power outages during a disaster without adding expensive communication infrastructures. This social sensor-based outage detection framework can be integrated with existing physical sensor-based monitoring systems to improve outage detection efficiency and accuracy.

In the future, we will leverage the Geographic Big Data Analytics platform, PlanetSense (Thakur et al. 2015) to harvest and analyze massive data from multiple social media (e.g. Twitter, Facebook, Instagram, Flicker) and news media to track the status of power outage in the US. Once fully developed, this methodology could be integrated with platforms designed to monitor near-real-time situational awareness of US energy infrastructures, such as the FNET/GridEye monitoring system (Chai et al. 2016) and EAGLE-I (Environment for Analysis of Geo-Located Energy Information) technology (U.S. Department of Energy 2015).

Acknowledgments

This manuscript has been authored by UT-Battelle, LLC under Contract No. DE-AC05-00OR22725 with the US Department of Energy. The United States Government retains and the publisher, by accepting the article for publication, acknowledges that the United States Government retains a non-exclusive, paid-up, irrevocable, worldwide license to publish or reproduce the published form of this manuscript, or allow others to do so, for United States Government purposes. The Department of Energy will provide public access to these results of federally sponsored research in accordance with the DOE Public Access Plan (http://energy.gov/downloads/doe-public-access-plan). Also, we would like to thank to Laurie Varma from ORNL for her diligent technical editing and proofreading of the manuscript.

Disclosure statement

No potential conflict of interest was reported by the authors.

Funding

The authors would like to acknowledge the financial support received from Oak Ridge National Laboratory (ORNL)'s Liane Russell Distinguished Early Career Fellowship and grant no. TG0100000.

ORCID

Jibonananda Sanyal http://orcid.org/0000-0002-7789-3199
Budhendra Bhaduri http://orcid.org/0000-0003-1555-1377

References

Bengio Y., P. Simard, and P. Frasconi. 1994. "Learning Long-Term Dependencies with Gradient Descent Is Difficult." *IEEE Transactions on Neural Networks* 5 (2): 157–166.

Boulton C. A., H. Shotton, and H. T. Williams. 2016. "Using Social Media to Detect and Locate Wildfires." *Tenth International AAAI Conference on Web and Social Media.*

Chai J., Y. Liu, J. Guo, L. Wu, D. Zhou, W. Yao, Y. Liu, T. King, J. R. Gracia, and M. Patel. 2016. "Wide-Area Measurement Data Analytics Using FNET/GridEye: A Review." *Power Systems Computation Conference (PSCC), 2016*, 1–6.

Chen S. J., T. S. Zhan, C. H. Huang, J. L. Chen, and C. H. Lin. 2015. "Nontechnical Loss and Outage Detection Using Fractional-Order Self-Synchronization Error-Based Fuzzy Petri Nets in Micro-Distribution Systems." *IEEE Transactions on Smart Grid* 6 (1): 411–420.

Collobert R., J. Weston, L. Bottou, M. Karlen, K. Kavukcuoglu, and P. Kuksa. 2011 August. Natural Language Processing (Almost) from Scratch. *Journal of Machine Learning Research* 12: 2493–2537.

Dernoncourt F., J. Y. Lee, and P. Szolovits. 2017. "NeuroNER: An Easy-to-Use Program for Named-Entity Recognition Based on Neural Networks." *Proceedings of the 2017 Conference on Empirical Methods in Natural Language Processing: System Demonstrations. Copenhagen: Association for Computational Linguistics*, 97–102.

Finkel J. R., T. Grenager, and C. Manning. 2005. "Incorporating Non-Local Information into Information Extraction Systems by Gibbs Sampling." *Proceedings of the 43rd Annual Meeting on Association for Computational Linguistics*, 363–370.

Gelernter Judith, and Shilp Balaji. 2013. "An Algorithm for Local Geoparsing of Microtext." *GeoInformatica* 17 (4): 635–667.

Gelernter Judith, and Nikola Mushegian. 2011. "Geo-Parsing Messages from Microtext." *Transactions in GIS* 15 (6): 753–773.

Goodchild Michael F. 2007. "Citizens as Sensors: The World of Volunteered Geography." *GeoJournal*69 (4): 211–221.

Graves A., and J. Schmidhuber. 2005. "Framewise Phoneme Classification with Bidirectional LSTM Networks." *Neural Networks, 2005. Proceedings of IJCNN'05.*, Vol. 4, 2047–2052.

Hochreiter S., and J. Schmidhuber. 1997. "Long Short-Term Memory." Neural Computation, 9 (8): 1735–1780.

Hu Yingjie, Huin. Mao, and Gran McKenzie. 2018. "A Natural Language Processing and Geospatial Clustering Framework for Harvesting Local Place Names from Geotagged Housing Advertisements." *International Journal of Geographical Information Science* 1–25.

Huang Q., and Y. Xiao. 2015. "Geographic Situational Awareness: Mining Tweets for Disaster Preparedness, Emergency Response, Impact, and Recovery." *ISPRS International Journal of Geo-Information* 4 (3): 1549–1568.

Huang Z., W. Xu, and K. Yu. 2015. "Bidirectional Lstm-Crf Models for Sequence Tagging." *arXiv preprint arXiv:1508.01991.*

Jones Christopher B., and Ross S. Purves. 2009. "Geographical Information Retrieval." *Encyclopedia of Database Systems*, 1227–1231. Springer.

Kinsella Sheila, Vaness Murdock, and Nei O'Hare. 2011. "I'm Eating a Sandwich in Glasgow: Modeling Locations with Tweets." *Proceedings of the 3rd International Workshop on Search and Mining User-generated Contents*, 61–68. ACM.

Lample G., M. Ballesteros, S. Subramanian, K. Kawakami, and C. Dyer. 2016. "Neural Architectures for Named Entity Recognition". *Proc. NAACL-HLT.*

LeCun Yann, Yoshu Bengio, and Geoffre Hinton. 2015. "Deep Learning." *Nature* 521 (7553): 436.

Lee K., J.y. Shin, and R. Zadeh. 2013. "Twitter's Effectiveness on Blackout Detection During Hurricane Sandy."

Leidner Jochen L., and Michael D. Lieberman. 2011. "Detecting Geographical References in the Form of Place Names and Associated Spatial Natural Language." *SIGSPATIAL Special* 3 (2): 5–11.

Li Rui, Shengji Wang, and Kevin Chen-Chua Chang. 2012. "Multiple Location Profiling for Users and Relationships from Social Network and Content." *Proceedings of the VLDB Endowment* 5 (11): 1603–1614.

Lingad J., S. Karimi, and J. Yin. 2013. "Location Extraction from Disaster-Related Microblogs." *Proceedings of the 22nd International Conference on World Wide Web*, 1017–1020.

Liu Y., and N. N. Schulz. 2002. "Knowledge-Based System for Distribution System Outage Locating Using Comprehensive Information." *IEEE Transactions on Power Systems* 17 (2): 451–456.

Malmasi S., and M. Dras. 2015. "Location Mention Detection in Tweets and Microblogs." *International Conference of the Pacific Association for Computational Linguistics*, 123–134.

Middleton S. E., L. Middleton, and S. Modafferi. 2014. "Real-Time Crisis Mapping of Natural Disasters Using Social Media." *IEEE Intelligent Systems* 29 (2): 9–17.

Molla D., and S. Karimi. 2014. "Overview of the 2014 ALTA Shared Task: Identifying Expressions of Locations in Tweets." *Australasian Language Technology Association Workshop 2014*, 151.

Monteiro Bruno R., Clodoveu A. Davis Jr, and Fre Fonseca. 2016. "A Survey on the Geographic Scope of Textual Documents." *Computers & Geosciences* 96: 23–34.

Pennington J., R. Socher, and C. D. Manning. 2014. "Glove: Global Vectors for Word Representation." *EMNLP* 14: 1532–1543.

Platt J. 1999. "Probabilistic Outputs for Support Vector Machines and Comparisons to Regularized Likelihood Methods." *Advances in Large Margin Classifiers* 10 (3): 61–74.

Sakaki T., M. Okazaki, and Y. Matsuo. 2010. "Earthquake Shakes Twitter Users: Real-Time Event Detection by Social Sensors." *Proceedings of the 19th International Conference on World Wide Web*, 851–860.

Santos C. D., and B. Zadrozny. 2014. "Learning Character-Level Representations for Part-of-Speech Tagging." *Proceedings of the 31st International Conference on Machine Learning (ICML-14)*, 1818–1826.

Sparck Jones K. 1972. "A Statistical Interpretation of Term Specificity and Its Application in Retrieval." *Journal of Documentation* 28 (1): 11–21.

Sun H., Z. Wang, J. Wang, Z. Huang, N. Carrington, and J. Liao. 2016. "Data-Driven Power Outage Detection by Social Sensors." *IEEE Transactions on Smart Grid* 7 (5): 2516–2524.

Takeichi Y., K. Sasahara, R. Suzuki, and T. Arita. 2014. "Twitter as Social Sensor: Dynamics and Structure in Major Sporting Events." *Artificial Life* 14: 778–784.

Thakur G. S., B. L. Bhaduri, J. O. Piburn, K. M. Sims, R. N. Stewart, and M. L. Urban. 2015. "PlanetSense: A Real-Time Streaming and Spatio-Temporal Analytics Platform for Gathering Geo-Spatial Intelligence from Open Source Data. *Proceedings of the 23rd SIGSPATIAL International Conference on Advances in Geographic Information Systems*, 11.

U.S. Department of Energy. 2014. "The Development of EAGLE-I: The First-Ever Technology to Track Power Outages Nationwide. Accessed February 15, 2017. https://energy.gov/articles/development-eagle-i-first-ever-technology-track-power-outages-nationwide.

Social and geographical disparities in Twitter use during Hurricane Harvey

Lei Zou, Nina S. N. Lam, Shayan Shams, Heng Cai, Michelle A. Meyer, Seungwon Yang, Kisung Lee, Seung-Jong Park and Margaret A. Reams

ABSTRACT

Social media such as Twitter is increasingly being used as an effective platform to observe human behaviors in disastrous events. However, uneven social media use among different groups of population in different regions could lead to biased consequences and affect disaster resilience. This paper studies the Twitter use during 2017 Hurricane Harvey in 76 counties in Texas and Louisiana. We seek to answer a fundamental question: did social-geographical disparities of Twitter use exist during the three phases of emergency management (preparedness, response, recovery)? We employed a Twitter data mining framework to process the data and calculate two indexes: Ratio and Sentiment. Regression analyses between the Ratio indexes and the social-geographical characteristics of the counties at the three phrases reveal significant social and geographical disparities in Twitter use during Hurricane Harvey. Communities with higher disaster-related Twitter use in Harvey generally were communities having better social and geographical conditions. These results of Twitter use patterns can be used to compare with future similar studies to see whether the Twitter use disparities have increased or decreased. Future research is also needed to examine the effects of Twitter use disparities on disaster resilience and to test whether Twitter use can predict community resilience.

1. Introduction

The ability of a community to prepare for, absorb, recover from, and more successfully adapt to disastrous events is one way to define disaster resilience (Lam et al. 2016; National Research Council 2012). Disaster resilience can be better understood by investigating human behaviors during the four phases of emergency management – preparedness, response, recovery, and mitigation (Federal Emergency Management Agency [FEMA] 2006). Given the same level of threat from a hazard, communities with different social and geographical characteristic have shown divergent experiences across emergency phases (Cai et al. 2016; Lam et al. 2015, 2016; Li et al. 2015; Mihunov et al. 2018; Muttarak and Pothisiri 2013; Petkova et al. 2018; Thieken et al. 2007). However, a major challenge is that data describing communities' behaviors in different phases of emergency management are difficult to access through traditional databases.

Social media, such as Twitter, provides an innovative approach to observe human behaviors under emergencies in real-time. Previous studies have investigated the utility of social media in emergency management and resilience after major disastrous events (Earle 2010; Kent and Capello 2013; Tsou 2015; Tsou and Leitner 2013), such as Hurricane Sandy in 2012 (Guan and Chen 2014; Kryvasheyeu et al. 2015, 2016; Zou et al. 2018), Colorado Flood in 2013 (Liu et al. 2018), and Hurricane Matthew in 2016 (Li et al. 2018; Martín, Li, and Cutter 2017). Their results demonstrate that information extracted from social media data could be used to detect affected areas (Assumpção et al. 2018), estimate potential damages (Kryvasheyeu et al. 2016), monitor evacuation and recovery process (Guan and Chen 2014; Martín, Li, and Cutter 2017), improve emergency management (Yin et al. 2012), and enhance disaster resilience (Dufty 2012).

However, these previous studies also show that utilizing social media data alone to derive scientific conclusions remains challenging. First of all, inaccurate or false information on location and user characteristics could invalidate the results. Developing a social media data-mining framework with standardized tools and indexes should help alleviate some of the problems and enable comparative analysis across events and regions (Zou et al. 2018). Secondly, social media use is known to vary across demographics groups, with younger, urban dwellers being more inclined to use social media than older, rural population (Li, Goodchild, and Xu 2013). This digital divide among disparate groups of population and across different regions could lead to biased conclusions if analysis is not carefully done. In the case of disasters, the social and geographical disparities of social media use could generate uneven respone and affect the long-term resilience of the communities. Thus, analyzing the patterns of social media use and its social and geographical characteristics across the different phases of emergency management is important, and findings from these types of studies should provide useful insights into reducing the disparities, reducing damage, and speeding recovery.

This paper studies Twitter use during the 2017 Hurricane Harvey in 76 counties in Texas and Louisiana. Hurricane Harvey made its first landfall in the United States on 25 August 2017 near Rockport, Texas as a category-4 hurricane. It lingered over the Houston area and produced over 50 inches of rainfall, causing widespread flooding and damages in the region. This unprecedented disastrous event is marked by controversies, especially on the issue of inadequate flood warning due to the unexpected amount of rainfall (Friedman and Schwartz 2018). At the same time, a new phenomenon emerged during the Harvey event: many residents in the Houston area resorted to social media to call for rescue from flooded homes when the 911 system was overloaded. This changing use of social media marks Harvey as one of the very first disastrous events in which social media has played an important role in fast-responding and rescue missions in the response phase of emergency management.

In this study, we seek to answer a fundamental question: did social and geographical disparities of Twitter use exist during three phases of emergency management (preparedness, response, recovery)? We hypothesize that communities with better socioeconomic conditions had more disaster-related Twitter use in all three phases of emergency management, implying that they are more likely to access disaster information and get the responses for rescues. We will employ the Twitter data-mining framework for resilience analysis developed in a previous study of Hurricane Sandy to process the data and calculate two indexes: Ratio and Sentiment (Zou et al. 2018). Regression analyses between thc Ratio indexes and the social and geographical characteristics of affected counties during three phrases of emergency management are conducted to identify the main social and geographical factors affecting Twitter use during the event. These results can be used to compare with future similar studies to evaluate whether Twitter use disparities have increased or decreased over time and whether such disparities may affect the resilience of the communities. Findings from this study will further elucidate potential applications and reveal the challenges of using Twitter in disaster resilience research and emergency management improvement.

2. Literature review

Social media use is rapidly changing and is increasingly infiltrating every sector of the society. Through social media, numerous users can share or exchange information at any time in any place. With the timestamps and geo-tags, every social media user can act as a smart sensor that collects information about their current locations or share feelings of any topics on social media in real-time. Analyzing social media data could provide useful information on understanding human behaviors during an event. Social media is also an effective way to measure public attitudes towards different global themes.

Because of the real-time nature and easy online access of the information, social media data such as Twitter data have increasingly been used to examine a variety of hazardous events and emergency management activities. Zou et al. (2018) provides a summary of the literature on social media use during disasters (Dufty 2012; Earle, Bowden, and Guy 2012; Kent and Capello 2013; Li, Goodchild, and Xu 2013; Lindsay 2011; Mandel et al. 2012; Merchant, Elmer, and Lurie 2011; Rizza and Pereira 2014; Shelton et al. 2014; Tsou 2015; Tsou and Leitner 2013). For instance, Dufty (2012) suggested that social media could play a significant role in building community resilience through risk reduction, emergency management, and post-hazard recovery. Earle, Bowden, and Guy (2012) demonstrated that Twitter activities could help identify affected areas faster than traditional monitoring methods during earthquakes. Kent and Capello (2013) analyzed the spatial patterns of user-generated contents on several social media platforms during the Horsethief Canyon fire of 2012. Their results show that regions with greater hazard threat levels, more young people, denser population and higher awareness levels produced more valuable disaster-related information on social media platforms. Several studies analyzed Twitter activities during Hurricane Sandy and concluded that monitoring activities on social media can help estimate damage and monitor post-disaster recovery status (Guan and Chen 2014; Kryvasheyeu et al. 2015, 2016; Shelton et al. 2014; Wang and Taylor 2014; Zou et al. 2018).

Most recently, Li et al. (2018) proposed a novel approach of using Twitter data for rapid flood mapping. By integrating Twitter data with stream gauge data and Digital Elevation Model data, their model could generate a reliable flood extend map in near real-time. Martín and others leveraged Twitter data to analyze residential evacuation behaviors during Hurricane Matthew in 2016 (Martín, Li, and Cutter 2017). Their research confirms that social media could be used to monitor public awareness and evacuation behavior during emergencies.

Despite the above successful studies, there are significant theoretical as well as technical issues related to the use of social media in emergency management and disaster resilience research. Significant technical issues include how to mine the huge volume of data which often have unrelated or false information, and how to develop efficient computer algorithms to visualize and identify social networks and relationships (Tsou 2015). As the amount of data created by social media grows, it is necessary to develop a social media data-mining framework to extract useful information from social media data and enable the comparison of social media activities across events and spaces. In terms of theoretical challenges, a major issue is the social and geographical disparities in social media use. Such disparities could have dire consequences in emergency management and disaster resilience (Li, Goodchild, and Xu 2013; Sloan et al. 2015). In an attempt to address the above two challenges, Zou et al. (2018) developed a geospatial data-mining framework with standardized indexes for analyzing Twitter use during disasters. The researchers applied the framework to study Hurricane Sandy and found significant social and geographical disparities in Twitter use existed in the event. Although social media use has become more popular nowadays, the digital divide in disaster-related social media use, which may affect the emergency management and disaster resilience building, needs more investigation.

3. Study area and data

3.1. Study area

Hurricane Harvey developed from a tropical wave to the east of the Lesser Antilles in the Caribbean Sea, reaching tropical storm status on 17 August 2017. On August 25 night, Harvey strengthened into a major hurricane at the intensity of Category-4, and made its first landfall near Rockport, Texas, at peak intensity (Figure 1). On August 26 morning, Harvey made its second landfall on the northeast of Copano Bay, Texas. Afterwards, Harvey had downgraded to a tropical storm as it stalled near the coastline of Texas, dropping unprecedented amounts of rainfall over Houston metropolitan areas. On August 28, it emerged back over the Gulf of Mexico, strengthening slightly before making a third and final landfall in Cameron, Louisiana on August 30. As Harvey drifted inland, it quickly weakened again and became extratropical storm on September 1, before dissipating on 2 September 2017 (Blake and Zelinsky 2018).

Harvey was the costliest tropical cyclone on record, inflicting approximately $125 billion in damage, primarily from widespread flooding in the Houston metropolitan area (National Oceanic and Atmospheric Administration [NOAA] 2018). Many areas received more than 40 inches of rain in less than one week as the system slowly meandered over eastern Texas and adjacent waters, causing catastrophic flooding. With peak accumulations of 60.58 inches, Harvey was the wettest tropical cyclone on record in the United States. The resulting floods inundated hundreds of thousands of homes, displaced more than 30,000 people, and prompted more than 17,000 rescues (Blake and Zelinsky 2018).

3.2. Twitter data

This study examines the disaster-related Twitter use during Hurricane Harvey. Twitter is a social media service that allows users to send and receive up to 140-character text messages (Twitter has

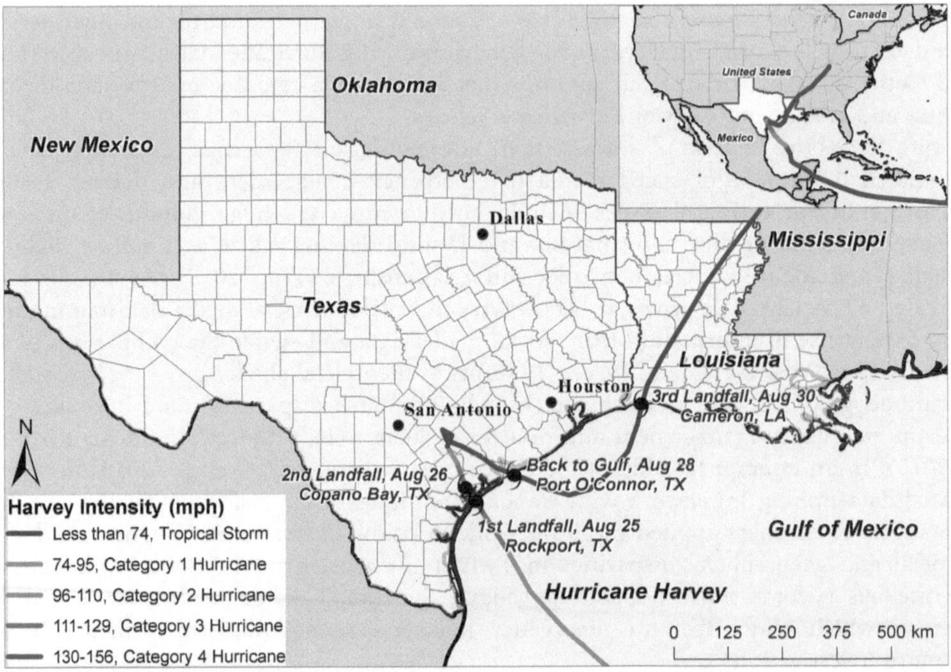

Figure 1. Track and timeline of Hurricane Harvey.

officially expanded its character count to 280 after November 2017), images, or website links, referred as 'tweets', through any online device. As one of the most popular social networking sites, more than 500 million tweets are posted on Twitter every day (based on authors' estimation), which makes it possible to obtain a large amount of information on human responses during any event. Twitter provides public Application Programming Interfaces (APIs) to help researchers collect data and analyze Twitter activities in near real-time. In addition to the tweet content and timestamps, the locational information of Twitter users could be obtained through different methods, which enables researchers to conduct spatial analysis.

In this research, Twitter data were obtained from Internet Archive (https://archive.org/), an online library which provides 1% (about 5 million tweets per day in 2017) randomly collected tweets from the full Twitter database since 2012 (Zou et al. 2018). Each tweet record contains information including the time when the tweet was created, tweet content, coordinates or places attached to the tweet, and information in the user profiles such as users' descriptions, addresses, account history, and follower/following statuses. We downloaded the Twitter data from 17 August to 12 September 2017 from this library. Five attributes derived from the Twitter data were used in this study: time when the tweet was created, text content, coordinates, place, and address in the user profile. Time of tweets was used to tabulate the Twitter data into the three stages of the emergency management cycle (preparedness, response, and recovery). Text contents were used for classification and sentiment estimation, while the latter three geographic elements were used to determine the location of each tweet. Twitter data during Hurricane Sandy (23 October – 12 November 2012) were also accessed from Internet Archive so that we could compare the disaster-related Twitter uses between the two hurricane events.

3.3. Geographical and social data

In addition to Twitter data, we collected 17 social-geographical variables (Table 1). Two hazard variables, including rainfall threat and wind threat, were derived from the rainfall observations (Figure 2A) and hurricane track data (Figure 2D) from the National Hurricane Center under NOAA (NOAA 2018). There are 3,085 rainfall observation stations in Texas and Louisiana, each recording daily rainfall depth from 25 August to 4 September 2017. Empirical Bayesian Kriging (EBK) interpolation in ArcGIS 10.6 was used to generate a continuous surface of rainfall depth in Texas and Louisiana based on the cumulative rainfall depth during that period (Zou et al. 2015; Figure 2B). The neighborhood searching number in EBK was defined as 12 neighboring observations with a grid size of 30m*30m. Rainfall threat in each county was computed as the averaged interpolated

Table 1. List of geographical and social variables.

Category	Abbreviation	Variable
Geographical (Hazard Threats)	Rainfall	Averaged interpolation of rainfall observations from Harvey, 2017
	Wind	Averaged kernel density by hurricane track and wind speed, 2017
Geographical (General)	Elevation	Mean elevation within the county, 2012
	Impervious	Mean impervious rates within the county, 2014
	WaterRate	% of water areas within the county, 2000
	PctUrban	% urban areas within the county, 2016
Social	MedAge	Median age of residents within the county, 2016
	Education	% population over 25 with a bachelor or a higher degree, 2016
	HIncome	Median household income, 2016
	FemaleH	% female headed households, 2016
	MobileH	% mobile households, 2016
	NoPhone	% housing units without telephone service available, 2016
	Poverty	% population living in poverty, 2016
	NoVehicle	% households without a vehicle, 2016
	EmployR	% civilian workforce that is employed, 2016
	PctYoung	% population 17 to 29 years old, 2016
	PopDensity	Population density, people per squared kilometers, 2016

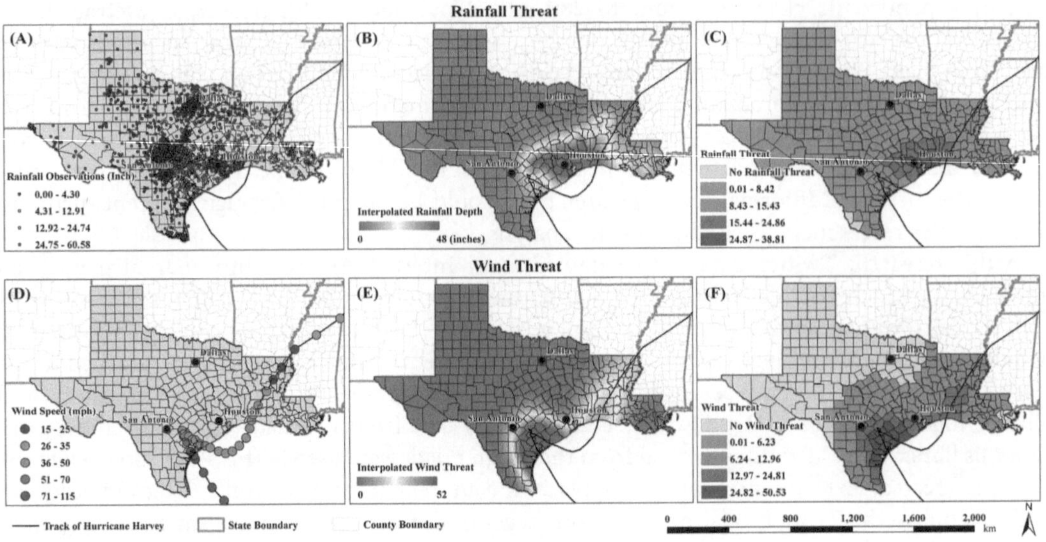

Figure 2. (A) Rainfall observations during Hurricane Harvey in Texas and Louisiana; (B) interpolated rainfall depth by Empirical Bayesian Kriging; (C) rainfall threat, averaged county-level rainfall depth; (D) track and wind speed of Hurricane Harvey; (E) kernel density surface by hurricane track and wind speed; (F) wind threat, county-level averaged kernel density.

rainfall depth of all grids within a county (Figure 2C). Wind threat in a county was calculated as the averaged kernel density based on the location of the hurricane track and its wind speed using ArcGIS 10.6 (Figure 2E&F). The grid size was the same (30m*30m), with a bandwidth of 500 km, which is the largest extent of Hurricane Harvey (NOAA 2018).

Four geographical variables, including mean elevation, impervious rate, water rate, and percent urban by county were obtained from Mihunov et al. (2018), who processed and tabulated the original data from the U.S. Geological Survey into county-level estimates. A total of 11 socioeconomic variables (Table 1) collected for the year 2016 (before Hurricane Harvey) were accessed from the U.S. Census (see https://www.census.gov/). The selection of these variables was based on three reasons. First, the same set of ten socioeconomic variables (except population density) and wind threat in Zou et al. (2018) were selected, since these eleven variables were found significant in indicating the resilience of communities and the results could be compared to previous studies. Second, we added population density, elevation, impervious surface rate, and percent urban to this study because they were found important in representing disaster resilience in previous studies (Cai et al. 2018; Mihunov et al. 2018; Reams, Lam, and Baker 2012; Zhao et al. 2018). Third, the rainfall threat was included because of the disastrous flooding caused by tremendous rainfalls during Hurricane Harvey. These 17 selected variables, together representing the social and geographical environments of the community, can be used to test the hypotheses that digital divide in disaster-related twitter use existed during the three phases of emergency management.

4. Methods

4.1. Twitter data mining

We followed the Twitter data-mining framework developed in Zou et al. (2018) for studying Hurricane Sandy and modified slightly for this study (Figure 3). First, we used 10% randomly selected tweets during the collection period as the background tweets. Similar to the Hurricane Sandy study, the unfiltered Harvey-related tweets were extracted based on four keywords: hurricane, Harvey, flood, and storm. These four keywords have been suggested by other studies as an effective set of

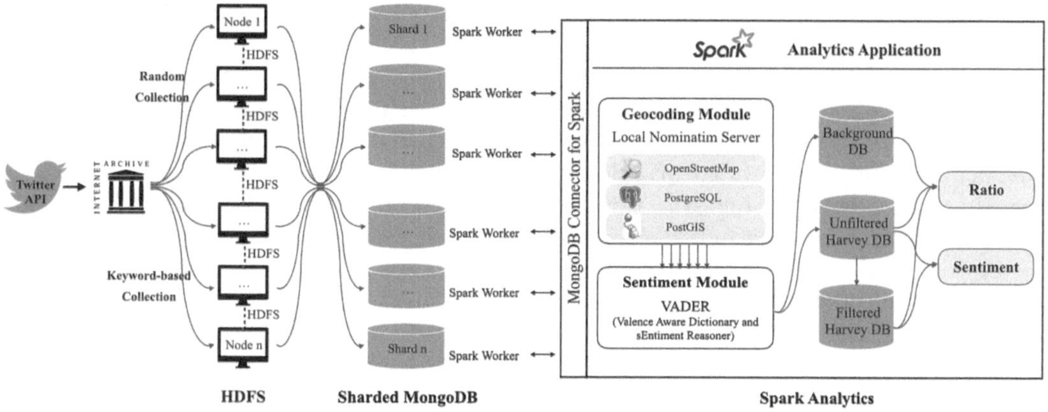

Figure 3. Twitter data mining framework (modified from Zou et al. 2018).

words to capture disaster-related tweets during hurricanes while avoiding too much unrelated information before and after events (Kryvasheyeu et al. 2015; Shelton et al. 2014). Any tweet containing one of the four keywords during the four weeks (17 August to 12 September 2017) was identified as an unfiltered Harvey-related tweet.

The data collection is implemented in a Hadoop Distributed File System (HDFS) with eight nodes. All selected background and unfiltered Harvey-related tweets were stored in a sharded MongoDB database (see https://www.mongodb.com/), which is well suited for social media data management due to its fast speed in data transmission (Walther and Kaisser 2013). Each Node, running MongoDB, is part of the Spark cluster and considered a spark worker. MongoDB connector for Spark maintains the connection between Spark environment and MongoDB. Machine learning applications and further analysis are being scheduled and distributed via Apache Spark and Yarn scheduler. Compared to the Twitter data mining framework in Zou et al. (2018), the framework used in this analysis with HDFS, Apache Spark, and local geocoding server is more suitable for data collection and processing of large-size dataset.

The second step was to determine the location associated with each tweet, which could be realized by three approaches: geo-tags embedded in tweets, addresses in user profiles, or locational information in the tweet contents. Only the first two methods were used in this study. Twitter data with geo-tags contain either a pair of latitude-longitude coordinates or a Twitter suggested place name. For cases without geo-tags, the addresses in the user profiles were used to represent their associated locations. We updated the geocoding module developed in Zou et al. (2018) (Figure 3) to accomplish the transformation of Twitter suggested place names or user addresses into pairs of coordinates. The geocoding module was based on a local Nominatim server, which uses OpenStreetMap planet data (https://planet.openstreetmap.org/, accessed in November 2017) as its data source and is managed by PostgreSQL and PostGIS databases. Compared with other commercial geocoding services, such as Google or ArcGIS geocoding APIs which have strict rate limits for free use by public users, the local Nominatim server provides unlimited geocoding services without rate restrictions, which makes it possible to handle large number of geocoding requests in a short time period (Singh 2017).

The third step was sentiment analysis. Sentiment analysis is the process to assign a sentiment score or category to each record based on its tweet content. The Sentiment module in this study utilizes the Valence Aware Dictionary and sEntiment Reasoner (VADER), a lexicon and rule-based python package for sentiment scoring, to evaluate the emotion status of each tweet (Hutto and Gilbert 2014). VADER assigns a sentiment score ranging from −1 (most negative) to 1 (most positive) to each input text message. Previous studies have successfully applied the VADER sentiment module to evaluate sentiment scores of Hurricane Sandy tweets (Zou et al. 2018), poll public attitude towards breast cancer screening (Wong et al. 2016) and predict election results from Twitter (Ramteke et al. 2016). After each selected tweet was geocoded and assigned a sentiment score, it was imported into the background and unfiltered Harvey databases, respectively.

The fourth step was filtration. Hurricane Irma was formed and named on 30 August 2017 when Hurricane Harvey was still devastating the Gulf of Mexico and made its third landfall on coastal Louisiana. The issue of collecting Irma-related tweets to Harvey database arises when users tweeted about Hurricane Irma after August 30 using one of the same keywords: hurricane, storm, or flood. We used an 'Irma' filter to exclude those cases. If one tweet mentioned 'Irma' but did not mention 'Harvey', then this tweet is identified as not related to Hurricane Harvey. Tweets not excluded by the 'Irma' filter in the unfiltered Harvey database were stored in the filtered Harvey database.

Fifth, we calculated two Twitter indexes, Ratio and Sentiment, from the Background and the two Harvey databases. Ratio index is the number of disaster-related tweets divided by the number of background tweets (equation 1), while Sentiment index is the averaged sentiment scores (equation 2). Both indexes can be tabulated into different spatial and temporal scales. Previous studies have demonstrated the uses of Ratio index in damage estimation (Zou et al. 2018), flood mapping (Li et al. 2018), and assessing citizen awareness on climate change (Kirilenko and Stepchenkova 2014, 2015), while Sentiment index has been applied to survey public mood and attitudes towards critical global themes (Bollen, Mao, and Zeng 2011; Cody et al. 2015) and emergency management (Kryvasheyeu et al. 2016; Zou et al. 2018).

$$\text{Ratio} = \frac{\#\ \text{Disaster} - \text{related Tweets}}{\#\ \text{Background Tweets}} \tag{1}$$

$$\text{Sentiment} = \text{Mean(sentiment scores)} \tag{2}$$

4.2. Data analysis

Twitter indexes (Ratio and Sentiment) were derived at multiple spatiotemporal scales. First, Ratio and Sentiment indexes for both Hurricanes Harvey and Sandy were calculated hourly and daily at the U.S. national scale to provide and compare the general trends of public awareness and sentiment toward these two events. To distinguish the spatial patterns of public awareness to Hurricane Harvey from Hurricane Irma, two U.S. county-level maps of Ratio index for the whole period were created using the unfiltered and filtered Harvey dataset, respectively.

Second, county-level Ratio indexes were tabulated into preparedness, response, and recovery phases in the severely affected counties in the states of Texas and Louisiana. If a county has very few background tweets, the small number problem might occur when computing the Ratio indexes. Therefore, as in Zou et al. (2018), we used a threshold of 50 background tweets to select counties with active Twitter use and 126 out of 191 counties were selected for more detailed analysis of Hurricane Sandy. Also, only counties that suffered rainfall and wind threats during Harvey were included in the analysis. A total of 76 counties in the two states were selected for Hurricane Harvey analysis based on these above criteria.

Ratio indexes for each selected county were tabulated into preparedness, response, and recovery phases according to FEMA's emergency management framework (Figure 4). By tabulating Twitter activities into the three phases, we can better understand the factors affecting disaster-related Twitter use in different phases. We defined a week before the landfall of Harvey as the preparedness phase (17–24 August), nine days during the landfall as the response phase (25 August – 02 September), and ten days after as the recovery phase (3–12 September; Figure 4). Since mitigation is a long-term recovery and adaptation process, information extracted from the post-disaster Twitter dataset (10 days after the event) may not be sufficient to provide a complete picture on mitigation. Therefore, this study only considered the first three phases of emergency management for subsequent analysis. Similarly, the preparedness, response, and recovery phases for Hurricane Sandy were defined as

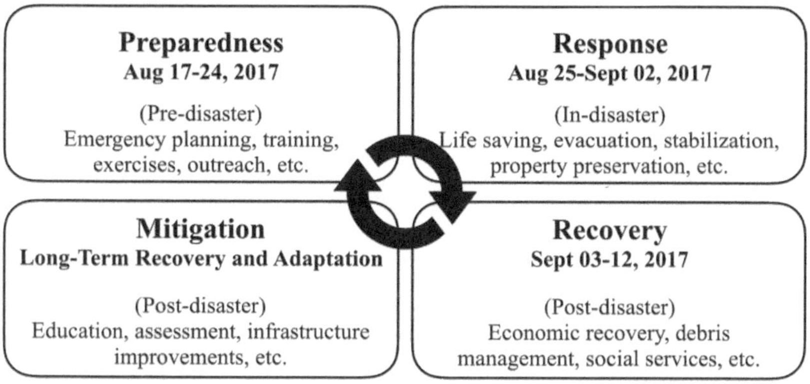

Figure 4. The four phases of emergency management.

23–28 October, 29–31 October, and 1–12 November 2012, according to Zou et al. (2018). Summary statistics of Ratio and Sentiment indexes in different phases during Harvey and Sandy at the regional level were computed and compared.

Third, to test whether geographical and social disparities in disaster-related Twitter use existed across different phases in severely affected counties, we first computed the Pearson correlations between the Ratio index at different phases and the 17 geographical-social variables (Table 1). Then, we conducted stepwise linear regressions to identify the geographical and social variables that are significant in affecting the Ratio index. Four models were derived for the preparedness, response, and recovery phases, and for the entire period.

5. Results and discussion

The Twitter data collected for this study are summarized in Table 2. During the study period (17 August–12 September 2017), a total of 458,290 out of 125 million (0.37%) tweets were collected into the unfiltered Harvey database. Among them, 268,464 (58.6%) tweets were successfully geocoded to a location (this number included 6,618 or 1.4% of tweets that had geo-tags). After the 'Irma' filtration, 385,213 tweets out of 125 million background tweets (0.31%) were collected into the filtered Harvey database. Among them, 218,694 (56.8%) tweets were successfully geocoded to a location based on users' addresses or attached places, while 5,633 (1.5%) tweets had geo-tags with coordinates. Meanwhile, 12,533,794 (10% of the total tweets) background tweets were randomly selected from the data library, and 4,077,447 (32.5%) of them were successfully geocoded or had geo-tags and used as a base layer to normalize the Harvey-related tweets. Because the selected background tweets are one tenth of the whole database, all calculated Ratio indexes were divided by ten to derive the true Ratio indexes. Results from the three analyses are summarized as follows.

5.1. The spatial-temporal patterns of the U.S.

Figure 5 depicts the temporal variations of Ratio and Sentiment indexes in the United States, which reflects the general trend of public awareness and sentiment towards Hurricanes Harvey (Figure 5A)

Table 2. Twitter data collected from internet archive (17 August-12 September 2017).

Databases	Total Tweets	Addresses/Places	Geo-tags	Geocoded
Background	12,533,794	3,939,518 (31.4%)	137,929 (1.1%)	4,077,447 (32.5%)
Unfiltered Harvey	458,290	261,846 (57.1%)	6,618 (1.4%)	268,464 (58.6%)
Filtered Harvey	385,213	218,694 (56.8%)	5,633 (1.5%)	224,327 (58.2%)

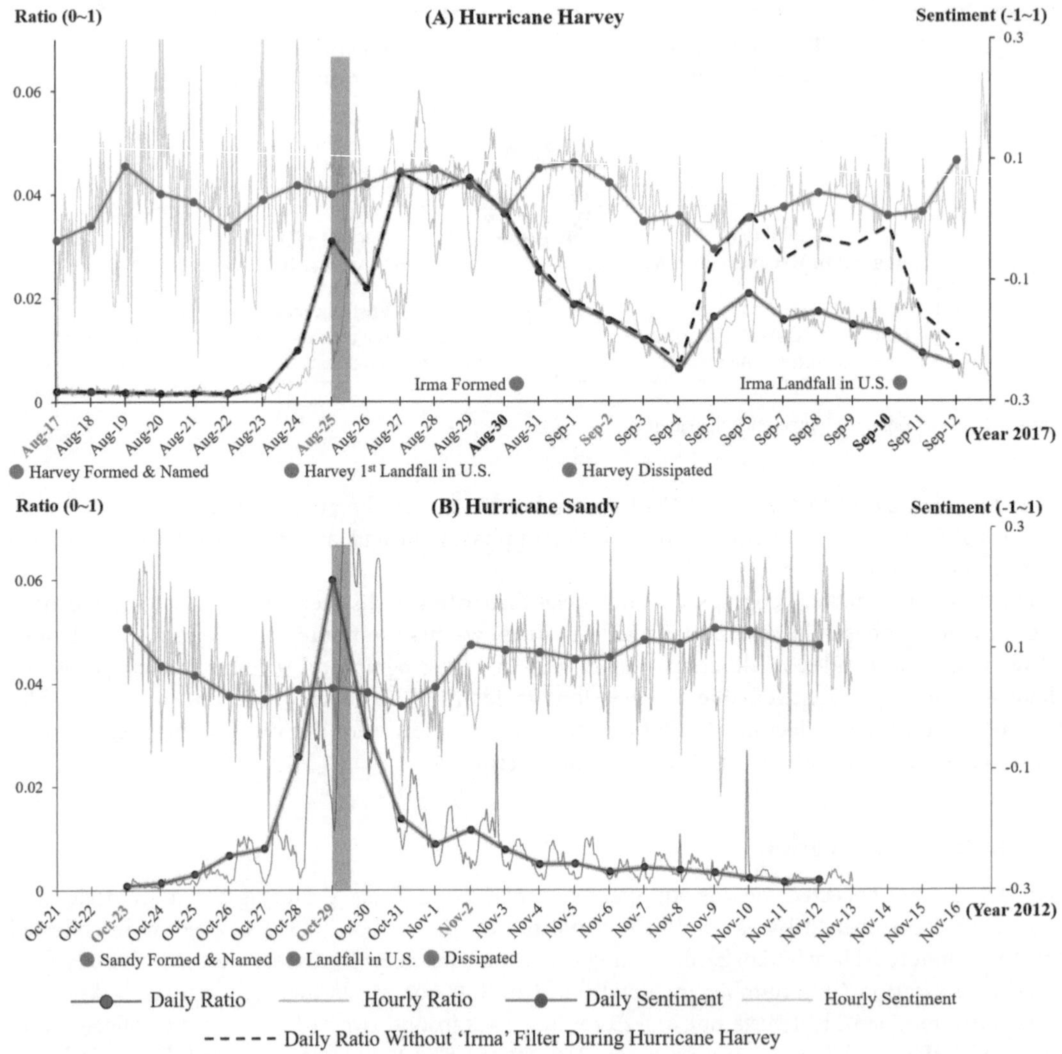

Figure 5. Temporal trends of hourly and daily Ratio and Sentiment indexes in the U.S. during (A) Hurricane Harvey in 2017 and (B) Hurricane Sandy in 2012.

and Sandy (Figure 5B) through time. The Ratio and Sentiment indexes using the filtered Harvey dataset were aggregated hourly and daily and represented as blue and red lines, respectively. The dashed black line represents the daily Ratio indexes using the unfiltered Harvey dataset. Compared to the unfiltered Harvey-related dataset, the filtered Harvey dataset shows significantly lower Ratio values after Hurricane Irma was formed, especially after 2 September 2017 when Harvey dissipated. This phenomenon is quite unique, as we seldom experienced two hurricanes hitting the U.S. within a short time span. It shows that keyword-based collection alone may collect a large amount of information not targeted to the specific event. Through a combination of keyword-based and rule-based collection, such as the 'Irma' filter in this case, we can effectively remove Irma-related tweets into the filtered Harvey database.

Table 3 lists the summary statistics of Ratio and Sentiment indexes during Harvey. We also list both indexes for Sandy to provide a context for comparison. For the Harvey event, the daily Ratio indexes ranged from 0.002 to 0.044, with an average value of 0.016. The highest hourly Ratio index value was at 1:00 pm. on 27 August 2017 when the rainfall was heaviest, whereas the

Table 3. Summary statistics of daily Ratio and Sentiment indexes at the national level during Harvey and Sandy.

Hurricane	Daily Twitter Indexes	Minimum	Maximum	Mean
Harvey (2017)	Ratio	0.002	0.044	0.016
	Sentiment	−0.05	0.098	0.036
Sandy (2012)	Ratio	0.001	0.059	0.010
	Sentiment	0.005	0.229	0.088

highest daily Ratio index value were during 27–29 August 2017 when the Houston metropolitan area suffered extensive flooding due to the heavy rainfall. These Ratios were very similar to that of Sandy (0.001 to 0.059, with an average of 0.010), suggesting that the degree of awareness in United States between the two hurricanes is very similar despite a five-year difference in time between the two events.

The Sentiment scores tabulated for the entire period at the national level between the two events differ slightly. For Harvey, the Sentiment scores ranged from −0.05 to 0.098 with a mean of 0.036, compared with a range of 0.005 to 0.229 and a mean of 0.088 for Sandy (Table 3). In general, Harvey has a narrower range and a more negative sentiment than Sandy. Figure 5 shows that there is no distinct pattern of Sentiment scores for Harvey at the national level, whereas for Sandy, a gentle dip in the Sentiment scores in the response phrase is observed.

To exemplify how people utilized Twitter during disasters and how tweets associated with the Sentiment scores, the top five saddest and happiest filtered Harvey-related tweets by Sentiment scores during the four weeks are listed in Table 4. The lowest and highest Sentiment scores of all tweets were −0.9482 and 0.9413, respectively. Two of the saddest Harvey-related tweets connected Hurricane Harvey with other tragic events around the world, one was worrying about the robbery caused by Hurricane Harvey, and the other two were simply depressed about this event. On the contrary, most of the happiest tweets were praying for or expressing their supports to people suffered from Hurricane Harvey.

The spatial patterns of county-level Ratio indexes in the United States during the entire period are displayed in Figure 6. A total of 1,877 counties with more than 25 background tweets were selected for the mapping to avoid the small number problem. Two maps were generated using the unfiltered and filtered Harvey datasets, both of which show very distinct patterns. The Ratio map using the unfiltered Harvey database shows two clusters of high Ratio values, one in coastal Texas and the other in Florida (Figure 6A), whereas the Ratio map using the filtered Harvey dataset shows that most of counties with high Ratio values were concentrated along the Texas coastline, including the metropolitan areas of Corpus Christi, Houston, and San Antonio, and no cluster of high

Table 4. The top 5 happiest and saddest filtered Harvey-related Tweets by Sentiment scores.

Rank	Tweet content	Scores
Top 5 saddest tweets		
1	#gorakhpur tragedy #railway accidents #floods so many innocent deaths but no one took to streets	−0.9482
2	not only do we have to worry about flooding we now have to worry about people robbing?! seriously? that shit is so sad	−0.9342
3	past few weeks news dominated by death & tragedy:(1. floods 2. swine-flu -250+ dead 3. gorakhpur-100+ babies dead 4. rail tragedy-40+ dead	−0.9313
4	feel awful for everyone effected by flooding. wouldn't wish it on my own worst enemy. out of our house for months when it happened to us.	−0.9162
5	seeing a hurricane on the news is scary but seeing a hurricane devastate ur hometown is heartbreaking	−0.9144
Top 5 happiest tweets		
1	to all my #texas friends may safety peace and comfort surround you. you are loved. #hurricaneharvey	0.9413
2	dear god comfort those affected by the flood provide their need restore hope and most of all bless them to see you at work in their life!	0.9194
3	god may not calm the storm, but his peace will definitely calm your heart. be at peace!	0.9180
4	wow thank you bow wow hurricane over houston is saved	0.9169
5	texas! i'm thinking of you. sending you strength light and love. please be safe. #hurricaneharvey	0.9168

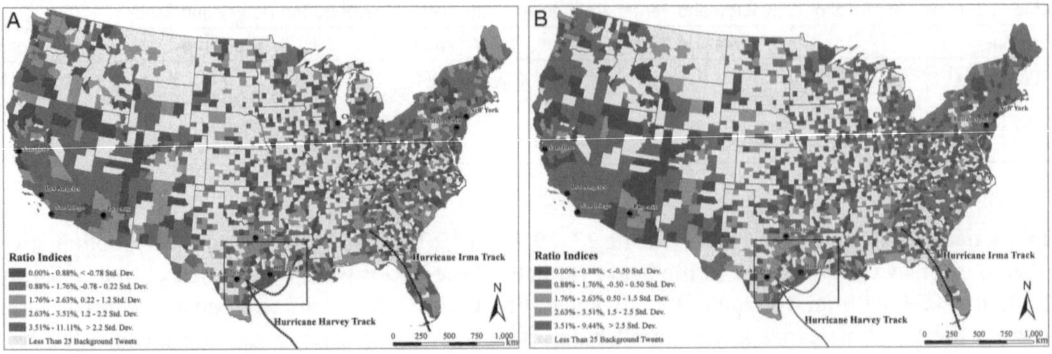

Figure 6. Comparison of county-level ratio indexes of Hurricane Harvey in the United States: (A) unfiltered Harvey dataset (B) filtered Harvey dataset.

Ratio values were found in Florida (Figure 6B). This is expected since many users in Florida had started to discuss Hurricane Irma during the overlapping period.

5.2. The spatial-temporal patterns in coastal Texas and Louisiana

Zooming into the severely affected area and tabulating the tweets into three phases reveals more information on how public awareness changed through the three phases. As mentioned above, a total of 76 counties in Texas and Louisiana were selected for the regional analysis because they had more than 50 background tweets and encountered threats from both wind and rainfall from Harvey. Figure 7 maps the spatial patterns of the Ratio indexes for the whole period and at the three phases using data of the 76 counties. These maps clearly show that higher Twitter use were

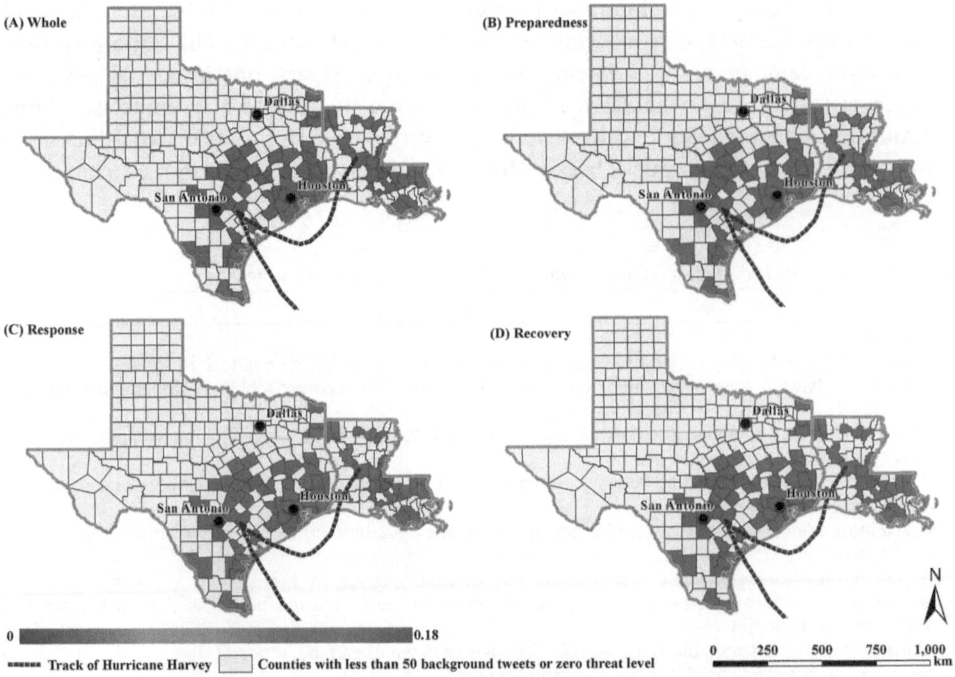

Figure 7. County-level ratio indexes for the entire period and at different phases of emergency management in affected areas.

Table 5. Summary statistics of Ratio and Sentiment indexes in different phases during Harvey, tabulated at the regional level.

Twitter Indexes	Phase	Minimum	Maximum	Mean	Mean (Sandy)
Ratio	Whole	0.0000	0.0572	0.0265	0.0185
	Preparedness	0.0000	0.0525	0.0081	0.0199
	Response	0.0000	0.1819	0.0718	0.0623
	Recovery	0.0000	0.0520	0.0133	0.0070
Sentiment	Whole	−0.2309	0.2842	0.0714	0.0398
	Preparedness	−0.5442	0.6125	0.0376	0.0363
	Response	−0.1964	0.3487	0.0833	0.0265
	Recovery	−0.5707	0.4588	0.0229	0.0906

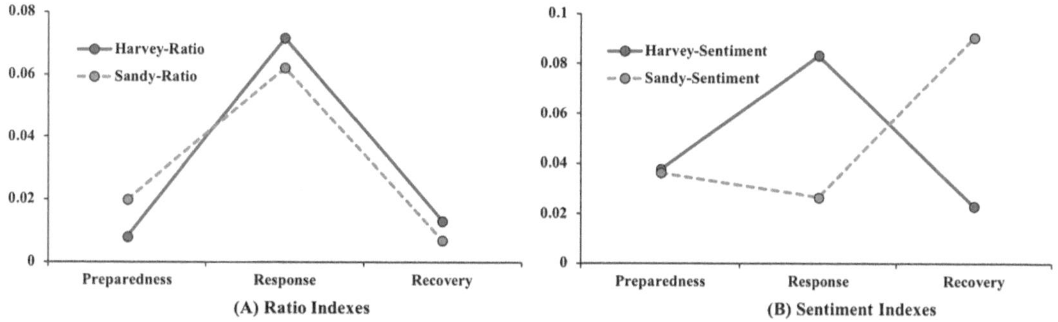

Figure 8. Comparison of the mean county-level Twitter indexes in the affected areas during Hurricanes Harvey and Sandy: (A) Ratio, (B) Sentiment.

concentrated in the Houston metropolitan area, with the highest Twitter activities occurring in the response phrase.

Table 5 lists the summary statistics of Ratio and Sentiment indexes in different phases during Harvey. The mean indexes of the 126 counties affected by Sandy are also included to enable comparison (Zou et al. 2018). Figure 8 provides a better visualization of Table 5 by plotting the mean values of the two indexes in the three phases for the two hurricanes. The results show that Twitter use patterns in the three phrases were very similar between the two hurricanes, with Harvey on average having a slightly higher Ratio than Sandy (0.0265 vs. 0.0185). Twitter use was the highest in the response phase, whereas the lowest Twitter use (i.e. Ratio index) was found in the preparedness phase for both hurricanes.

However, there is a marked difference in the Sentiment scores between Harvey and Sandy (Figure 8B). For Harvey, the highest Sentiment scores were in the response phrase and the lowest Sentiment scores were in the recovery phase. For Sandy, the pattern was reversed, with the highest and the lowest Ratio indexes being in the recovery and response phases, respectively (Zou et al. 2018). This contrast could be explained by the following three reasons. First, there was insufficient flood warning for Harvey because of the unexpected amount of rainfall. Residents in the impacted areas were underprepared for this event. During the response phrase, more Twitter users started to share disaster-related information, ask for help, or help victims in the rescue operation. Since more tweets were used to convey disaster information and they generally had fewer emotional words, this could lead to higher Sentiment scores during the response phase. However, in the recovery phrase, the catastrophic flooding caused by Hurricane Harvey had a lingering effect. The flood water hindered residents' daily lives and reminded them of the damage caused by Harvey, their tweets were more likely to be negative and emotional, thus leading to a lower Sentiment score. Third, Hurricane Irma was formed and made its landfall in the U.S. during the recovery phase of Hurricane Harvey, which could make many users to connect these two events together. All three reasons could result in higher Sentiment scores in the response phase and lower Sentiment scores in the recovery phase for Harvey.

5.3. Social and geographical disparities

Pearson correlations and stepwise linear regression analyses between Ratio indexes and the 17 social-geographical variables at the three phases and for the entire period were conducted to test the main hypothesis that communities with better socioeconomic conditions had more disaster-related Twitter use in all three phases of emergency management. This analysis resulted in four regression models with R^2 of 0.42, 0.40, 0.48, and 0.16 for the whole, preparedness, response, and recovery phases, respectively (Table 6). The standardized regression coefficients listed in each model can be used to evaluate the relative importance of the selected variables on the Ratio indexes. The model for the whole period provides a general view of the relationships between disaster-related Twitter use and social-geographical conditions. Rainfall threat, wind threat, household income, and percentage of young population between 17 and 29 years were found to be positively correlated with the Ratio index ($Ratio_{Whole}$).

An examination of the three models for the different phases further reveal how social-geographical conditions affect disaster-related Twitter use during the cycle of emergency management. In the preparedness phase (the second model), Ratio index was positively associated with wind threat while negatively associated with percentage of households without a vehicle and median age. In the response phase (the third model), the social-geographical disparities were most significant, and the model yielded the highest R^2 of all models (0.48). In this model, rainfall threat, instead of wind threat, was selected as a positive indicator, while percentages of mobile homes, female-headed households, and households without telephone services were negatively associated with disaster-related Twitter use. In the recovery phase, neither rainfall threat nor wind threat were selected in the model, and only household income (positively associated) and elevation (negatively associated) were selected, though this last model yielded the lowest R^2 (0.16). All four models support the hypothesis that communities with better social-geographical conditions had higher Twitter use.

Two additional observations are made here. First, the changing hazard threats through the three phases during Harvey have been adequately reflected by the Twitter use, which also reflects residents' perception of risks. In the preparedness phase, residents' perceptions of disaster risk were based on hurricane forecasts with projected hurricane track information. Therefore, as residents along the hurricane track are more aware of the risks, they are more likely to tweet. As a result, the wind threat variable was selected in this phase. In the response phase, communities with high rainfall threat were at high risk of intensive flooding that residents payed more attention to hazard information, asked for help, or helped others on Twitter. All these reasons led to increased Twitter use, and the rainfall threat variable was selected. In the recovery phase, neither wind threat nor rainfall threat were significant in affecting disaster-related Twitter use. The Pearson correlations (Table 7) further illustrate the changing perception of residents on disaster risk, with the highest correlations between Ratio index and wind and rainfall threats occurred in the preparedness and response phases.

From a socioeconomic perspective, percentages of households without a vehicle, mobile homes, female-headed households, and households without telephone services are common indicators of low socioeconomic conditions, while average household income and percentage of young population are considered positive indicators of community resilience. The relationships found in all four models and the correlation analysis show positive and significant correlations between socioeconomic conditions and disaster-related Twitter use in all phases and the entire period (Table 7). Variables positively correlated with the Ratio indexes were positive indicators of disaster resilience

Table 6. Stepwise regression models (p-value < 0.05) between Ratio indexes and the 17 variables.

Phases	Model	R^2
Whole	$Ratio_{Whole} = -0.014 + 0.374 * Rainfall + 0.232 * Wind + 0.364 * HIncome + 0.202 * PctYoung$	0.42
Preparedness	$Ratio_{Preparedness} = 0.026 + 0.564 * Wind - 0.254 * NoVehicle - 0.243 * MedAge$	0.40
Response	$Ratio_{Response} = 1.033 + 0.587 * Rainfall - 0.357 MobileH - 0.232 * FemaleH - 0.180 * NoPhone$	0.48
Recovery	$Ratio_{Recovery} = 0.002 + 0.389 * HIncome - 0.232 * Elevation$	0.16

Table 7. Pearson correlations between Ratio indexes and the 17 variables.

Pearson Correlation *r*	$Ratio_{Whole}$	$Ratio_{Preparedness}$	$Ratio_{Response}$	$Ratio_{Recovery}$
Rainfall Threat	0.425**	0.077	0.508**	0.230*
Wind Threat	0.321**	0.564**	0.250*	0.043
MedAge	−0.170	−0.279*	−0.149	−0.037
Education	0.272*	0.073	0.278*	0.144
HIncome	0.404**	0.242*	0.359**	0.312**
PctUrban	0.280*	0.131	0.265*	0.199
FemaleH	−0.239	−0.011	−0.282*	−0.035
MobileH	−0.292*	−0.296**	−0.277*	−0.125
NoPhone	0.035	−0.114	0.024	0.159
Poverty	−0.268*	−0.139	−0.216	−0.262*
NoVehicle	−0.325**	−0.116	−0.308**	−0.178
EmployR	0.320**	0.292*	0.250*	0.254*
PctYoung	0.114	0.112	0.134	−0.039
Elevation	−0.102	−0.081	−0.099	−0.158
Impervious	0.276*	0.124	0.273*	0.203
waterrate	0.032	0.039	0.038	0.007
PopulationDensity	0.272*	0.129	0.273*	0.176

*p-value < 0.05.
**p-value < 0.01.

(Education, HIncome, PctUrban, EmployR, PctYoung, Impervious, and PopulationDensity). Negative indicators of disaster resilience, including MedAge, FemaleH, MobileH, NoPhone, Poverty, NoVehicle, and Elevation, were negatively associated with the Ratio indexes. A total of nine socioeconomic factors were found to correlate significantly with Ratio indexes in the response phase, while only four and three were significantly associated with Ratio indexes in the preparedness and recovery phases, respectively. The top three most significant socioeconomic variables were averaged household income, percent of households without a vehicle, and percentage of employment. These communities with higher economic vitality have better accessibility to transportation tools; they are likely to have better access to disaster-related information through Twitter during Harvey, which may help them better respond to and recover from disasters.

These regression and correlation results demonstrate that the digital divide in disaster-related Twitter use existed during Hurricane Harvey. Communities with better socioeconomic conditions had higher disaster-related Twitter use in all three phases of emergency management during Hurricane Harvey. These results are consistent with previous studies of Hurricane Sandy and Horsethief Canyon fire in 2012 (Kent and Capello 2013; Zou et al. 2018). Although social media has become more ubiquitous in 2017, the digital divide in social media use during disasters still existed in Harvey.

5.4. Limitations and future research

There are some limitations of this study, which necessitate future research. First, the keyword/rule-based method to extract Harvey-related tweets could be improved by including a set of location keywords to identify if a tweet is related to Harvey or Irma. For example, tweets mentioning 'Houston' or 'Texas' are more likely to be associated with Harvey, while tweets containing 'Florida' or 'Miami' are related to Irma and should be excluded from the filtered Harvey database.

Second, the social and geographical analysis indicates that the selected 17 geographical and social factors cannot fully explain the disaster-related Twitter activities. Other factors, such as damage caused by disasters, recovery status, and hazard experiences, may also affect disaster-related Twitter activities and could be considered in future research.

Third, we conducted a preliminary correlation analysis between sentiment indexes and socioeconomic variables, but no clear correlation was detected. Previous studies indicate that damage may impact public sentiment towards disasters (Kryvasheyeu et al. 2016; Zou et al. 2018). How public sentiment towards Hurricane Harvey is associated with different topics as well as local socioeconomic conditions is unknown. More future investigations on the relationships among sentiment indexes, social-geographical factors, and popular Twitter topics at different emergency phases could be conducted. Twitter data could be categorized as different disaster-related usage types, such as sharing hazard status, comments of disaster impacts, rescue-related information, and communications. The results may lead to new research findings on how social-geographical disparities affect public emotions and concerns during disasters.

Fourth, the Twitter data mining framework could be improved by collecting real-time data using Twitter streaming API and the Spark Streaming extension. The data used in this study were accessed and analyzed after Hurricane Harvey. In future events, however, it is necessary to collect and process social media data in real-time to enable monitoring as well as applications of social media activities during disasters to enhance emergency management.

Finally, although this study reveals the social and geographical disparities in disaster-related Twitter use during Hurricane Harvey, how such digital divide affects the long-term disaster resilience of different communities and how to reduce such digital divide so that hazard victims could easily access disaster- or rescue-related information are important questions for future research to address.

6. Conclusion

This study analyzed the Twitter activities during Hurricane Harvey from 17 August to 12 September 2017. The study sought to answer a fundamental question: did social and geographical disparities of Twitter use exist during the three phases of emergency management (preparedness, response, recovery). The study employed a Twitter data-mining framework modified from a previous Hurricane Sandy study and computed two indexes: Ratio and Sentiment. Regression analyses between the Ratio indexes and the social-geographical variables confirm our hypothesis and reveals significant social and geographical disparities in Twitter use, with the highest R^2 (0.48) found in the response phase. Communities with higher disaster-related Twitter use generally were communities of better social-geographical conditions. Compared with Hurricane Sandy, the sentiment in Harvey was tied to the prolonged flooding effect, with the lowest sentiment scores occurring in the recovery phase instead of the response phase.

There are important implications of the framework and research findings. First, this study improved the utility of the previously developed Twitter data mining framework. The updated framework is more efficient for fast data collection and processing of large amounts of Twitter dataset than the framework in Zou et al. (2018). Second, the computed Twitter indexes provide useful baseline information on Twitter activity in Hurricane Harvey, which can be used to compare with similar disaster events across space and through time. Third, the social and geographical disparities in Twitter use found in this study imply that communities with better socioeconomic conditions are more likely to have access to hazard information and receive rescue responses during emergencies. This digital divide of Twitter use could further deepen the divide in recovery and resilience among communities. Future research is needed to examine the effects of Twitter use and its disparities on the disaster resilience of communities, and to test whether Twitter use can serve as an indicator for predicting the resilience of different communities.

Disclosure statement

No potential conflict of interest was reported by the authors.

Funding

This article is based on work supported by two research grants from the U.S. National Science Foundation: one under the SBE Office of Multidisciplinary Activities (SMA) organization in Interdisciplinary Behavioral and Social Science Research (IBSS) Program (Award No. 1620451), and the other under the NSF Social and Economic Sciences Division

(SES) Hurricane Harvey 2017 Program (Award No. 1762600). Any opinions, findings, and conclusions or recommendations expressed in this material are those of the authors and do not necessarily reflect the views of the funding agencies.

ORCID

Lei Zou http://orcid.org/0000-0001-6206-3558
Nina S. N. Lam http://orcid.org/0000-0002-5344-9368
Heng Cai http://orcid.org/0000-0001-6068-5150
Michelle A. Meyer http://orcid.org/0000-0002-8750-8178
Margaret A. Reams http://orcid.org/0000-0002-5636-0318

References

Assumpção, T. H., I. Popescu, A. Jonoski, and D. P. Solomatine. 2018. "Citizen Observations Contributing to Flood Modelling: Opportunities and Challenges." *Hydrology and Earth System Sciences* 22 (2): 1473–1489. doi:10.5194/hess-22-1473-2018.

Blake, Eric S., and David A. Zelinsky. 2018. *Hurricane Harvey*. AL092017. National Hurricane Center Tropical Cyclone Report. National Hurricane Center. https://www.nhc.noaa.gov/data/tcr/AL092017_Harvey.pdf.

Bollen, Johan, Huina Mao, and Xiaojun Zeng. 2011. "Twitter Mood Predicts the Stock Market." *Journal of Computational Science* 2 (1): 1–8. doi:10.1016/j.jocs.2010.12.007.

Cai, Heng, Nina S. N. Lam, Lei Zou, and Yi Qiang. 2018. "Modeling the Dynamics of Community Resilience to Coastal Hazards Using a Bayesian Network." *Annals of the American Association of Geographers* 108 (5): 1260–1279. doi:10.1080/24694452.2017.1421896.

Cai, Heng, Nina S.-N. Lam, Lei Zou, Yi Qiang, and Kenan Li. 2016. "Assessing Community Resilience to Coastal Hazards in the Lower Mississippi River Basin." *Water* 8 (2): 46. doi:10.3390/w8020046.

Cody, Emily M., Andrew J. Reagan, Lewis Mitchell, Peter Sheridan Dodds, and Christopher M. Danforth. 2015. "Climate Change Sentiment on Twitter: An Unsolicited Public Opinion Poll." *PLOS ONE* 10 (8): e0136092. doi:10.1371/journal.pone.0136092.

Dufty, Neil. 2012. "Using Social Media to Build Community Disaster Resilience." *Australian Journal of Emergency Management, The* 27 (1): 40.

Earle, Paul. 2010. "Earthquake Twitter." *Nature Geoscience* 3 (April): 221–222. doi:10.1038/ngeo832.

Earle, Paul S., Daniel C. Bowden, and Michelle Guy. 2012. "Twitter Earthquake Detection: Earthquake Monitoring in a Social World." *Annals of Geophysics* 54 (6), doi:10.4401/ag-5364.

FEMA (Federal Emergency Mnaagment Agency). 2006. *Principles of Emergency Management: Independent Study Manual*. https://www.hsdl.org/?abstract&did=485174.

Friedman, Lisa, and John Schwartz. 2018. "How Hurricane Harvey Became So Destructive." *The New York Times*, January 20, sec. Climate. https://www.nytimes.com/2017/08/28/climate/how-hurricane-harvey-became-so-destructive.html.

Guan, Xiangyang, and Cynthia Chen. 2014. "Using Social Media Data to Understand and Assess Disasters." *Natural Hazards* 74 (2): 837–850. doi:10.1007/s11069-014-1217-1.

Hutto, C. J., and Eric Gilbert. 2014. "VADER: A Parsimonious Rule-Based Model for Sentiment Analysis of Social Media Text." 10

Kent, Joshua D., and Henry T. Jr Capello. 2013. "Spatial Patterns and Demographic Indicators of Effective Social Media Content During TheHorsethief Canyon Fire of 2012." *Cartography and Geographic Information Science* 40 (2): 78–89. doi:10.1080/15230406.2013.776727.

Kirilenko, Andrei P., Tatiana Molodtsova, and Svetlana O. Stepchenkova. 2015. "People as Sensors: Mass Media and Local Temperature Influence Climate Change Discussion on Twitter." *Global Environmental Change* 30 (January): 92–100. doi:10.1016/j.gloenvcha.2014.11.003.

Kirilenko, Andrei P., and Svetlana O. Stepchenkova. 2014. "Public Microblogging on Climate Change: One Year of Twitter Worldwide." *Global Environmental Change* 26 (May): 171–182. doi:10.1016/j.gloenvcha.2014.02.008.

Kryvasheyeu, Yury, Haohui Chen, Esteban Moro, Pascal Van Hentenryck, and Manuel Cebrian. 2015. "Performance of Social Network Sensors During Hurricane Sandy." *PLOS ONE* 10 (2): e0117288. doi:10.1371/journal.pone.0117288.

Kryvasheyeu, Yury, Haohui Chen, Nick Obradovich, Esteban Moro, Pascal Van Hentenryck, James Fowler, and Manuel Cebrian. 2016. "Rapid Assessment of Disaster Damage Using Social Media Activity." *Science Advances* 2 (3): e1500779. doi:10.1126/sciadv.1500779.

Lam, Nina S.-N., Yi Qiang, Helbert Arenas, Patricia Brito, and Kam-biu Liu. 2015. "Mapping and Assessing Coastal Resilience in the Caribbean Region." *Cartography and Geographic Information Science* 42 (4): 315–322. doi:10.1080/15230406.2015.1040999.

Lam, Nina S.-N., Margaret Reams, Kenan Li, Chi Li, and Lillian P. Mata. 2016. "Measuring Community Resilience to Coastal Hazards Along the Northern Gulf of Mexico." *Natural Hazards Review* 17 (1): 04015013. doi:10.1061/(ASCE)NH.1527-6996.0000193.

Li, Linna, Michael F. Goodchild, and Bo Xu. 2013. "Spatial, Temporal, and Socioeconomic Patterns in the Use of Twitter and Flickr." *Cartography and Geographic Information Science* 40 (2): 61–77. doi:10.1080/15230406.2013.777139.

Li, Kenan, Nina S. N. Lam, Yi Qiang, Lei Zou, and Heng Cai. 2015. "A Cyberinfrastructure for Community Resilience Assessment and Visualization." *Cartography and Geographic Information Science* 42 (sup1): 34–39. doi:10.1080/15230406.2015.1060113.

Li, Zhenlong, Cuizhen Wang, Christopher T. Emrich, and Diansheng Guo. 2018. "A Novel Approach to Leveraging Social Media for Rapid Flood Mapping: A Case Study of the 2015 South Carolina Floods." *Cartography and Geographic Information Science* 45 (2): 97–110. doi:10.1080/15230406.2016.1271356.

Lindsay, Bruce R. 2011. "Social Media and Disasters: Current Uses, Future Options and Policy Considerations." *Journal of Current Issues in Media & Telecommunications* 2 (4): 287.

Liu, Xiaohui, Bandana Kar, Chaoyang Zhang, and David M. Cochran. 2018. "Assessing Relevance of Tweets for Risk Communication." *International Journal of Digital Earth*, June. https://www.tandfonline.com/doi/abs/10.1080/17538947.2018.1480670.

Mandel, Benjamin, Culotta Aron, Boulahanis John, Stark Danielle, Lewis Bonnie, and Rodrigue Jeremy. 2012. "A Demographic Analysis of Online Sentiment During Hurricane Irene." Proceedings of the Second Workshop on Language in Social Media, 27–36. LSM '12. Stroudsburg, PA: Association for Computational Linguistics. http://dl.acm.org/citation.cfm?id=2390374.2390378.

Martín, Yago, Zhenlong Li, and Susan L. Cutter. 2017. "Leveraging Twitter to Gauge Evacuation Compliance: Spatiotemporal Analysis of Hurricane Matthew." *PLOS ONE* 12 (7): e0181701. doi:10.1371/journal.pone.0181701.

Merchant, Raina M., Stacy Elmer, and Nicole Lurie. 2011. "Integrating Social Media into Emergency-Preparedness Efforts." *New England Journal of Medicine* 365 (4): 289–291. doi:10.1056/NEJMp1103591.

Mihunov, Volodymyr V., Nina S. N. Lam, Lei Zou, Robert V. Rohli, Nazla Bushra, Margaret A. Reams, and Jennifer E. Argote. 2018. "Community Resilience to Drought Hazard in the South-Central United States." *Annals of the American Association of Geographers* 108 (3): 739–755. doi:10.1080/24694452.2017.1372177.

Muttarak, Raya, and Wiraporn Pothisiri. 2013. "The Role of Education on Disaster Preparedness: Case Study of 2012 Indian Ocean Earthquakes on Thailand's Andaman Coast." *Ecology and Society* 18 (4), http://www.jstor.org/stable/26269420.

National Research Council. 2012. *Disaster Resilience: A National Imperative*. doi:10.17226/13457.

NOAA (National Oceanic and Atmospheric Administration). 2018. "2017 Atlantic Hurricane Season." Accessed July 7, 2018. https://www.nhc.noaa.gov/data/tcr/.

Petkova, Elisaveta P., Jaishree Beedasy, Eun Jeong Oh, Jonathan J. Sury, Erin M. Sehnert, Wei-Yann Tsai, and Michael J. Reilly. 2018. "Long-Term Recovery from Hurricane Sandy: Evidence From a Survey in New York City." *Disaster Medicine and Public Health Preparedness* 12 (2): 172–175. doi:10.1017/dmp.2017.57.

Ramteke, J., S. Shah, D. Godhia, and A. Shaikh. 2016. "*Election Result Prediction Using Twitter Sentiment Analysis.*" 2016 international conference on inventive computation technologies (ICICT), 1:1–5. doi:10.1109/INVENTIVE.2016.7823280.

Reams, Margaret A., Nina S. N. Lam, and Ariele Baker. 2012. "Measuring Capacity for Resilience among Coastal Counties of the U. S. Northern Gulf of Mexico Region." *American Journal of Climate Change* 01 (December): 194–204. doi:10.4236/ajcc.2012.14016.

Rizza, Caroline, and Ângela Guimarães Pereira. 2014. "Building a Resilient Community through Social Network: Ethical Considerations about the 2011 Genoa Floods." 6.

Shelton, Taylor, Ate Poorthuis, Mark Graham, and Matthew Zook. 2014. "Mapping the Data Shadows of Hurricane Sandy: Uncovering the Sociospatial Dimensions of 'Big Data'." *Geoforum* 52 (March): 167–179. doi:10.1016/j.geoforum.2014.01.006.

Singh, Sushant K. 2017. "Evaluating Two Freely Available Geocoding Tools for Geographical Inconsistencies and Geocoding Errors." *Open Geospatial Data, Software and Standards* 2 (1): 11. doi:10.1186/s40965-017-0026-3.

Sloan, Luke, Jeffrey Morgan, Pete Burnap, and Matthew Williams. 2015. "Who Tweets? Deriving the Demographic Characteristics of Age, Occupation and Social Class from Twitter User Meta-Data." *PLOS ONE* 10 (3): e0115545. doi:10.1371/journal.pone.0115545.

Thieken, Annegret H., Heidi Kreibich, Meike Müller, and Bruno Merz. 2007. "Coping with Floods: Preparedness, Response and Recovery of Flood-Affected Residents in Germany in 2002." *Hydrological Sciences Journal* 52 (5): 1016–1037. doi:10.1623/hysj.52.5.1016.

Tsou, Ming-Hsiang. 2015. "Research Challenges and Opportunities in Mapping Social Media and Big Data." *Cartography and Geographic Information Science* 42 (sup1): 70–74. doi:10.1080/15230406.2015.1059251.

Tsou, Ming-Hsiang, and Michael Leitner. 2013. "Visualization of Social Media: Seeing a Mirage or a Message?" *Cartography and Geographic Information Science* 40 (2): 55–60. doi:10.1080/15230406.2013.776754.

Walther, Maximilian, and Michael Kaisser. 2013. "Geo-Spatial Event Detection in the Twitter Stream." In *Advances in Information Retrieval*, 356–367. Lecture Notes in Computer Science. Berlin, Heidelberg: Springer. doi:10.1007/978-3-642-36973-5_30.

Wang, Qi, and John E. Taylor. 2014. "Quantifying Human Mobility Perturbation and Resilience in Hurricane Sandy." *PLOS ONE* 9 (11): e112608. doi:10.1371/journal.pone.0112608.

Wong, Kai O., Faith G. Davis, Osmar R. Zaïane, and Yutaka Yasui. 2016. "Sentiment Analysis of Breast Cancer Screening in the United States Using Twitter - Semantic Scholar." In Porto, Portugal. https://pdfs.semanticscholar.org/10a6/16f71a8f2f2ba9d61091612bfd91e7da9d84.-pdf.

Yin, J., A. Lampert, M. Cameron, B. Robinson, and R. Power. 2012. "Using Social Media to Enhance Emergency Situation Awareness." *IEEE Intelligent Systems* 27 (6): 52–59. doi:10.1109/MIS.2012.6.

Zhao, Qunshan, Heather Fischer, Wei Luo, and Elizabeth Wentz. 2018. "Community Resilience in Maricopa County, Arizona, USA: The Analysis of Indoor Heat-Related Death and Urban Thermal Environment." In Madison, Wisconsin, USA. https://www.researchgate.net/publication/325247226_Community_Resilience_in_Maricopa_County_Arizona_USA_The_Analysis_of_Indoor_Heat-related_Death_and_Urban_Thermal_Environment.

Zou, Lei, Joshua Kent, Nina S.-N. Lam, Heng Cai, Yi Qiang, and Kenan Li. 2015. "Evaluating Land Subsidence Rates and Their Implications for Land Loss in the Lower Mississippi River Basin." *Water* 8 (1): 10. doi:10.3390/w8010010.

Zou, Lei, Nina S. N. Lam, Heng Cai, and Yi Qiang. 2018. "Mining Twitter Data for Improved Understanding of Disaster Resilience." *Annals of the American Association of Geographers* 108: 1422–1441. doi:10.1080/24694452.2017.1421897.

Population distribution modelling at fine spatio-temporal scale based on mobile phone data

Petr Kubíček, Milan Konečný, Zdeněk Stachoň, Jie Shen, Lukáš Herman, Tomáš Řezník, Karel Staněk, Radim Štampach and Šimon Leitgeb

ABSTRACT

Population distribution modelling can benefit many different domains, for example, transportation, urban planning, ecology or emergency management. Information about the location and number of people in an affected area is crucial for decision-makers during emergencies and crises. Mobile phone data represents relatively reliable and time accurate information on real-time population distribution, movement and behaviour. In this study, we evaluate the spatio-temporal distribution of population derived from phone data of the selected pilot area (City of Brno, Czech Republic). Analysis is based on the dataset describing the estimated human presence (EHP) with two values – visitors and transiting persons. The temporal change of data is first analysed and further processed using two methodological approaches. First, the dasymetric method is used where the building geometry and technical attributes served as a target layer. Second, the results of building level analysis are transformed into a regular grid zone of both visitors and the general EHP. Resulting spatio-temporal patterns are compared to the census data. Results demonstrate how the proposed building level dasymetric approach can improve the spatial granularity of EHP. Potential use of proposed methodology within selected emergency situations is further discussed.

1. Introduction

Information about the location and number of people in an affected area is crucial for decision-makers in different fields such as transportation, urban planning, ecology or emergency management. Mobile phone data is relatively reliable and time accurate information about real-time and routine population distribution, movement and behaviour and can provide updated demographic data for risk assessments. This kind of information can also be of benefit in areas previously observed only infrequently and indirectly (Data-Pop Alliance 2017). Emergency management is generally understood as comprising several phases that roughly correspond to 'before' (preparedness), 'during' (response), and 'after' (recovery) phases of any event. The use of geographic knowledge and cartographic visualization during all these emergency phases requires high flexibility and depends on the dynamic nature of the situation, scope of decision-making and other user involvement (Kubíček et al. 2011).

Use of accurate data and information is crucial for minimizing loss of property and life during a crisis (Konečný and Reinhardt 2010). This study evaluates the population information derived from the phone data of the selected pilot area (city centre Brno, Czech Republic). The data are analysed and visualised using different methodological approaches. Results are compared to traditional demographic data sources and discussed.

Several research questions arise:

- What are the differences between the number of inhabitants declared by the official census and the number provided by the mobile phone data?
- What are the daily rhythms of mobile phone users in different administrative units?
- What is the spatio-temporal pattern of mobile phone users in these administrative units?

These research questions will be examined and discussed in relation to the mobile phone data available for the pilot area.

To maintain consistency with existing studies, the term Estimated Human Presence (EHP, Järv, Tenkanen, and Toivonen 2017) or population present (PP, Martin, Cockings, and Leung 2015) is used, which is a general term that can be defined as the sum of three categories:

- resident population (PR)
- non-resident population (PNR)
- transiting population (PT)

The term Building Estimated Human Presence (BEHP) is used in our paper for the sum of two categories that stay more than 30 minutes at the observed administrative unit:

- resident population (PR)
- non-resident population (PNR)

The word 'building' used in term symbolizes exclusion of PT category. These people are unlikely to be staying inside buildings but moving through the administrative unit on streets.

2. Related work

A review of existing research is presented in two separate discussions – one dealing with the technological background of mobile phone localization studies and the other analysing the potential of mobile phone data localization under emergency management.

2.1. Mobile phone data localization methods

Three elementary sets of mobile phone localization methods have been identified:

- Network -driven, non-intrusive. This method requires administration rights for the mobile network to localize a mobile phone. The operator of the mobile network then uses/sells aggregated data on user location, or even more typically, sells information and/or services based on user location and additional attributes rather than the obtained data.
- Device-driven, non-intrusive. In this method, a user agrees to the location of their mobile phone being tracked. It is typical when a user installs an application and agrees to the proposed permissions, such as localization. The application's administrator then receives information about the localization of a mobile phone. The business model is very similar to the first method and

according to Verkasalo and Hämmäinen (2007) is based on the hardware of individual phones (GNSS, WiFi, Bluetooth, or similar technologies).

- Third-party driven, intrusive. This method is based on tracking a mobile phone without the network administrator's or the mobile phone owner/user's permission. A false BTS (Base Transceiver Station) approach is used under this method. The third-party driven method is illegal in many countries around the world and should only be used with the permission of the relevant national authority.

Starting with the network-based solutions, we can distinguish two categories – passive and active localization. The first category is based on georeferencing and aggregation of mobile network operational data (Pucci, Manfredini, and Tagliolato 2015). These are basically logs of communications between phones and mobile networks connected by calls, SMS, or data connection. Each activity is then recorded and can be further listed for users or base stations (BTS) as so-called Call Detail Records (CDR). Because of Cell ID (CI) parameters, each communication is positioned to a certain BTS (Dufková et al. 2008; Jonge, Pelt, and Roos 2012). GSM cells vary in spatial extent, being smaller and more detailed in urbanized areas and bigger in rural areas (Steenbruggen, Tranos, and Nijkamp 2015; Lískovec 2016, Figure 1).

Figure 1. GSM cells in the Brno city area. The spatial resolution of cells is highly variable, the most detailed being in the city centre (GSMWEB 2018).

Passive localization methods have recently opened new horizons for mobility studies or travel behaviour (Ahas et al. 2010; Gao 2015; Kang et al. 2012). Silm and Ahas (2010) used this localization method to study seasonal variability of population in Estonian municipalities. Pucci, Manfredini, and Tagliolato (2015) analysed everyday mobility (rhythmicity) of citizens in Lombardy (Italy). Similarly, Novák and Temelová (2012) studied the everyday mobility of young people in Prague (Czech Republic). Čtyroký (2017), also in Prague, attempted to optimize opening hours of public institutions, and Ouředníček et al. (2018) delimited Prague's metropolitan area.

Extensive research based on mobile phone call detail records (CDR) was conducted in China. Mobile phone data is mostly used to analyse the patterns of the user distribution and behaviour. Through the ratio of users in the crowd, Kang et al. (2012) use mobile phone data to estimate urban population. The mobile phone location data in Shenzhen has been used to reveal the diversity and heterogeneity of spatio-temporal distribution of human aggregated activities in urban space (Cao et al. 2017; Xu et al. 2015). Li and Zhou (2015) explored how to obtain that data in the indoor situation.

Liu, Kang, and Wang (2014) presents a generic workflow for big-data-driven (including mobile phone data) human mobility analyses and summaries major movement measures. Kang et al. (2010) compute statistical characteristic of mobile phone usage, derive aggregated mobility patterns for millions of mobile phone users and individual mobility patterns for different groups which are divided by some selected social factors, such as gender and age. Ahas et al. (2015) compare the time and space differences in daily activities in different cities, Yuan, Raubal, and Liu (2012) try to use this pattern to analyse travel behaviour. Zhou et al. (2015) use the understanding of user's behaviour to reverse the user's location. Gao (2015) presented an analytical framework to explore human mobility patterns focusing on quantitative analytics to identify spatio-temporal patterns (hotspots). Also emphasizing the limitations of CDR data, he suggested the integration of phone calls with other more holistic data sources on human activity. In addition, based on mobile phone data and user behaviour patterns, conducted traffic flow analysis (Kang, Liu, and Wu 2017; Wu et al. 2012) and urban land use function recognition (Peng et al. 2018; Zhou et al. 2014).

These studies varied in space (municipalities, counties), time (day, week, and season) and resolution. Passive and active approaches have also been used to study different social granularity, ranging from personal sensing (following a personal SIM card), social/group sensing (Dufková et al. 2008; Novák and Temelová 2012), public/community sensing (Lane et al. 2010). Yuan and Raubal (2016) used mobile phone data to analyse individual human behaviour, such as the dimensions of human activity space. They also focused on age group and gender data (of nine big cities in China) and attempted to correlate them with the urban environment and individual urban context.

2.2. Analysis of demands for mobile phone localization in emergency management

The potential use of mobile phone data for emergency management could be divided into three successive phases: emergency preparedness, emergency response, and disaster recovery. Emergency preparedness and disaster recovery units use passive techniques to detect mobile phones, for example, analysing patterns based on regular data measurements from mobile service providers, while emergency response requires ad-hoc data that are efficiently obtained through so-called active localization methods. Active methods are considered a 'man-in-the-middle-attack' and therefore cannot be generally used in most countries around the world without special permission. However, through this type of connection, emergency messages (e.g. SMS) may be sent to any or all mobile phone(s) in an affected area during an emergency. Řezník, Horáková, and Szturc (2015) described an active method example for mobile phone localization using an International Mobile Subscriber Identity (IMSI) catcher and verified the methods in a field test.

Several case studies have already been published referencing the use of mobile phone data during the recovery phase of the recent earthquake in Haiti (Bengtsson et al. 2011; Řezník, Horáková, and Szturc 2015), flooding (Pastor-Escuredo et al. 2014), or dynamic population mapping (Deville et al.

2014). Oxendine et al. (2014) discussed the use of non-authoritative data (including mobile phones and social media) for evacuation purposes and stressed the importance of such data for increasing situational awareness and the common operational picture. For instance, the example from Haiti demonstrated that an active method through an IMSI catcher may be installed and used within two hours of its delivery to the desired place (Řezník, Horáková, and Janiurek 2011; Řezník, Horáková, and Szturc 2015). Rescue work was coordinated according to the information on mobile phone positions; the ruins of buildings with mobile phones discovered within were explored for survivors first.

Emergency managers want population data at a finer spatial resolution than the level of administrative units. Fine resolution can be produced by various means using dasymetric modelling (Wright 1936; Martin 1996; Slocum 1999; Nagle et al. 2014). Dasymetric models increase the spatial resolution of population data by using the relationships between population distribution and ancillary geographic layers. Traditionally, this method further divides a homogenous administrative unit into separate units weighted by the probability of occurrence of citizens. Ancillary data commonly used for spatial pattern refinement include land cover (Mennis 2003), road density (Reibel and Bufalino 2005), LiDAR-based building heights (Xie 2006; Kressler and Steinnocher 2008), Corine Land Cover and Urban Atlas (Gallego 2010). A commonly mentioned study is the modelling and integration of population and ancillary data sources. However, these do not consider the temporal and other societal contexts of human presence in space.

Mobile phone data and its temporality can further address population modelling in this direction. Recently, Järv, Tenkanen, and Toivonen (2017) presented a generic conceptual framework of a multi-temporal function-based dasymetric (MFD) interpolation method for mobile phone data. These authors introduced a raster-based stepwise methodology combining land use data for spatial disaggregation of target zones, time-dependant human activities for estimating human presence, and the integration of passive mobile phone data for final spatio-temporal reallocation by target zones.

Recently, a group of emergency management experts studied the emergency threats and vulnerabilities of the Czech Republic (Paulus et al. 2016) and identified and categorized the most topical emergency situations. From this analysis, 22 typological emergency situations were pinpointed. A detailed and typified plan for each emergency was defined, including the responsible public administrative organization and the spatial level on which the plans are used (central, regional, local). Challenges for mobile phone data were further discussed from both academic and practical perspectives.

The possibilities and limitations of using mobile phone data will be presented in examples from the Czech Republic. The number of mobile phones per household member aged 6+ in the Czech Republic in 2015 was 0.99 (CZSO 2017, 28). This means that mobile phone data could provide extremely fine and dynamic sources about large populations. The information available could be useful for designing evacuation routes, providing exposure maps sensitive to regular population movements such as daily or seasonal migrations, evaluating the effectiveness of early warning systems, or assessing the risk of disease outbreak following a disaster.

The demand for mobile phone data use in crisis management was initiated during ZONA Integrated Rescue System training (in 2015) when an emergency situation's context dealt with evacuating inhabitants from municipalities affected by a simulated nuclear power plant failure. Both paper and digital versions of the evacuation maps were prepared in advance (Stachoň et al. 2016). However, in planning the number of evacuation vehicles to be sent to various locations, a problem arose concerning the actual number of inhabitants in need of evacuation as some of these municipalities were traditional sleep-over areas with minimum population during working hours, while others were collocations of industrial zones, shopping centres, and educational facilities with a high concentration of inhabitants during working hours (Russnák and Herman 2017). Thus, census data used to roughly estimate numbers of inhabitants to be evacuated is inadequate and can underestimate or overestimate the actual spatial distribution of inhabitants.

3. Research design

3.1. Pilot area, data and methods

The central part of the City of Brno, with a high population density and distinct rhythmicity, was selected for the pilot study. The pilot area embraces several transportation hubs and includes all the main railway and coach stations as well as the most important city public transport stations. Most of the city's historical monuments are also concentrated here. It is a transit zone for many daily commuters and other visitors. The big retail centres are in the southern and western parts of the pilot area (Binek et al. 2017, 6).

Brno is dealing with the problem of an underestimated population size in the official statistical data sources. An available research report (Seidenglanz, Toušek, and Chvátal 2013) shows the differences between official census data and the estimated current population at the city level (Table 1). The non-registered population in Brno comprises Czech citizens and foreigners living in Brno without registration, university students and homeless persons (Seidenglanz, Toušek, and Chvátal 2013, 3). Indicators from the Ministry of Education, Youth and Sports of the Czech Republic show that nearly 65,000 students were attending universities in Brno in 2018 (https://data.brno.cz/graf/pocet-studentu-vs/). Most of them were part of the non-registered population of Brno. A large percentage of the current population comprises daily commuters (to work, school, services etc.) and visitors.

For a more detailed spatial and demographic granularity, one source of mobile phone data is freely available in the Czech Republic. Mobile operator O2 has recently published two alternative mobile phone datasets through the Liberty API (Application Programming Interface). While MAPI ('the mobility API') gives information about the number of people moving between two locations (basic settlement units), SDAPI ('the socio-demographic API') shows the gender and age structure of people who were present at the basic settlement unit. The source of this data is the O2 mobile network, but data is further extrapolated to the entire population of the Czech Republic considering O2's market share. SDAPI aggregates the number of mobile phone users into specific gender groups, provides gender differentiation, and defines two basic types of mobile phone user:

- visitors (PR+PNR) – those who stay more than 30 minutes at the observed administrative unit). This type is equal to BEHP value in this paper.
- transiting persons (PT) – those staying less than 30 minutes

The availability of SDAPI mobile phone data is limited to just one day (24 hours) a week (Sunday 21:00–Monday 21:00) in segments of one hour. Values falling under 100 mobile phone users in an age group are omitted to maintain the differential privacy of users.

We used the SDAPI to harvest the data of four basic administrative units in the centre of Brno (Figure 2), which was selected as the pilot area. The pilot area comprised four spatial units designated as 1 – 'Náměstí Svobody', 2 – 'Janáčkovo divadlo', 3 – 'Zelný trh', and 4 – 'Přízová'. Selection was based on the fact that each day has a strict limitation on database queries. The units in the city centre are characteristic for a high temporal variability of citizens, as they contain a lot of commerce and connection hubs for public transport, and a high concentration of BTS assuring a relatively higher data spatial accuracy (e.g. Mulíček, Osman, and Seidenglanz 2015).

Table 1. Differences of the population between census data and current estimated population. The numbers for 2020 were estimated in a demographic analysis. Adapted from Seidenglanz, Toušek, and Chvátal (2013, 26).

Population in Brno	2011	%	Est. 2020	%
Population registered in Brno (census data)	3,789,765	70.37	379,372	68.47
Non-registered population living in Brno	Approx. 50,000	9.26	Approx. 55,000	9.91
Current population (only daily commuters and visitors)	Approx. 110,000	20.37	Approx. 120,000	21.62
Total	Approx. 540,000	100	Approx. 555,000	100

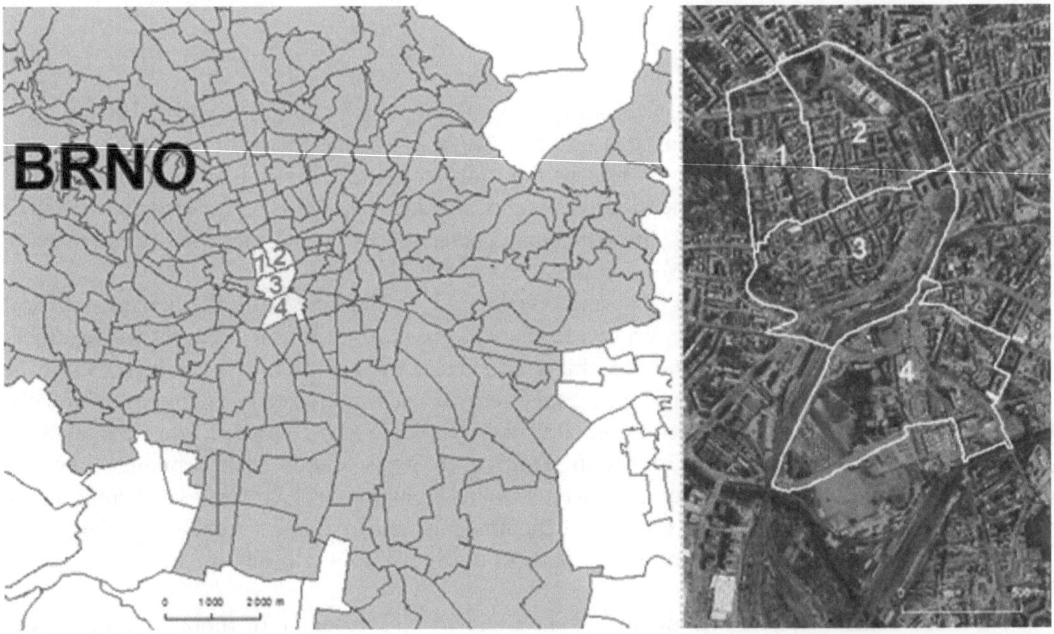

Figure 2. Localization of the pilot area (yellow) within the Brno. Selected spatial units: 1 – Náměstí Svobody', 2 – 'Janáčkovo divadlo', 3 – 'Zelný trh', and 4 – 'Přízová'.

Additionally, dasymetric spatial interpolation was applied to divide the administrative units (source zone) into target zones (buildings). Building footprints were used as the limiting variable. We presumed that visitors (PR+PNR) stay inside buildings (indoor personal activity). The footprint geometry of buildings originating from the INSPIRE Building theme (European Commission Joint Research Centre 2013) was combined with statistical attributes from the Czech Statistical Office (CZSO). For grid statistics, we modified the GEOSTAT grid dataset. The original GEOSTAT grid is produced with a cell dimension of 1 × 1 km using ETRS89 / LAEA coordinate reference system (Eurostat 2011). To improve spatial granularity, the original cell size was reduced to 100 × 100 m (each GEOSTAT cell was divided into 100 cells).

The target layer was further modelled using the following ancillary data sources:

- Temporal distribution of visitors based on the SDAPI analysis with a two-hour step – time dependant human activity (Figure 3).
- Aggregation of functional attributes of buildings into three categories – administrative, commercial, and residential. Categories were used for the time weighting of individual buildings (Figure 4).
- The vertical extent of buildings based on the number of floors attributed in CZSO data – volumetric extent (Figure 4).

3.2. Data processing and analysis

At the beginning, a temporal profile for mobile phone SDAPI data was created with a two-hour step based on the data of twelve consecutive Mondays (November 2017–January 2018). Both PR+PNR and PT were analysed and compared with the latest available census data (CZSO 2011) of selected spatial units (Figure 2). The data were processed in R software and simple arithmetic averages were used to show the temporal changes of both variables.

All buildings were aggregated into the three groups:

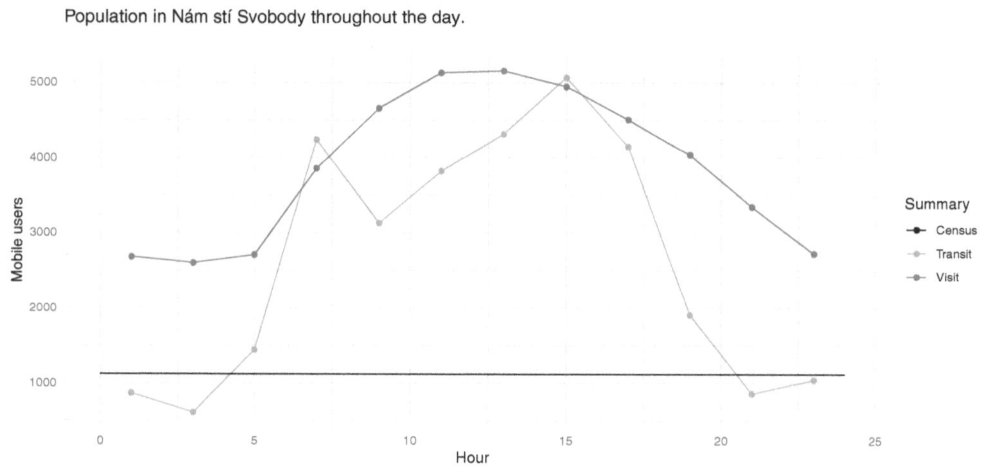

Figure 3. Temporal variability of estimated human presence (spatial unit no. 1 – 'Náměstí Svobody').

- residential buildings (individual housing, block of flats, dormitories)
- administrative buildings (public administration, schools and universities, post offices)
- commercial buildings (restaurants, cafes, libraries, markets, industry and services, banks, and others) using the functional delimitation of the of the construction classification CZ-CC (CZSO 2009, Figure 4).

The total floor space (A_p) for each building is calculated as:

$$A\_p = A\_b * P/n \quad (1)$$

where A_b is the area (m^2) of the building footprint, P is the number of floors above the ground, and n is the number of database records (if more addresses are assigned to one building footprint)

Each database record (address point) is calculated separately and then aggregated for the whole building. This approach enabled buildings even with mixed functional categories to be handled.

Different weights were used to differentiate the probability of human presence in individual building categories. Values and timing were consulted with social scientists and used a very simplified approach based on expert judgement and knowledge of existing working hours. This is particularly

Figure 4. Functional delimitation of buildings in Brno centre (a – building type; and b – number of floors).

valid for administrative and commercial buildings. The values for administrative buildings took into account the specificity of the day of the week, weight being set to 1 for business hours (8:00–17:00) and 0 for the rest of the day. Commercial buildings have higher variability of opening hours, so the weight was set to 1 for the hours between 7:00 and 21:00. Even more detailed segmentation of commercial buildings is possible based on the opening hours of individual commercial premises, but for the beginning, only one period was used. The weight for residential buildings is lowered to 0.5 between 7:00 and 21:00, when a lower presence of permanent inhabitants at home is assumed.

The authors are aware of the potential consequences of this simplified approach. The weights have only been given as very rough approximations to verify the concept, not to strictly determine reliable information about the population. An important part of using the method will be surveying spatio-temporal human activity for the available age categories of mobile phone data. Valuable insight may be also gained also when more mobile phone data is acquired by analysing zones with homogenous building types (e.g. residential areas, campuses, science parks or groups of office buildings). The former or latter method of deriving human activity data can be taken as the primary source and its accuracy confirmed using other method (e.g. sociological surveys, detailed field measurements in a limited area, etc.). If an accurate enough method of deriving human activity data from mobile phone data can be established and confirmed with a survey, the proposed method would also become useful for areas where human activity surveys are not available or impractical to conduct.

The BEHP (PR+PNR, excluding PT) for certain times is then calculated in the following way. The BEHP for each building is calculated using mobile phone data, the functional category of the building, relative functional weights (changes in time, see Table 2), and footprint. The weighted footprint (R_p) is then calculated using the following formula:

$$R\_p = A\_p * (W\_k/sum(W)) \tag{2}$$

where A_p is the total floor space (see Formula (1)), W_k is the weight for the functional building category (Table 1), and $sum(W)$ is the sum of all weights for a particular hour.

Based on this formula and the results, the BEHP (P_o) is calculated as:

$$P\_o = (R\_p/sum(R\_p)) * P\_m \tag{3}$$

where P_m is the mobile phone data sum for the whole administrative unit and particular hour, R_p is the weighted footprint (formula (2)), and $sum(R\_p)$ – sum of the weighted footprints

Employing the aforementioned steps and formulas, all records for a particular building and time sessions are aggregated and the BEHP is calculated for each building (Figure 5, part a). This process is then repeated for available temporal slices.

To be consistent with the grid statistics used by other authors (e.g. Järv, Tenkanen, and Toivonen 2017) an alternative visualization method of BEHP using a grid with 100 m spatial resolution has been prepared. The grid is spatially compliant with the population grid dataset (GEOSTAT, see above). The BEHP is then estimated for grid cells as follows.

All buildings in a cell are identified using the 'Union' function; those with multiple membership are divided between more cells. The potential building footprint fragment areas (R_a) are calculated

Table 2. Weights per functional delimitation and day/night.

	1	3	5	7	9	11	13	15	17	19	21	23
Administrative	0	0	0	0	1	1	1	1	1	0	0	0
Commercial	0	0	0	1	1	1	1	1	1	1	1	0
Residential	1	1	1	0.5	0.5	0.5	0.5	0.5	0.5	0.5	0.5	1

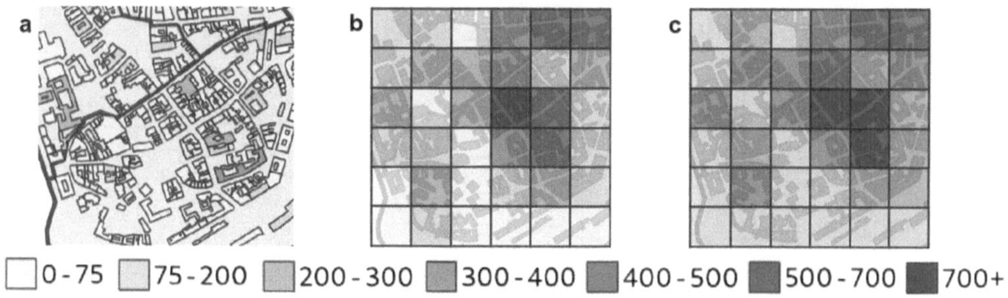

Figure 5. Details of analytical maps (a – PR+PNR; b – PR+PNR; and c – PR+PNR+PT).

with the formula:

$$R\_a = A/A\_b \tag{4}$$

where A is the fragment area and A_b – building area.

The BEHP per particular building fragment and hour (P_fh) is then recalculated for particular fragments and multiplied by the relative area of the fragment:

$$P\_fh = P\_bh * R\_a \tag{5}$$

where R_a is the relative footprint of the building and P_bh – BEHP in the whole building and time period.

All the buildings and building fragments with the same cell ID are aggregated into the BEHP per cell using the 'Spatial Join' function (Figure 5, part b).

So far, the PT has not been considered. These mobile phone users are unlikely to be staying inside buildings but moving through the administrative unit on streets by walking, driving cars, or using municipal public transport. However, even their short presence influences the total number of EHP within an administrative unit. The second grid output, therefore, calculates the general EHP taking into account PR+PNR and PT. The calculation is based on the aforementioned grid results, namely, the data set of buildings and building fragments assigned to particular grid cells. The area of built-up zones for each cell is already known, and the rest of the cell is called a non-built zone, where PT are expected. The relative area of built-up zones is then multiplied by P_fh (Formula 5) for a particular cell and the remaining fragments of non-built zones is multiplied by relative number of PT for particular time span. Both values are further aggregated (summed) using the cell ID, and the resulting value is the general EHP per cell (Figure 5, part c).

The entire procedure of building level dasymetric population modelling using mobile phone data is documented in Figure 6.

4. Results

4.1. Analysis of estimated human presence in the centre of Brno

The resulting spatio-temporal patterns of BEHP were compared to the 2011 census data (CZSO 2011). The results of temporal analysis based on mobile phone data (Figure 7) clearly show a higher number of inhabitants in most buildings during both day and night compared to the census data.

Figure 8 compares the day and night BEHP populations to census data. Census data are relatively accurate for buildings with smaller populations (less than 100 inhabitants). However, the values from census data are much lower than the values from mobile phone data in categories of buildings with the highest populations. The accurate estimation of buildings occupied by greater numbers of people is even more important in emergency situations.

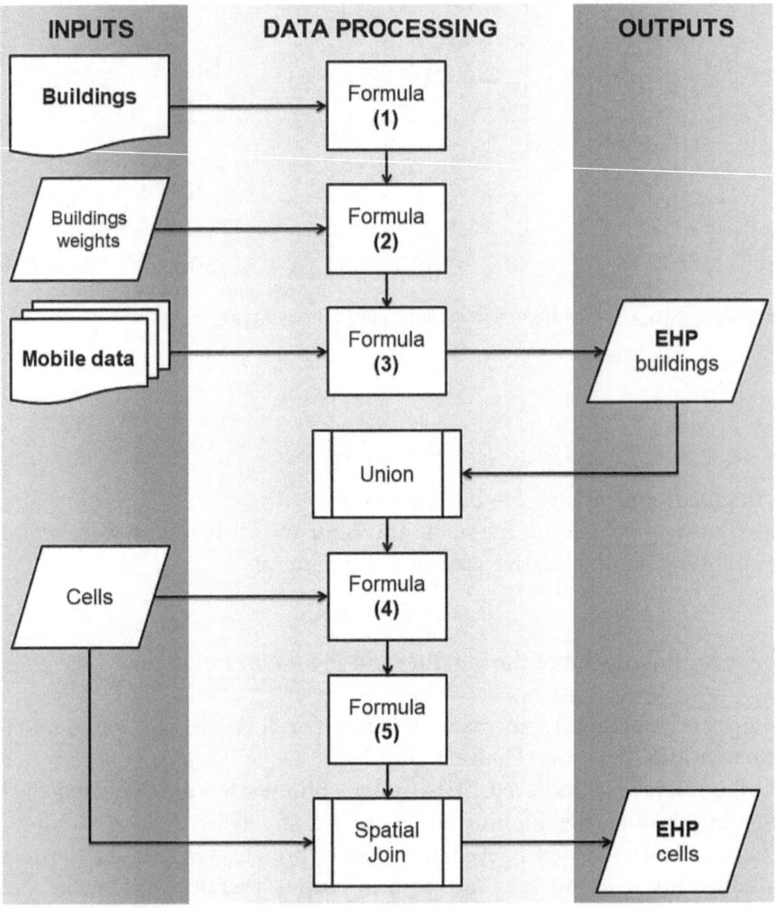

Figure 6. Schematic description of the procedure used for building level dasymetric population.

Transiting inhabitants (PT, less than 30 minutes in an administrative unit) actively moved through the pilot area starting at 5:00 and ending at 21:00 with two visible peaks, one at 7:00 and the other at 15:00. The number of visiting people (PR+PNR) starts increasing at 5:00, culminates between 13:00 and 15:00, and stabilizes at 20:00. The maximum average EHP was reached at Přízová district at 15:00, peaking at twenty thousand mobile phone users.

Analytical maps for all three methods and selected times (5:00, 15:00, and 21:00) are presented below (Figure 9).

A visual comparison was made between all three methods and the census level of inhabitants. The population distribution and its spatial patterns were separately analysed for night time (measurements 23:00–5:00) and day time (7:00–21:00).

Taking into account only residential buildings, the average BEHP for both night and day was compared to the census data per building (Figure 10). A visual comparison of both day and night data confirmed the general trend of higher BEHP values than in the census data for a substantial number of buildings (87% for the night and 70% for the day). This trend is visible despite the lowered weight for residential buildings during the day. The only exceptions are buildings with a very small footprint with 20–50 regular inhabitants. Their BEHP values are lower than the census data values. The resulting patterns are strongly influenced by the methodology and weights used.

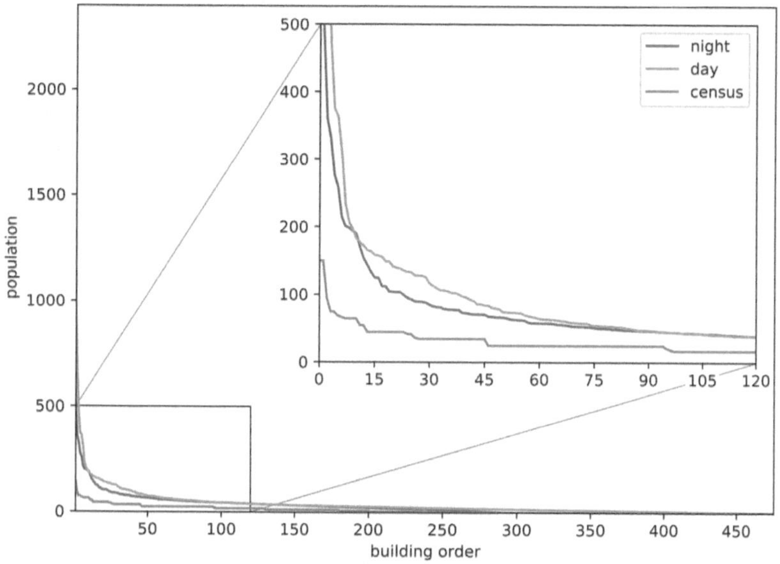

Figure 7. Census population compared to night and day averages of BEHP population (mobile phone data, PR+PNR) in individual buildings. Buildings are ordered by the value of BEHP population. Building order *n* denotes *n*-th building with the largest population in a given series.

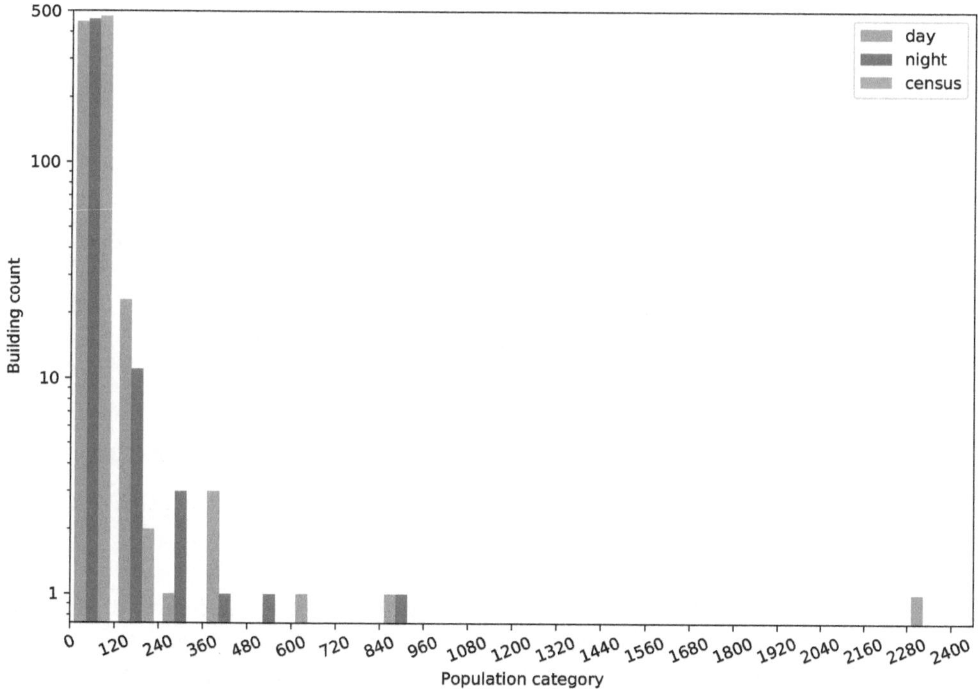

Figure 8. Frequency distribution of buildings from census data compared to day and night averages of BEHP population (mobile phone data). Axis X = category of buildings by population.

The building level data results are also available as an interactive 3D visualization (http://olli.wz.cz/ehp/) which enables selected time periods to be explored and documents the overall spatial context of analysed data, including the background topographic map.

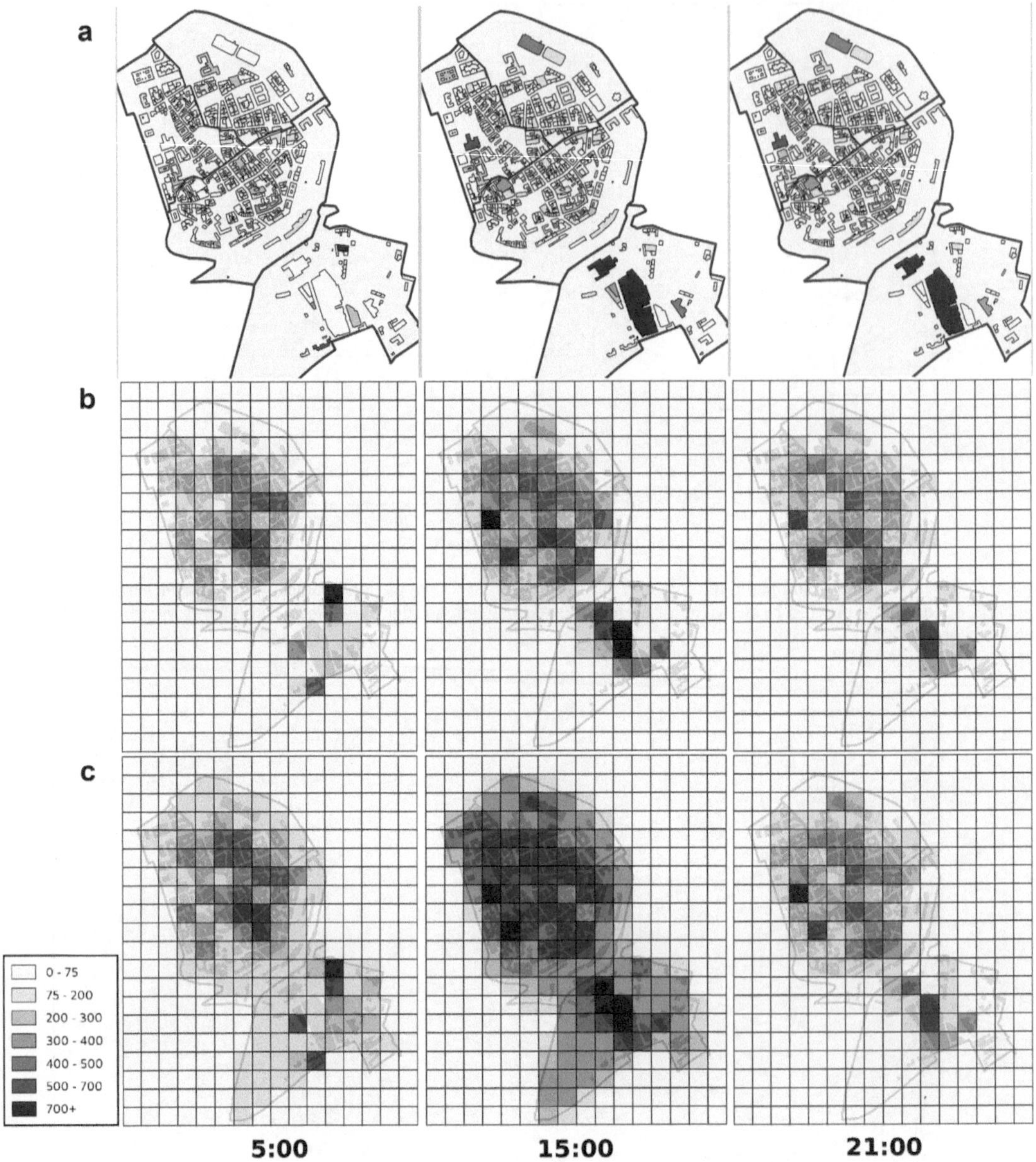

Figure 9. Analytical maps for three methods (a – PR+PNR on building level; b – PR+PNR on grid level; and c – PR+PNR+PT on grid level) and selected hours.

The average BEHP per night or day (Figure 11) clearly distinguishes the centres of high human presence at different times of the day. While the large residential buildings (night) and commercial centres such as 'Tesco' and 'Vaňkovka' stores (day) are 'expected champions', other buildings would be indistinguishable without BEHP evaluation attributes. The analysis refines the spatial pattern and enables not only the primary BEHP concentration to be identified but also the order of relevance of buildings.

The biggest difference in spatial patterns is visible between the general EHP (PR+PNR+PT) averages per day and night (Figure 12).

The average spatial distribution of visiting humans (PR+PNR) is shown in Figure 13. While night time visits are more concentrated in the city centre, daytime peak visits are more distributed with a

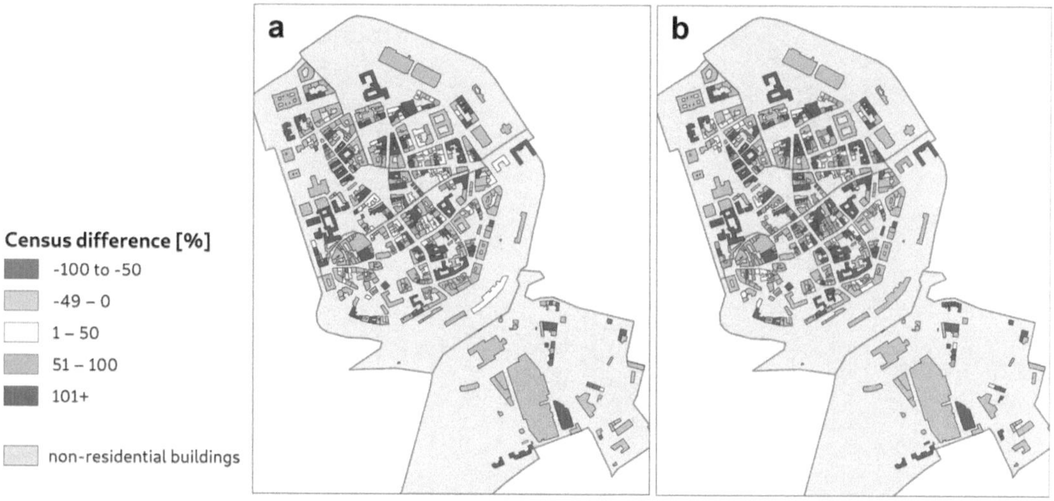

Figure 10. The difference for residential buildings between building level average BEHP (PR+PNR) and census data for day (a) and night (b).

higher concentration of inhabitants in the Přízová spatial unit in the south-eastern part of the study area where large retail shops are situated.

The patterns of grid data for residential zones which only take into account visitors were also compared (Figure 14). The comparison clearly shows a significantly higher number of visitors over residents in the pilot area. This situation is consistent with existing social and demographic research (Seidenglanz, Toušek, and Chvátal 2013), as described in Section 3.1.

4.2. Using the method for the purposes of emergency management

The necessity of identifying a detailed distribution of inhabitants in space and time was also established during the Integrated Rescue System (IRS) training 'Drinking Water' held in 2017, which

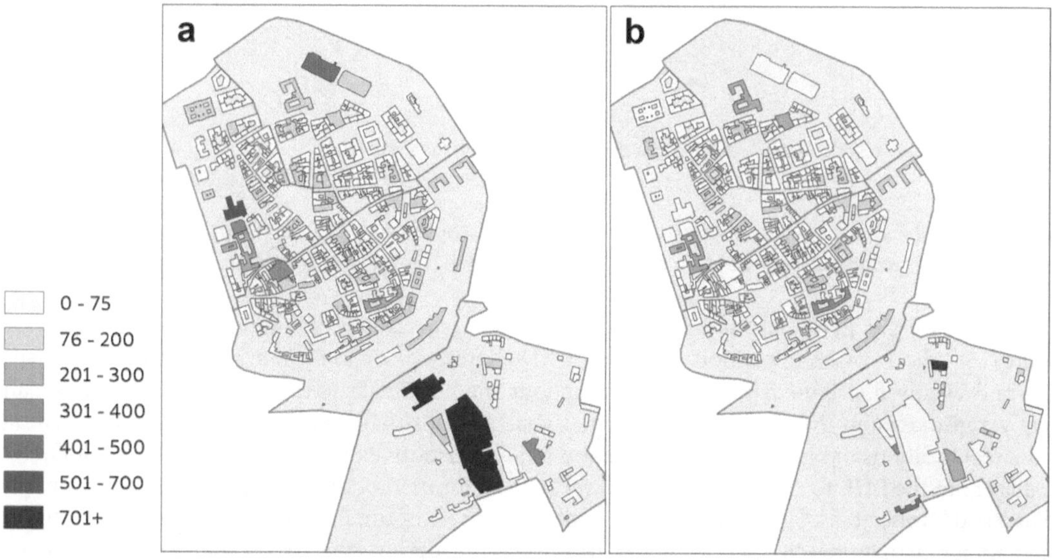

Figure 11. Building level BEHP (PR+PNR) average for day (a) and night (b).

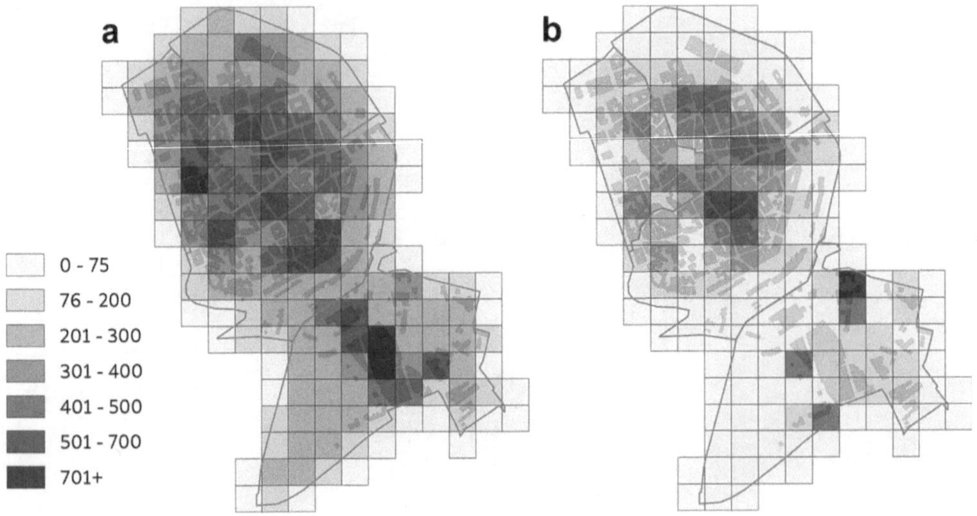

Figure 12. Average number of PR+PNR+PT for day (a) and night (b) for a grid data.

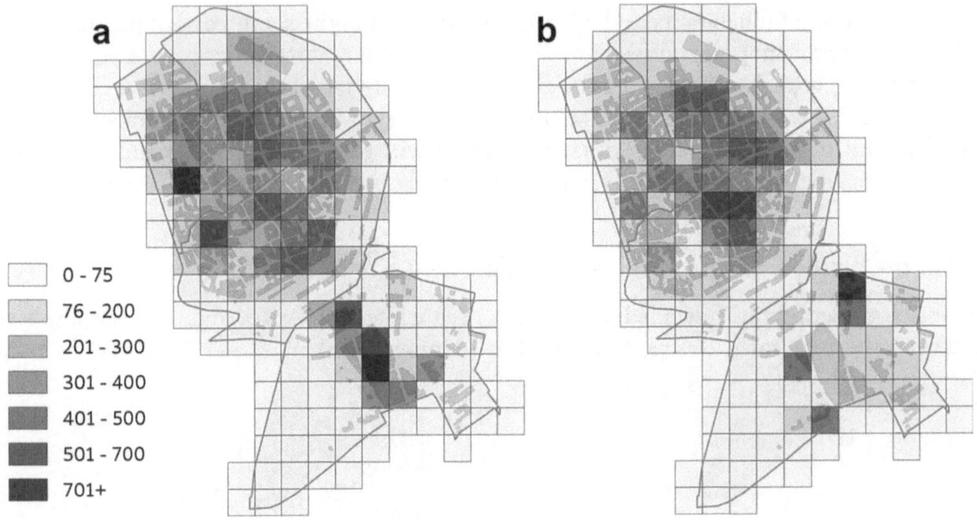

Figure 13. Average number of visitors (PR+PNR) for day (a) and night (b) for a grid data.

focused on solving a crisis involving contamination of the drinking water supply with a chemical substance. This training was held in the city of Brno where the pilot area of our analysis is also situated.

The use of the EHP dasymetric approach for this emergency situation is demonstrated in Figure 15. The location-allocation analysis on the leftmost side only takes into account census data and evenly distributed population throughout the administrative unit. Each water tank can supply approximately 2000 people. The second analysis adjusts the water tank locations according to the real locations of buildings and population in administrative units. The third and fourth analysis quantifies the EHP for working days and weekends. Our methodology proposes a greater number of necessary water tanks in administrative units, and their optimal locations change according to population fluctuations. In such cases, an approach considering time as a variable plays a crucial role in the location-allocation analysis of drinking water sources in particular administrative

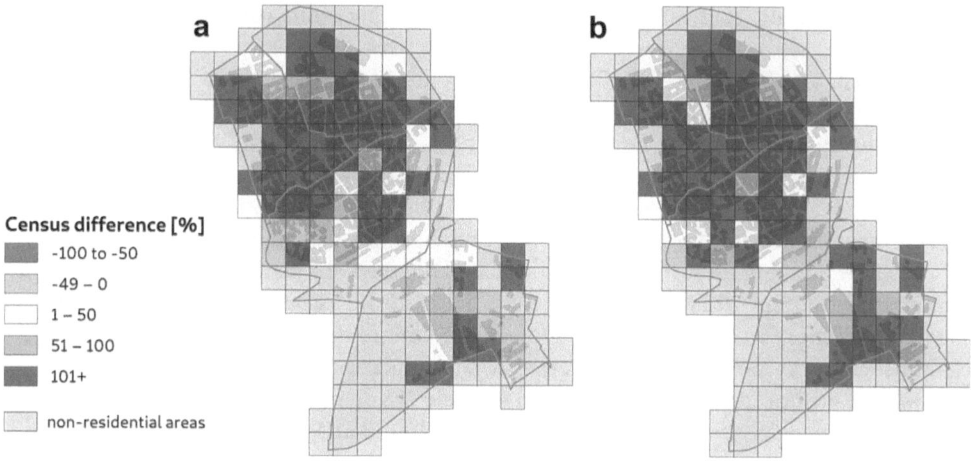

Figure 14. Comparison of PR (residents) and PN+PNR (visitors) data for day (a) and night (b) for a grid data.

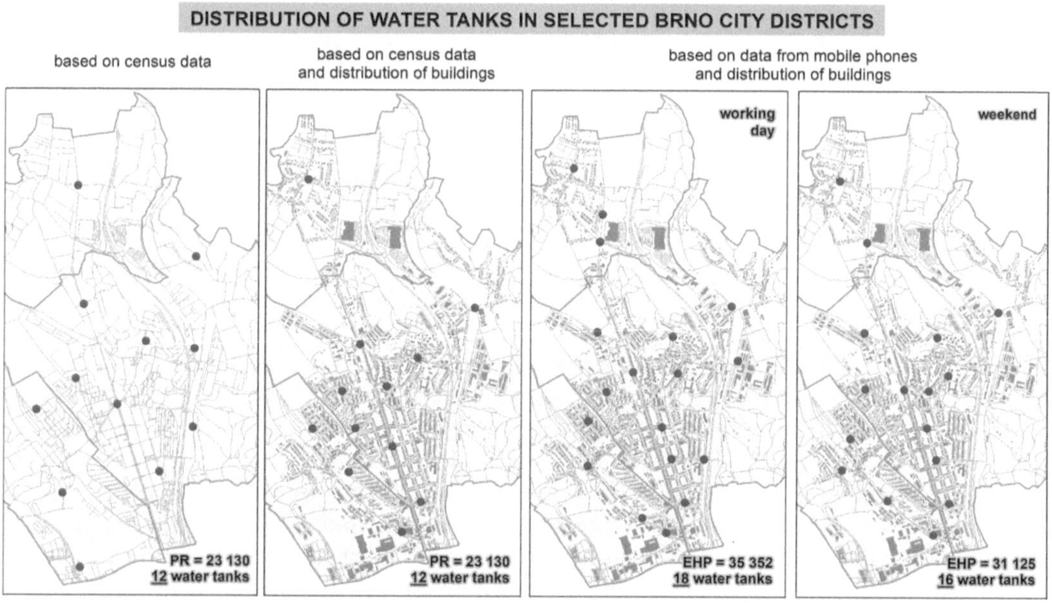

Figure 15. The role of spatio-temporal distribution of population in the case of water shortage (see the text for a detailed description).

units, and a more detailed description of human presence influences the success of emergency measures.

5. Discussion and conclusion

The availability of spatial data from various sources, including mobile phone data and social media, for example, is rising. The variety of data sources also allows data from numerous sources to be used and combined. This study has used available mobile phone data to provide information about the population and its distribution. The authors compared the obtained data to recent and available traditional demographic data sources (CZSO 2011) and also developed the general framework for estimating a more realistic spatio-temporal distribution of inhabitants using a dasymetric and regular grid method.

Järv, Tenkanen, and Toivonen (2017) broadly introduced the dasymetric interpolation method and the refinement of the population between target zones (buildings) for risk assessment at the micro level. Our approach can be considered as an alternative when CDR data are available for administrative/spatial units but not for GSM cells. Specifically, we differ in the following points:

- We further developed the use of building level administrative data. Järv, Tenkanen, and Toivonen (2017) used the building footprints and associated 3D space/volume to better represent the potential human distribution within the regular grid. We used the associated census technical data on individual buildings (building footprints) to achieve a more detailed spatial granularity and connected this data with the functional use of particular buildings. By contrast, we did not use numbers of inhabitants for individual buildings in our calculation of BEHP. These numbers were only used to compare census data with our results (calculated BEHP) at the level of buildings.
- The analysis was performed with more detailed mobile phone data. While the original authors used normalized mobile phone data to represent the relative population distribution, we were able to distinguish visitors, transiting persons, gender and age groups to achieve a better spatio-temporal differentiation of demographic/gender attributes and specifics of activity (PR +PNR/PT).
- Both studies differ also in the overall period and sampling method. Discontinuous data (day and night samples from a working day) for a one-month period were used in the original study (Järv, Tenkanen, and Toivonen 2017). We used longer period (12 weeks) and continuous data for 24 hours representing both weekends and the working days. The pilot study revealed high spatio-temporal differences in human population distribution with implications for crisis management and decision-making processes.

The analysed data confirmed the general assumption of overestimated mobile phone data measurements within city centres (Järv, Tenkanen, and Toivonen 2017). However, temporal analysis also the revealed differing rhythmicity between visitors and transiting people and fluctuations during day and night times. Spatial patterns at the level of individual buildings detailed the granularity of analysis and allowed other administrative and economic data to be linked.

The results showed that census data significantly underestimates both the number of visitors and transiting persons for day and night EHP. However, this situation is consistent with existing social and demographic research (Seidenglanz, Toušek, and Chvátal 2013), as described in Section 3.1.

The data from mobile operator O2 used for the study provide similar values which can be corroborated by using the alternative data from mobile operator T-Mobile currently available from the Brno city administration. T-Mobile statistics of human presence variability for basic administrative units can be checked at http://operator.brno.ml. The average numbers of EHP for selected spatial units in our pilot area (Figure 2) provide a similar underestimation of census data.

Table 1 in Section 3.1 shows the difference between census data and estimated presence of population for the city of Brno. However, the differences in the city centre can be expected to be even greater because of the concentration of transportation hubs, retail centres and historical monuments as mentioned above. The other reason for greater differences in the pilot area is a population increase in the suburbs and increase of job opportunities in the centre of Brno (Sýkora and Mulíček 2014). All these reasons caused that census data to significantly underestimate the real population in the pilot area. The comparison with the census data emphasized the necessity of further developing the methodology for EHP calculation.

The important part of the suggested method is its easy repeatability and adjustment. The EHP calculation and entire analytical process have been automated using Python code. Individual input (building footprints, mobile phone user values per hour, administrative units), as well as coefficients (weights) or formulas, can be adjusted, and the general model further calibrated. An interactive example of the workflow for one of the pilot areas can be found at https://nbviewer.jupyter.org/github/gis4dis/MFD-Brno/blob/master/MFD%20example.ipynb or on GitHub at https://github.com/gis4dis/MFD-Brno. The code is described in a step-by-step fashion with map plots, table examples and graphs.

These developments in the analytical and visualization rules are also closely linked to validation and the real use of results. One potential validation process is consultation with social scientists focusing on human behaviour and daily rhythms. EHP can be partially validated for commercial buildings and large malls via Google popular times graphs. More detailed discussion was performed by Martin, Cockings, and Leung (2015).

The disaggregation of the GEOSTAT population grid to a target resolution of 100 m opens new possibilities of combining EHP with other statistical data and also possible cross validation of results (Batista e Silva, Gallego, and Lavalle 2013; Scholz, Andorfer, and Mittlboeck 2013). A detailed grid also enables integration with other spatial data such as crime and traffic offence data (Russnák et al. 2016; Herman et al. 2018), city climate data (Dobrovolný and Krahula 2015; Geletič, Lehnert, and Dobrovolný 2016), retail economic data (Mulíček, Osman, and Seidenglanz 2015), or public transportation analysis (Mesarec and Lep 2009).

The proposed approach can provide more accurate dynamic population distribution mapping to assess object-oriented risks e.g. for purposes of Integrated Rescue System (IRS). Although the pilot area is relatively small, several emergencies have occurred directly in the pilot area, including fires (most recently in 2016), in some cases also with casualties (in 2002), and demonstrations (in 2007, 2015, 2017). The approach can be applied to a larger spatial extent, such as during a flood, if the data are available. For example, Smith, Martin, and Cockings (2016) mention that the estimated population exposure to hazards discovered by adopting an innovative integrated modelling approach that explicitly includes both space and time can improve flood risk management.

The challenges and opportunities for more extensive use of mobile phone data exist not only for demographic and social sciences, but also for a range of other application domains in crisis management. This study presents an extensible methodology of dasymetric mapping and encourages further development in big spatial data application.

Challenges and limitations to this study still exist. Firstly, the city centre was selected as the pilot area to reveal the highest differences, but even a small area can influence the results in general (city centres are generally considered to be overestimated – Järv, Tenkanen, and Toivonen 2017). Secondly, only a relatively small amount of mobile phone data were available for the study. More detailed and long-term mobile phone data (not only for daily rhythms, but for weekly, monthly, and seasonal changes) enable buildings to be differentiated in more detail and their functional specifications described (opening hours for commercial and administrative buildings, refinement of functional classification of buildings) in combination with other ancillary data (traffic data, spatial planning data). Thirdly, the population grids need to be discussed and validated in terms of suitable grid size and potential combination with other statistical data from population grids.

The presented study and methods can be expanded in several directions. The authors are working on extending the spatial and temporal parameters and target populations (different age and gender groups) and better representation of other administrative units outside the city centre which potentially serve other functions (see Mulíček, Osman, and Seidenglanz 2015 for comparison). Extension of the analysed time frame will improve the data validity and can also help identify peculiarities and irregularities (e.g. Easter Monday), when the behaviour of inhabitants differs from usual patterns.

Disclosure statement

No potential conflict of interest was reported by the authors.

Funding

This research was funded by the grant of the Ministry of Education, Youth and Sports of the Czech Republic [grant number LTACH-17002] Dynamic mapping methods oriented to risk and disaster management in the era of big data, by the National Key R&D Program of China [project number 2016YFE0131600], National Natural Science Foundation of China [project number 41871371, 41671457], by the grant of the Czech Science Foundation [number 17-02827S] Mapping everydayness: representation of routine spaces and by the grant of the Masaryk University "Integrated research on environmental changes in the landscape sphere of Earth III" [grant number MUNI/A/1251/2017].

ORCID

Petr Kubíček http://orcid.org/0000-0003-4613-1068
Zdeněk Stachoň http://orcid.org/0000-0003-1807-1459
Radim Štampach http://orcid.org/0000-0002-0480-8862

References

Ahas, R., A. Aasa, Y. Yuan, M. Raubal, Z. Smoreda, Y. Liu, C. Ziemlicki, M. Tiru, and M. Zook. 2015. "Everyday Space–Time Geographies: Using Mobile Phone-based Sensor Data to Monitor Urban Activity in Harbin, Paris, and Tallinn." *International Journal of Geographical Information Science* 29 (11): 2017–2039. doi:10.1080/13658816.2015.1063151.

Ahas, R., S. Silm, O. Järv, E. Saluveer, and M. Tiru. 2010. "Using Mobile Positioning Data to Model Locations Meaningful to Users of Mobile Phones." *Journal of Urban Technology* 17 (1): 3–27. doi:10.1080/10630731003597306.

Batista e Silva, F., J. Gallego, and C. Lavalle. 2013. "A High-resolution Population Grid Map for Europe." *Journal of Maps* 9 (1): 16–28. doi:10.1080/17445647.2013.764830.

Bengtsson, L., X. Lu, A. Thorson, R. Garfield, and J. von Schreeb. 2011. "Improved Response to Disasters and Outbreaks by Tracking Population Movements with Mobile Phone Network Data: A Post-Earthquake Geospatial Study in Haiti." *Plos Medicine* 8 (8): 1–9. doi:10.1371/journal.pmed.1001083.

Binek, J., O. Šerý, H. Svobodová, and Z. Šilhan. 2017. *Strategický plán rozvoje městské části Brno-střed na období 2018–2025* [Strategic plan for the development of the Brno-střed for the period 2018–2025]. http://www.brno-stred.cz/uploads/Strategicky_plan/SP_MC_Brno_stred_pro_tisk_9.6.17_.pdf.

Cao, J., W. Tu, Q. Li, and R. Cao. 2017. "Spatio-temporal Analysis of Aggregated Human Activities Based on Massive Mobile Phone Tracking Data." *Journal of Geo-Information Science* 19 (4): 467–474. doi:10.3724/SP.J.1047.2017.00467.

CZSO. 2009. *Klasifikace stavebních děl [Construction Classification]*. Accessed February 14, 2018. https://www.czso.cz/csu/czso/klasifikace_stavebnich_del_-cz_cc-_platna_od_1_10_2009.

CZSO. 2011. *Census 2011 Results*. Accessed February 24, 2018. https://www.czso.cz/csu/sldb/home; https://www.czso.cz/csu/rso/-registr-scitacich-obvodu-a-budov.

CZSO. 2017. *Information Society in Figures 2017*. Accessed February 17, 2018. https://www.czso.cz/csu/czso/information-society-in-figures-2014-2016.

Data-Pop Alliance. 2017. "Data-Pop Alliance." Accessed February 14, 2018. http://datapopalliance.org/.

Deville, P., C. Linard, S. Martine, M. Gilbert, F. R. Stevens, A. E. Gaughan, V. D. Blondel, and A. J. Tatem. 2014. "Dynamic Population Mapping Using Mobile Phone Data." *PNAS* 111 (45): 1–6. doi:10.1073/pnas.1408439111.

Dobrovolný, P., and L. Krahula. 2015. "The Spatial Variability of air Temperature and Nocturnal Urban Heat Island Intensity in the City of Brno, Czech Republic." *Moravian Geographical Reports* 23 (3): 8–16. doi:10.1515/mgr-2015-0013.

Dufková, K., M. Ficek, L. Kencl, J. Novák, J. Kouba, I. Gregor, and J. Danihelka. 2008. "Active GSM Cell-ID Tracking: 'Where Did You Disappear?'" In *Proceedings of the first ACM international workshop on Mobile entity localization and tracking in GPS-less environments*, 7–12. doi:10.1145/1410012.1410015.

European Commission Joint Research Centre. 2013. "*INSPIRE Data Specification on Buildings – Technical Guidelines*." 323 p. Accessed February 19, 2018. https://inspire.ec.europa.eu/file/1533/download?token=ouzQkBuP.

Eurostat. 2011. *Reference Data*. Accessed February 12, 2018. http://ec.europa.eu/eurostat/web/gisco/geodata/reference-data.

Gallego, F. J. 2010. "A Population Density Grid of the European Union." *Population and Environment* 31 (6): 460–473. doi:10.1007/s11111-010-0108-y.

Gao, S. 2015. "Spatio-temporal Analytics for Exploring Human Mobility Patterns and Urban Dynamics in the Mobile Age." *Spatial Cognition & Computation: An Interdisciplinary Journal* 15 (2): 86–114. doi:10.1080/13875868.2014.984300.

Geletič, J., M. Lehnert, and P. Dobrovolný. 2016. "Modelled Spatio-temporal Variability of air Temperature in an Urban Climate and its Validation: A Case Study of Brno, Czech Republic." *Hungarian Geographical Bulletin* 65 (2): 169–180. doi:10.15201/hungeobull.65.2.7.

GSMWEB. 2018. *Non-commercial Server about BTSs and Mobile Operators*. Accessed March 25, 2018. http://www.gsmweb.cz/.

Herman, L., J. Russnák, R. Stuchlík, and J. Hladík. 2018. "Visualization of Traffic Offences in the City of Brno (Czech Republic): Achieving 3D Thematic Cartography Through Open Source and Open Data." In *Proceedings of 25th Central European Conference*, edited by H. Svobodová, 270–280. Brno: Masaryk University. doi:10.5817/CZ.MUNI.P210-8908-2018.

Järv, O., H. Tenkanen, and T. Toivonen. 2017. "Enhancing Spatial Accuracy of Mobile Phone Data Using Multi-temporal Dasymetric Interpolation." *International Journal of Geographical Information Science* 31 (8): 1630–1651. doi:10.1080/13658816.2017.1287369.

Jonge, E., M. Pelt, and M. Roos. 2012. *Time Patterns, Geospatial Clustering and Mobility Statistics Based on Mobile Phone Network Data*. Hague: Statistics Netherlands. 26 p. Accessed February 21, 2018. https://www.cbs.nl/NR/rdonlyres/4EDB51ED-927A-4A69-B8F3-4DC57A44DDE4/0/Timepatternsgeospatialclusteringandmobilitystatistics.pdf.

Kang, C., S. Gao, X. Lin, Y. Xiao, Y. Yuan, Y. Liu, and X. Ma. 2010. "Analyzing and Geo-visualizing Individual Human Mobility Patterns Using Mobile Call Records." In *2010 18th International Conference on Geoinformatics*, 1–7. doi:10.1109/GEOINFORMATICS.2010.5567857.

Kang, C., Y. Liu, X. Ma, and L. Wu. 2012. "Towards Estimating Urban Population Distributions from Mobile Call Data." *Journal of Urban Technology* 19 (4): 3–21. doi:10.1080/10630732.2012.715479.

Kang, C., Y. Liu, and L. Wu. 2017. "An Analysis of Entropy of Human Mobility from Mobile Phone Data." *Wuhan Daxue Xuebao (Xinxi Kexue Ban)/Geomatics and Information Science of Wuhan University* 42 (1): 63–69. doi:10.1320.

Konečný, M., and W. Reinhardt. 2010. "Early Warning and Disaster Management: The Importance of Geographic Information (Part A)." *International Journal of Digital Earth* 3 (3): 217–220. doi:10.1080/17538947.2010.508884.

Kressler, F., and K. Steinnocher. 2008. "Object-oriented Analysis of Image and LiDAR Data and Its Potential for a Dasymetric Mapping Application." In *Object-Based Image Analysis*, edited by T. Blaschke, S. Lang, and G. J. Hay, 611–624. Berlin, Heidelberg: Springer.

Kubíček, P., E. Mulíčková, M. Konečný, and J. Kučerová. 2011. "Flood Management and Geoinformation Support Within the Emergency Cycle (EU Example)." In *Environmental Software Systems. Frameworks of Environment*, edited by J. Hřebíček, G. Schimak, and R. Denzer, 77–86. Berlin: Springer. doi:10.1007/978-3-642-22285-6_9.

Lane, N. D., E. Miluzzo, H. Lu, D. Peebles, T. Choudhury, and A. T. Campbell. 2010. "A Survey of Mobile Phone Sensing." *IEEE Communications Magazine* 48 (9): 140–150. doi:10.1109/MCOM.2010.5560598.

Li, Q., and B. Zhou. 2015. "Smartphone-based Individual Indoor Spatiotemporal Behavior Analysis." *Progress in Geography* 34 (4): 457–465. doi:10.11820/dlkxjz.2015.04.007.

Lískovec, R. 2016. "*Výzkum časoprostorových rytmů prostřednictvím geolokalizačních technologií [Exploration of Timespace Rhythms Using Geolocalization Technologies]*." Bachelor thesis. Brno: Masaryk University. 51 p.

Liu, Y., C. Kang, and F. Wang. 2014. "Towards Big Data-driven Human Mobility Patterns and Models." *Wuhan Daxue Xuebao (Xinxi Kexue Ban)/Geomatics and Information Science of Wuhan University* 39 (6): 660–666. doi:10.13203/j.whugis20140149.

Martin, D. 1996. "An Assessment of Surface and Zonal Models of Population." *International Journal of Geographic Information Systems* 10 (8): 973–989. doi:10.1080/02693799608902120.

Martin, D., S. Cockings, and S. Leung. 2015. "Developing a Flexible Framework for Spatiotemporal Population Modeling." *Annals of the Association of American Geographers* 105 (4): 754–772. doi:10.1080/00045608.2015.1022089.

Mennis, J. 2003. "Generating Surface Models of Population Using Dasymetric Mapping." *The Professional Geographer* 55 (1): 31–42. doi:10.1111/0033-0124.10042.

Mesarec, B., and M. Lep. 2009. "Combining the Grid-based Spatial Planning and Network-based Transport Planning." *Technological and Economic Development of Economy* 15 (1): 60–77. doi:10.3846/1392-8619.2009.15.60-77.

Mulíček, O., R. Osman, and D. Seidenglanz. 2015. "Urban Rhythms: A Chronotopic Approach to Urban Timespace." *Time & Society* 24 (3): 304–325. doi:10.1177/0961463X14535905.

Nagle, N. N., B. P. Buttenfield, S. Leyk, and S. Spielman. 2014. "Dasymetric Modeling and Uncertainty." *Annals of the Association of American Geographers* 104 (1): 80–95. doi:10.1080/00045608.2013.843439.

Novák, J., and J. Temelová. 2012. "Každodenní život a Prostorová Mobilita Mladých Pražanů: Pilotní Studie Využití Lokalizačních dat Mobilních Telefonů [Everyday Life and Spatial Mobility of Young People in Prague: A Pilot Study Using Mobile Phone Location Data]." *Sociologický Časopis / Czech Sociological Review* 48 (5): 911–938.

Ouředníček, M., J. Nemeškal, P. Špačková, M. Hampl, and J. Novák. 2018. "A Synthetic Approach to the Delimitation of the Prague Metropolitan Area." *Journal of Maps* 14 (1): 26–33. doi:10.1080/17445647.2017.1422446.

Oxendine, C. E., E. Schnebele, G. Cervone, and N. Waters. 2014. "Fusing Non-authoritative Data to Improve Situational Awareness in Emergencies." In *Proceedings of the 11th International ISCRAM Conference*, edited by S. R. Hiltz, et al., 760–764. University Park: ISCRAM. doi:10.13140/2.1.4980.5443.

Pastor-Escuredo, D., A. Morales-Guzmán, et al. 2014. "Flooding Through the Lens of Mobile Phone Activity." In *IEEE Global Humanitarian Technology Conference*, 1–8. San Jose: IEEE. doi:10.1109/GHTC.2014.6970293.

Paulus, F., A. Krömer, J. Petr, and J. Černý. 2016. *Analýza hrozeb pro Českou republiku* [Threats Analysis for the Czech Republic]. Lázně Bohdaneč: IOOL. 9 p. Accessed February 20, 2018. www.hzscr.cz/soubor/analyza-hrozeb-zprava-pdf.aspx.

Peng, Z., Z. Sun, Q. Cheng, H. Jiao, and W. Chen. 2018. "Urban Land Use Function Recognition Method Using Sequential Mobile Phone Data." *Wuhan Daxue Xuebao (Xinxi Kexue Ban)/Geomatics and Information Science of Wuhan University* 43 (9): 1399–1407. 1437. doi:10.13203/j.whugis20170329.

Pucci, P., F. Manfredini, and P. Tagliolato. 2015. *Mapping Urban Practices Through Mobile Phone Data*. 1st ed. New York: Springer International Publishing. 90 p. doi:10.1007/978-3-319-14833-5.

Reibel, M., and M. E. Bufalino. 2005. "Street-weighted Interpolation Techniques for Demographic Count Estimation in Incompatible Zone Systems." *Environment and Planning A* 37 (1): 127–139. doi:10.1068/a36202.

Řezník, T., B. Horáková, and D. Janiurek. 2011. "Emergency Support System: Actionable Real-time Intelligence with Fusion Capabilities and Cartographic Displays." *Advances in Military Technology* 6 (2): 83–97.

Řezník, T., B. Horáková, and R. Szturc. 2015. "Advanced Methods of Cell Phone Localization for Crisis and Emergency Management Applications." *International Journal of Digital Earth* 8 (4): 259–272. doi:10.1080/17538947.2013.860197.

Russnák, J., and L. Herman. 2017. "Visualization and Spatial Analysis of Open Data for Evacuation Planning - Pilot Study of the Dukovany Nuclear Power Plant (Czech Republic)." In *17th International Multidisciplinary Scientific GeoConference SGEM 2017, Informatics, Geoinformatics and Remote Sensing* (Vol. 17, Is. 23), 853–860. doi:10.5593/sgem2017/23/S11.107.

Russnák, J., P. Ondrejka, L. Herman, P. Kubíček, and A. Mertel. 2016. "Visualization and Spatial Analysis of Police Open Data as a Part of Community Policing in the City of Pardubice (Czech Republic)." *Annals of GIS* 22 (3): 187–201. doi:10.1080/19475683.2016.1200670.

Scholz, J., M. Andorfer, and M. Mittlboeck. 2013. "Spatial Accuracy Evaluation of Population Density Grid Disaggregations with Corine Landcover." In *Geographic Information Science at the Heart of Europe*, edited by D. Vandenbroucke, B. Bucher, and J. Crompvoets, 267–283. Cham: Springer. doi:10.1007/978-3-319-00615-4_15.

Seidenglanz, D., V. Toušek, and F. Chvátal. 2013. *Odborná analýza vývoje počtu osob oficiálně neregistrovaných ve městě Brně* [Analysis of the Number of Persons Officially Unregistered in the City of Brno]. https://www.brno.cz/fileadmin/user_upload/sprava_mesta/Strategie_pro_Brno/dokumenty/Odhad_13.pdf.

Silm, S., and R. Ahas. 2010. "The Seasonal Variability of Population in Estonian Municipalities." *Environment and Planning A* 42 (10): 2527–2546. doi:10.1068/a43139.

Slocum, T. 1999. *Thematic Cartography and Visualization*. 1st ed, 293. Upper Saddle River, NJ: Prentice Hall.

Smith, A., D. Martin, and S. Cockings. 2016. "Spatio-Temporal Population Modelling for Enhanced Assessment of Urban Exposure to Flood Risk." *Applied Spatial Analysis and Policy* 9 (2): 145–163. doi:10.1007/s12061-014-9110-6.

Stachoň, Z., P. Kubíček, R. Štampach, L. Herman, J. Russnák, and M. Konečný. 2016. "Cartographic Principles for Standardized Cartographic Visualization for Crisis Management Community." In *Proceedings, 6th International Conference on Cartography and GIS (Vol. 1 and Vol. 2)*, edited by T. Bandrova and M. Konečný, 781–788. Sofia: Bulgarian Cartographic Association.

Steenbruggen, J., E. Tranos, and P. Nijkamp. 2015. "Data from Mobile Phone Operators: A Tool for Smarter Cities?" *Telecommunications Policy* 39 (3–4): 335–346. doi:10.1016/j.telpol.2014.04.001.

Sýkora, L., and O. Mulíček. 2014. "Prague: Urban Growth and Regional Sprawl." In *Confronting Suburbanization: Urban Decentralization in Postsocialist Central and Eastern Europe*, edited by Kiril Stanilov and Luděk Sýkora, 133–162. Chichester: Wiley Blackwell.

Čtyroký, J. 2017. "Kde má Praha centrum? [Where is Prague Center?]." *Sborník příspěvků - Konference GIS Esri v ČR*. p. 4. Accessed February 21, 2018. http://download.arcdata.cz/konf/2017/sbornik_konference_2017.pdf.

Verkasalo, H., and H. Hämmäinen. 2007. "A Handset-based Platform for Measuring Mobile Service Usage." *Info* 9 (1): 80–96. doi:10.1108/14636690710725085.

Wright, J. K. 1936. "A Method of Mapping Densities of Population: With Cape Cod as an Example." *Geographical Review* 26 (1): 103–110.

Wu, J., L. Huang, Y. Liu, J. Peng, W. Li, S. Gao, and C. Kang. 2012. "Traffic Flow Simulation Based on Call Detail Records." *Dili Xuebao/Acta Geographica Sinica* 67 (12): 1657–1665.

Xie, Z. 2006. "A Framework for Interpolating the Population Surface at Residential-housing-unit Level." *GIScience and Remote Sensing* 43 (3): 1–19. doi:10.2747/1548-1603.43.3.233.

Xu, J., Z. Fang, S. Xiao, and L. Yin. 2015. "The Spatio-temporal Heterogeneity Analysis of Massive Urban Mobile Phone Users' Stay Behavior: A Case Study of Shenzhen City." *Journal of Geo-Information Science* 17 (2): 197–205. doi:10.3724/SP.J.1047.2015.00197.

Yuan, Y., and M. Raubal. 2016. "Analyzing the Distribution of Human Activity Space from Mobile Phone Usage: an Individual and Urban-Oriented Study." *International Journal of Geographical Information Science* 30 (8): 1594–1621. doi:10.1080/13658816.2016.1143555.

Yuan, Y., M. Raubal, and Y. Liu. 2012. "Correlating Mobile Phone Usage and Travel Behavior – A Case Study of Harbin, China." *Computers, Environment and Urban Systems* 36 (2): 118–130. doi:10.1016/j.compenvurbsys.2011.07.003.

Zhou, B., Q. Li, Q. Mao, W. Tu, and X. Zhang. 2015. "Activity Sequence-Based Indoor Pedestrian Localization Using Smartphones." *IEEE Transactions on Human-Machine Systems* 45 (5): 562–574. doi:10.1109/THMS.2014.2368092.

Zhou, X., Y. Yue, A. G. O. Yeh, H. Wang, and T. Zhong. 2014. "Uncertainty in Spatial Analysis of Dynamic Data – Identifying City Center." *Wuhan Daxue Xuebao (Xinxi Kexue Ban)/Geomatics and Information Science of Wuhan University* 39 (6): 701–705. doi:10.13203/j.whugis20140074.

Discovering the relationship of disasters from big scholar and social media news datasets

Liang Zheng, Fei Wang, Xiaocui Zheng and Binbin Liu

ABSTRACT

The construction method for chains of disasters or events is still one of the core scientific questions in studying the common rules of disaster's evolution. Especially when dealing with the complexity and diversity of disasters, it is critical to make a further investigation on reducing the dependency of prior knowledge and supporting the comprehensive chains of disasters. This paper tries to propose a novel approach, through collecting the big scholar and social news data with disaster-related keywords, analysing the strength of their relationships with the co-word analysis method, and constructing a complex network of all defined disaster types, in order to finally intelligently extract the unique disaster chain of a specific disaster type. Google Scholar, Baidu Scholar and Sina News search engines are employed to acquire the needed data, and the respectively obtained disaster chains are compared with each other to show the robustness of our proposed approach. The achieved disaster chains are also compared with the ones concluded from existing research methods, and the very reasonable result is demonstrated. There is a great potential to apply this novel method in disaster management domain to find more secrets about disasters.

1. Introduction

The dynamic process of emergency response determines that the emergency management system is an open giant system, which has complex characteristics such as multi-agency, multi-factor, multi-scale, variability and so on (Fan 2007). There are many profound scientific problems in this research field. It is of great significance to master the regular patterns of the occurrence and development of disasters in order to control and reduce their negative impacts (Xue, Peng, and Zhang 2002). Many historical cases show that a disaster is not an isolated and static event. Under certain conditions, the secondary and derivative disasters will occur along with the development of the original one. The initial disaster and a series of derived disasters may constitute a chain, and each one is likely to lead to another one (Wang and Yuan 2010). The mutual coupling of various disaster-causing factors, such as natural factors, biological factors, physical factors, chemical factors, human factors and so on, further lead to the complex correlation along the chain of disasters (Yu et al. 2014). The in-depth study of disaster chain is one of the key scientific issues in finding the regularities during disasters' evolution processes. Especially in the face of complex and frequently occurred emergencies, how to discover the relationships among latent disasters and build up the comprehensive disaster chain is an urgent challenge in the field of public safety science (Li 2017).

The devastating hurricane Katrina in 2005 caused a series of storm surges and floods in Gulfport, Mississippi, New Orleans, Louisiana, and areas in between, and resulted in thousands of fatalities and billions of US dollars of economic losses (CNN 2017). In 2008, the freezing rain and snow disaster in the south of China triggered a serious of hazards, including roads blocking, power grids collapse and so on, which brought the economic losses of about 150 billion yuan (Chen et al. 2011). In 2011, the earthquake and tsunami in the east of Japan not only took the lives of more than twenty thousands of people, but also destroyed the Fukushima nuclear plant and led to a serious radioactive material leakage accident, which brought a worldwide and long-term impact (Forward 2017). The heavy casualties and losses of these catastrophes are caused by the complex temporal and spatial interactions of both the original and secondary disasters. In many cases, the secondary and derivative disasters could have even more destructive power than the original one (Ha et al. 2016).

Therefore, we should not regard each disaster as an independent individual, but consider multi-hazard as a chain-cycle system that has the mutual coupling relationships. Due to the chain relationships among disasters, it is meaningful to investigate the disaster chain and find out the major secondary disasters from an original one, which can help the emergency manager to better prepare in advance and minimum the losses by eliminating the key triggered factors and cutting off the chain (Wang and Yuan 2010).

The typical methods for studying disaster chain mainly include probability analysis method based on simulation data and experimental data (Helbing and Kühnert 2003; Li et al. 2010; Wang, Gao, and Xin 2010), geospatial analysis method based on remote sensing images (Gitis, Petrova, and Pirogov 1994; Zhou et al. 2014; Hallegatte 2008), and disaster evolution analysis based on complex network theory (Liu and Wu 2015; Lin et al. 2012; Chen, Li, and Sun 2008; Li and Chen 2014). The above research approaches mainly apply qualitative researches, and rely heavily on the historical cases and the prior knowledge of experts. Most of the research results are the disaster chain of one or several typical disasters in certain local areas, which could not provide comprehensive expressions for ubiquitous disaster types (Ha et al. 2016).

With the rapid development of big data theory, the techniques and methods for mining the explicit and implicit information from massive data have been increasingly becoming mature (Sánchez, Craglia, and Bregt 2016). Currently, in many scholar data repositories and news portal websites, such as Baidu Scholar, Google Scholar, Sina News and CNN News, there are tremendous amounts of information related to disasters. It can be speculated that there must be some useful scholar papers in the search results to help us finding the secrets of the relationships among various disasters. Therefore, it is worthy of trying the brand-new attempt to apply the big data methodology in the disaster chain research domain.

From the new perspective, this paper intends to explore the big data mining method to analyse the latent relationships of the disasters classified in four major categories and more than 300 sub-categories proposed by China (China 2007), based on the massive scholar data and social media news data from different sources. By means of scholar and news big data analysis, this methodology no longer depends on the domain expert's priori knowledge or limited historical cases, and is able to provide a universal and intelligent method to find out the patterns of disaster chains for all defined types of disasters. Furthermore, the degree of relationships among disaster types can be evaluated quantitatively by the co-occurrence analysis and a certain chain of disasters can be extracted through our method.

The remainder of the paper is organized as follows. Section 2 introduces several major research methods for studying disaster chain so far. Section 3 elaborates on the big datasets, the major theories and the technology roadmap adopted in our research. Section 4 explains the proposed methodology and gives a detailed description about the computational procedures. Section 5 compares the research results of three search engines (Google, Baidu and Sina), and as well compares the results with the ones achieved by existing methods. Then the last section gives our conclusions and discusses the further work.

2. State of art in the field of disaster chain

There are many scholars concerning about the disaster chain researches, one hotspot of the disaster management or emergency management domain. But due to the numerous types and the complex coupling relationships of all possible disasters, it is hard to launch experiments to simulate them all in the lab. Also, the historical cases data is relatively inadequate. The researches in this direction mainly focus on several specific disasters, such as earthquake, typhoon and flood. Based on our extensive literature work, three typical approaches for studying disaster chain are summarized as follows.

2.1. Probability analysis method based on simulation data and experimental data

The disaster chain involves the triggering relationship between disasters, and this triggering relationship is not inevitable in reality. Almost any disaster has the probability, but not certainly, to trigger some other secondary disasters. The probability analysis method tries to build an event tree to analyse and filter the secondary disasters that may be triggered by an original one, and then calculate the conditional probabilities of the occurrence for those secondary disasters (Yu 2010; Dong 2009).

There are several ways to calculate the conditional probabilities for the disaster chain. Some scholars use the Bayesian network model, which can infer the conditional probabilities of events in the same network based on the limited known information (Wang, Gu, and Huang 2013). In addition, other methods, such as neural networks (Wei et al. 2000), and expert scoring systems (Liu et al. 2006), can be used to determine the conditional probabilities of disasters in the chain as well. Wang, Gao, and Xin (2010) combined field data with artificial neural networks to analyse the probability of building collapse caused by the earthquake. Li et al. (2010) proposed a mathematical function model that described the interactions among disaster-causing factors and disaster-bearing bodies, and then constructed a disaster chain of the earthquake and performed some experimental verification. Helbing, Dirk, and Christian Kühnert (2003) applied the main equations of statistical physics to construct the influence matrix which reflected the correlation between disasters, and then considered a semi-quantitative analysis of the time-dependence of catastrophes for the purpose of anticipation.

The current research based on probabilistic analysis method is mostly focused on the interactions among disaster-causing factors. Other factors such as the hazard-inducing environment have not received enough considerations (Han et al. 2014). Meanwhile, the commonly used method for calculating conditional probabilities is the expert scoring method at present, which mainly depends on the prior knowledge of experts and may affect the objectivity and universality of the result.

2.2. Geospatial analysis method based on remote sensing images

Geospatial technology of Digital Earth has entered into many scientific applications and been proven crucial for effective disaster management, including traditional GIS-based technologies (Mccarthy, Graniero, and Rozic 2008), human–GIS interaction issues (Cai 2008), 3D GIS solutions (Kwan and Lee 2005) and many others. For the disaster chain researches, the large-scale and widespread disasters such as the earthquake and geological disasters are often analysed by the means of geospatial technology such as remote sensing image processing and GIS spatial analysis methods.

After analysing thousands of satellite images about the Sumatran earthquake in Indonesia and the great flood in the West River, Yao and Qiang (2006) proposed the mechanism of the earthquake disaster chain and the typical phenomenon of seismic-flood. Liu (2013) used multi-source and multi-temporal satellite images to extract debris flow information and studied the characteristics of typical channels in depth, and then analysed the disaster chain of debris flows in Parlung Zangbo basin in Tibet.

Meanwhile, with the application of multiple remote sensing sensor platforms and the improvement of interpretation methods, it can provide a reliable access to discover the secondary disasters from high-resolution remote sensing images. Especially after the 2008 Wenchuan earthquake, many scholars started to study the disaster chain of the earthquake. Fan et al. (2008) used multi-source remote sensing data to obtain the information of barrier lakes induced by the earthquake. Xu et al. (2012) used the remote sensing images to study the consequences of the chain of the earthquake, including the debris flow, the drastic river bed movements and the ecological changes. Liang, Tang, and Wang (2012) fused and compared the surveying data and remote sensing images of Redstone river basin of Qingchuan city before and after the earthquake to analyse the characteristics of earthquake-geological disaster chain.

Obviously, the geospatial analysis method has outstanding features to support scientists to study the relationships between disasters, especially for geological, and meteorological disasters. Many good research results have been achieved and are beneficial for the disaster management. However, datasets with enough temporal scale, spatial scale and good quality are relied on, and only limited types of disaster chains can be investigated. Thus, it is difficult to extend this method to many other types of disasters.

2.3. Disaster evolution analysis method based on complex network theory

Complex network is a network with the complicated topological features, which has been used in many fields such as life sciences, engineering sciences, information sciences, social sciences and so on (Miller, Vandome, and Mcbrewster 2010). It has begun to attract the attention of some scholars to study the dynamic process of disaster chain using the complex network theory (Zheng et al. 2017).

For disasters such as snow storm, rain storm and typhoon, the preliminary applications of disaster evolution analysis method have been carried out. Liu and Wu (2015) proposed a risk evaluation model for disaster chain based on the complex network structure, and gave the mathematical description about the chaining mechanism. Lin et al. (2012) used it to analyse the evolution mechanism of college campus fire hazards. Li and Chen (2014) expressed the process of the disaster chain of city blackouts by constructing a complex network. Chen, Li, and Sun (2008) adopted it to construct the network of the freezing rain and snow disasters, and analysed its evolution process and characteristics.

However, at present, the applications of complex network theory in the disaster chain research field are still underway, and generally are only adopted for demonstrating the visual process. The scholars have not yet proceeded with the further research on the ‘degree’ of network nodes and the semantic assignment of network edges. The overall features of the complex networks have not been explored in depth. The mining and analysis of the networks are still absent, and the structural advantages of complex networks have not yet been fully utilized.

The above three approaches demonstrating the researches on disaster chain are nowadays popular. However, a unified and mature theory and methodology system has not been established (Yu et al. 2014). Three main problems still remain.

(1) The existing researches are still limited to several typical disasters in certain study areas, such as the typhoon disaster chain in the coastal areas and the seismic geological disaster chain in the mountainous areas.
(2) Disaster evolution analysis methods and geospatial analysis methods are mostly adopting qualitative researches about the evolution process and the propagation mechanisms. Only a few studies have quantified the disaster chain by establishing mathematical and physical models.
(3) Most probability analysis methods and disaster evolution analysis methods are confined to one or several historical cases, and the results usually depend on a large amount of prior knowledge and expert experience. Consequently, it is lack of universality and feasibility to explore many other type of disasters.

Aiming at solving the exposed problems, this paper intends to discover the relationships of all types of disasters through mining the big scholar and social news datasets and construct one complex network of disasters, whose nodes are the disaster types and edges are the relationships among disasters. What's more, the subnet division theory would be investigated to extract a specific disaster chain from the disaster complex network.

3. Data and methodology

This research tries to employ the pure big data method to reveal the relationships among disasters, and therewith the primary task is to determine the data sources. People discuss the reasons, developments, caused damages and all other aspects about disasters in many places. The massive data directly or indirectly illuminate the complex relationships among the disasters. For instance, some explicit descriptions about the relationships occur in the academic articles, like the disaster chain of typhoon (Chen, Chen, and Wang 2007), the secondary disasters of the earthquake (Tang 2008; Men 2007). Some descriptions are not focusing on the disaster chains, but mention some of their relationships, and these cases appear in both academic articles and news web pages, like the derivative damages caused by the earthquake (CHINA Daily 2011) and typhoon (CNN 2013). Therefore, the scholar data and Internet news data are chosen in our research work.

Google Scholar and Baidu Scholar data repository, as well as Sina News website, are selected as the data sources. Google Scholar (https://scholar.google.com/) is a freely accessible web search engine that indexes the full text or metadata of scholarly literature all over the world, including most online academic journals and books, conference papers, theses and dissertations, preprints, technical reports, and others. Most of the articles published in English can be searched there. Baidu Scholar (http://xueshu.baidu.com/) originating from Baidu search engine which is the biggest Chinese search engine in the world, covers most of the articles published in Chinese and is superior to Google Scholar in Chinese search (Xie, Guan, and Yang 2015). Google and Baidu both provide a convenient one-stop search service allowing us to easily access the detailed information of articles with developed software tool. Other scholar search engines like Oalib, Library Genesis and many others, only specialize in certain subjects and are inferior to Google Scholar and Baidu Scholar in search amount and range. In this way, Google Scholar is taken as the English literature data search engine and Baidu Scholar is regarded as the Chinese literature search engine.

News articles capture a variety of topics about our society, which reflect not only the socioeconomic activities that happened in our physical world, but also some of the cultures, human interests, and public concerns that exist only in the perceptions of people (Hu, Ye, and Shaw 2017). Disasters are frequently discussed in news articles, and the content of news articles always aggregate all related information about each disaster, such as spatial, temporal and hazard-related information (Wang and Stewart 2015). Two or more disasters may co-occur in the same article. Such co-occurrence often suggests certain relatedness among the disasters. From the Internet news, we could find the whole course of the events, including the reasons, the situational developments, the statistical data and other information. Comparing with scholar data, news data may provide wider coverage and more details.

Sina News (http://news.sina.com.cn/) is one of the most famous Chinese news websites, and it provides a lot of latest feature news. A simple test has been done to show the popularity of different Internet news websites, including Sina News, Baidu News, Tencent News and Ifeng News, who are the most leading news portals in China. Table 1 shows the search results returned from those websites when using different disaster-related keywords: typhoon, fire hazard, earthquake, and tsunami (in Chinese Pinyin separately called 'taifeng', 'huozai', 'dizhen', 'haixiao'). Obviously, Sina News have much more amounts of data information than others. It also offers advanced search options for performing search and filter assignments. Therewith, Sina News is selected as the news repository in our research.

Table 1. Comparison of the number of search results between news websites (accessed March 9, 2018. Similarly hereinafter for all searched datasets).

	Disaster type			
News websites	Taifeng (typhoon)	Huozai (fire hazard)	Dizhen (earthquake)	Haixiao (tsunami)
Sina News	264,383	438,981	920,245	212,490
Baidu News	42,800	44,500	53,300	36,300
Tencent News	72,123	52,292	171,224	19,163
Ifeng News	33,957	56,555	60,975	8916

Because news data is considered as the supplement of scholar data to enhance the volumes of the disaster-related datasets in our research, the search keywords of news are only restricted to Chinese. Although more amounts of experimental data can be acquired through other language's news websites, only search results of Sina News with Chinese keywords are considered due to the balance between effectiveness and quantity.

Aiming at finding the latent relationships among various disasters from the massive data, our research technical route is illustrated in Figure 1. Google Scholar and Baidu Scholar as the scholar experimental datasets, and Sina News as the social news database are taken into account in this research. In order to reduce the invalid data or noisy data from the original data sources, data

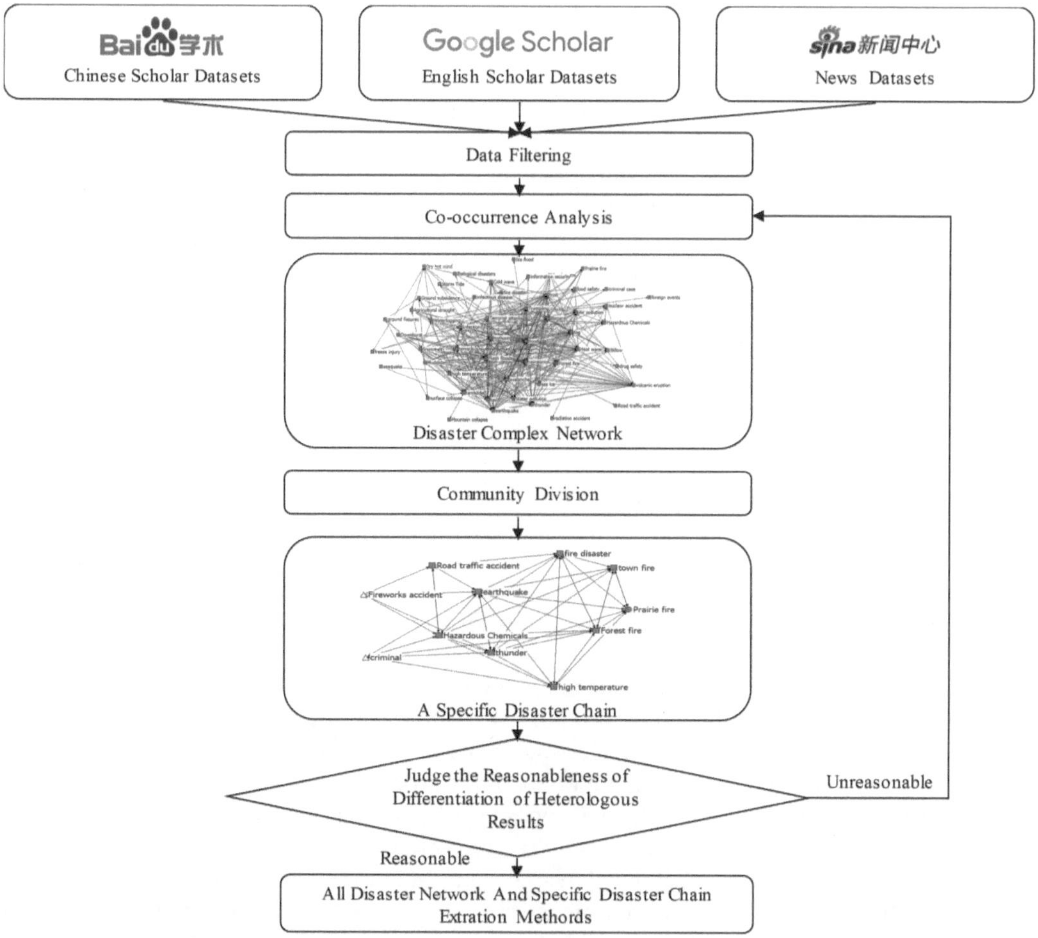

Figure 1. Research technical route.

filtering method is applied to improve the quality of the analysing datasets. All relevant datasets about disasters from scholar papers and news are achieved through this step.

Co-word analysis is a form of co-occurrence analysis method, which is one kind of content analysis methods that have been widely employed to conduct a variety of researches for exploring the hot topics and development trends of some specific domain (He 1999; Liu et al. 2014). It does not rely on any priori definition of research terms, and thus can objectively detect the dynamics of science. Indexes based on the co-word frequency of items can be used to measure the strength of relationships between items.

Enlightened by the productive applications of co-word analysis method in other domains, we adopt it to compute the co-occurrence of disaster type related terms within the experimental datasets, and thus construct the complex network of those terms in order to represent the degree of relationships. Complex network method as a graph with more advanced topological features is recently studied and becomes an active area of scientific research inspired by the empirical study of real-world networks such as brain networks, social networks and computer networks. The related researches are developing at a brisk pace and become a multidisciplinary field including mathematics, physics, sociology, computer science, biology, engineering science and many others. The chain of hundreds of disaster types actually involves relationships between any of two disaster types. The weight of relationship can be graphed and represented as the connection between the two disaster types according to their closeness. Disaster chain can be naturally constructed and more importantly be analysed by the means of complex network method, and thus to establish the primary relationship network. Based on that, different results achieved from different data sources, including Google Scholar, Baidu Scholar and Sina News are compared and discussed. The overall analysing results will also be compared with the traditional studies.

The refined network of all hazards consists of all possible relationships among them. But in practical applications, the decision maker of emergency management always cares for a specific chain of a disaster. For example, the disaster chain of typhoon and the disaster chain of earthquake cause different attention in coastal regions or in mountainous regions. Therefore, an intelligent extraction method to get any chain of interest from the refined complex network is investigated. Community division method is a well-known approach to divide the complex network into different subnets, and it includes different algorithms such as GN algorithm (Girvan and Newman 2001), Label propagation algorithm (Kothari and Jain 2002), and Maximum flow algorithm (Ford and Fulkerson 1956). One of the algorithms is adopted in our research to fulfil the purpose of extracting the wished chain of a specific disaster.

4. Implementation of the proposed methodology

4.1. Determination of search mode

Co-word analysis method is employed in this research to measure the importance of the relationships among different disaster types. Basically, co-occurrence of two disaster types in the same article is counted when this matched article is found in scholar datasets or news datasets. By means of search engine, the amount of search results matching the co-occurrence rule is recorded.

Searching the keywords at different locations of an article may bring about different search results. Baidu Scholar, Google Scholar and Sina News all provide similar search modes including 'search words occur in the title of the article' or 'anywhere in the article' within their advanced search function. Obviously, disaster type keywords may appear many times in one article if it really concerns many aspects of the disaster types. Therewith, 'search words occur anywhere in the article' mode is taken into account as the fundamental search rule. Moreover, 'Find articles with all of the words' and 'with the exact phrase' search modes are optional in all of the selected search engines and the research results may differ. Particularly in Chinese words, a long term may be divided into more than one semantic word even no space character between the words if the first option is selected, and thus research results are not same as expectation. Therefore, 'Find articles with the exact phrase' mode is employed in our research as the fundamental search rule as well. For instance, if applying the co-occurrence of 'typhoon' and 'flood' keywords through Google Scholar search engine according to the above-determined search rules, the amount of searched results is 34,700.

For the same disaster type, different semantic expressions of search keywords certainly bring about different search results, no matter in Chinese or English. For example, 'fog' keyword achieves 1,100,000 search records and 'fog disaster' keyword achieves only 90,800 search records through the Google Scholar search engine. The search results with 'fog' keyword contain huge amounts of papers studying on the weather, climate or meteorology fields, and they contain disaster-related researches as well since it has the inclusion relationship with 'fog disaster' keyword. But the search results with 'fog disaster' keyword may omit scholar papers only with 'fog accident', 'fog incident' or other semantic expressions relating to disaster researches. Therefore, the data filtering assignment is firstly performed every time to single out the papers in the disaster research domain with either of the keywords of 'accident', 'incident', 'emergency' and 'disaster'. In this way, the consequent secondary search will not take those disasters semantically related keywords again.

For the search mode in Chinese, traditional Chinese and simplified Chinese characters with completely same semantic meaning in the search engine will receive totally different search results, due to the complexity and diversity of Chinese. For example, 'typhoon' keyword in simplified Chinese achieves 155,000 records and in traditional Chinese achieves 27,100 records, and the search results do not have any duplication. Therefore, we will consider both simplified and traditional Chinese keywords when facing with one semantic search term. However, this case doesn't apply in English search work.

For the search mode in English, singular and plural issue matters the search results. For example, 'sandstorm' keyword in Google Scholar achieves 20,600 search results, 'sandstorms' achieves 21,300 search results, and their overlapping 'sandstorm sandstorms' achieves 3360 search results. That is, the papers discussing the semantic meaning relating to sandstorm have the amount of 38,540 (20,600 + 21,300 − 3,360) search records, which are far more than the ones using only singular or plural keyword. Therefore, we need to search both singular and plural keywords and integrate the results to avoid the omission. Interestingly, this case doesn't apply in Chinese search work.

4.2. Determination of disaster-related keywords filtering

To better understand the nature of disasters and set up the suitable strategies, China explicitly classified disasters into four major categories and more than 300 sub-categories, including natural disasters, calamitous accidents, public health incidents and social safety incidents (China 2007). Each kind of major disaster is again classified into detailed subclasses in several levels in order to prevent the ambiguity of the general classification. For example, a landslide event is one of the subclasses of geological disasters, and a geological disaster is one of the subclasses of natural disasters.

Many of the subclass disaster types seldom occur in the reality or are not often discussed in the scholar papers and news articles. Considering over 300 subclasses of all possible disaster types into the search work will bring complicated processing data, and more importantly not all of them are interesting for the consequent disaster chain research. Therefore, first and foremost, all of the disaster type subclasses are separately searched through Baidu Scholar and the number of search results for each keyword is tallied up. Web crawler is used to help to get the data. Web crawler can be regarded as an Internet bot that systematically browses the World Wide Web normally for the purpose of web indexing based on certain defined rules. It can be a program or script developed in different computer languages (Guo 2017), and Python programming language is adopted in our research. Some of the results are picked and listed in Table 2.

The left column of Table 2 lists the top 10 disaster types of the largest number of search results, and the right column lists the other 10 disaster types of around 1000 search results. More search results occur, more scholars or public concern about the disaster types. The statistics show that there are 73 disaster types with the search results over 1000 items, 50 disaster types with the search results between 100 and 1000 items, and the remaining over 200 disaster types with only search results less than 100 items. When searching the same 73 disaster types via Google Scholar, the amount of search results for each of 73 gets the items over 1000 as well. When the complexity of computation and the importance of disaster type are considered together, only 73 disaster types with over 1000 search results are determined to be selected as the keywords in the consequent co-word analysis. The expression in both Chinese Pinyin and English of all 73 disaster types are listed in Appendix.

4.3. Computation of co-occurrence matrix

Based on the determination of search mode and selected keywords, three search engines are used to find the matched records. The co-occurrence frequencies between each of the 73 candidates (see Appendix) are able to be calculated, and a matrix can be created. Since the total amount of co-occurrence relationships between each pair of 73 candidates is 2701, it is hardly possible to search them one by one manually.

Web crawler is adopted in our research to implement the task of finding the matched keywords, automatic updating keywords, recording and summarizing search results, and finally constructing the 73*73 matrix of co-occurrence frequencies. Partial results from Google Scholar search engine are presented in Table 3.

In Table 3, the numbers on the diagonal can be read as the amount of articles containing the keyword, for example, there are 1,854,000 articles mention the term ‘flood’. The rest of the numbers mean the amount of articles that talk about two terms, for example, there are 9560 articles mention both the ‘flood’ and ‘waterlogging’.

When measuring the degree of relationship between two disaster types, the occurrence of each disaster type have also to be considered. The more the matched articles of one disaster type are found, the more the matched articles of its co-word with other disaster types are found, and vice versa. It is needed to set up a parameter for considering the impact, and well qualify the strength of the correlation between different disaster types. Therefore, the method of calculating the co-occurrence degree is introduced to handle with this problem. To get the co-occurrence degree, recent studies of Dice index, Jaccard index, Salton index and others are proposed and applied by many scholars (Yin and Yasuda 2005). Egghe (2009) compared those index algorithms and concluded that Salton index has correlation relation with most of other index algorithms, except Jaccard index. Both Jaccard index and Salton index can make two closely related keywords become closer, and two distantly related keywords more distant. Loet (2008) made a detailed comparison between Jaccard index and Salton index, and found that if the only option is to generate a co-occurrence matrix especially applying in webometric research, the Jaccard index is the best basis for the normalization. Therewith, we adopt the Jaccard index in the next computation step.

Table 2. Some of the search results through Baidu Scholar.

Disaster type	Number	Disaster type	Number
High temperature	2,630,000	Freeze injury	6,750
Earthquake	2,130,000	Oil spill at sea	5,830
Infectious disease	1,040,000	Electricity accident	5,660
Food safety accident	636,000	Water pollution	5,270
Information security	525,000	Marine accident	4,460
Flood	379,000	Crash	993
Landslide	316,000	National conflict	958
Rainstorm	315,000	Bluetongue	883
Typhoon	182,100	Acute occupational poisoning	850
Thunder	154,000	Agricultural pests	839

Table 3. Partial disaster co-occurrence frequency matrix obtained using Google Scholar.

	Disaster type					
Disaster type	Flood	Waterlogging	Agricultural drought	Typhoon	Tornado	Rainstorm
Flood	1,854,000	9560	4270	23,900	27,900	17,300
Waterlogging	9560	14,800	826	1140	181	1860
Agricultural drought	4270	824	7820	509	222	502
Typhoon	23,800	1140	509	51,100	4090	6200
Tornado	27,900	181	222	4090	68,800	1980
Rainstorm	17,300	1860	502	6210	1980	30,900

The Jaccard Index formula is

$$J_{ij} = \frac{c_{ij}}{c_i + c_j - c_{ij}}. \tag{1}$$

In Equation (1), J_{ij} represents the co-occurrence index of disaster type i and disaster type j. c_{ij} represents the co-occurrence frequency of disaster type i and disaster type j, and c_i and c_j separately represent the frequency of disaster type i and disaster type j, as searched through various scholar or news search engine. Table 4 illustrates the co-occurrence index matrix calculated through Jaccard index.

4.4. Construction of disaster complex network

With the calculated co-occurrence matrix, the disaster complex network can be constructed based on the principal complex network theory. Figure 2 represents a complex network, where the search results accessing from Google Scholar, the co-occurrence matrix calculated by Jaccard index, and the disaster types defined in Appendix. In this complex network graph, every disaster type of 73 kinds represents a node, and the co-occurrence rate of any of two disaster types denotes an edge if the co-occurrence value is above 0.

Figure 2 consists of 73 nodes and 1567 edges. The number of nodes connecting to each node is called the node degree, and the average of all nodes' degree is named the network average degree, which are used to represent the complexity of a complex network. Thus, the network average degree in Figure 2 equals $1567 * 2/73 \approx 43$, which means that averagely each node has the connection with 43 other nodes. In order to reduce the complexity of the network and extract the prominent information, it is necessary to optimize the graph. The co-occurrence shows the strength of correlation relation between two disaster types, that is, the smaller the value is, the weak the correlation is. Appropriate threshold can be set to eliminate the interference of the unimportant connections, and to simplify the network. For example, if the value of 0.03 is set to the threshold, the number of network edges become 536, the network average degree is around 15, and 2 isolated nodes occur (the nodes not connecting with others). If the value of 0.04 is applied, the number of network edges become 363, the network average degree is around 10, and 5 isolated nodes occur. The selection of different threshold values is meaningful and may be altered in respective researches. The threshold decision rule will be discussed in next subsection. Figure 3 shows the optimized complex network graph with the threshold value of 0.03.

Table 4. Jaccard index-based disaster co-occurrence frequency matrix (partly).

	Disaster type					
Disaster type	Flood	Waterlogging	Agricultural drought	Typhoon	Tornado	Rainstorm
Flood	1	0.005142	0.002274	0.012705	0.014724	0.009263
Waterlogging	0.005142	1	0.019493	0.017603	0.00217	0.042427
Agricultural drought	0.002274	0.019445	1	0.006444	0.002289	0.008538
Typhoon	0.012651	0.017603	0.006444	1	0.035316	0.081794
Tornado	0.014724	0.00217	0.002289	0.035316	1	0.020262
Rainstorm	0.009263	0.042427	0.008538	0.081937	0.020262	1

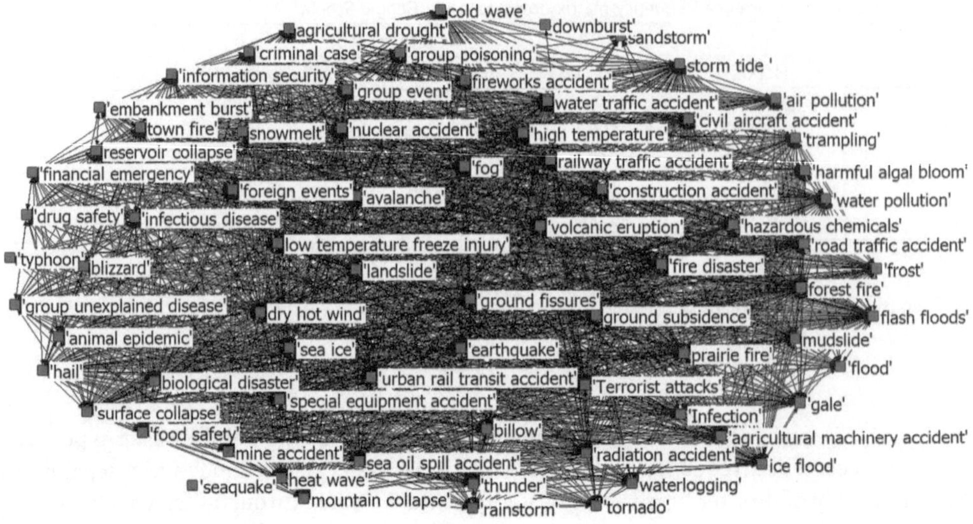

Figure 2. The constructed complex network graph of 73 disaster types.

4.5. Extracting a special disaster chain from the complex network

Although the complex network has been simplified with proper threshold settings, however, it is still too inconvenience to be used directly in the practical applications. The decision maker of disaster management always cares for a specific chain of a disaster. Therefore, an intelligent subnet division method is investigated to extract a disaster chain of interest from the refined complex network.

Community division method is a well-known approach to divide a whole complex network into several individual communities. Each community can be regarded as a subnet. There are two kinds of community structures. One is disjoint communities, and the other is overlapping communities, as presented in Figure 4 (a,b). As for disjoint communities, the intersection of vertex sets of any two communities is empty, and overlapping communities have common vertex sets.

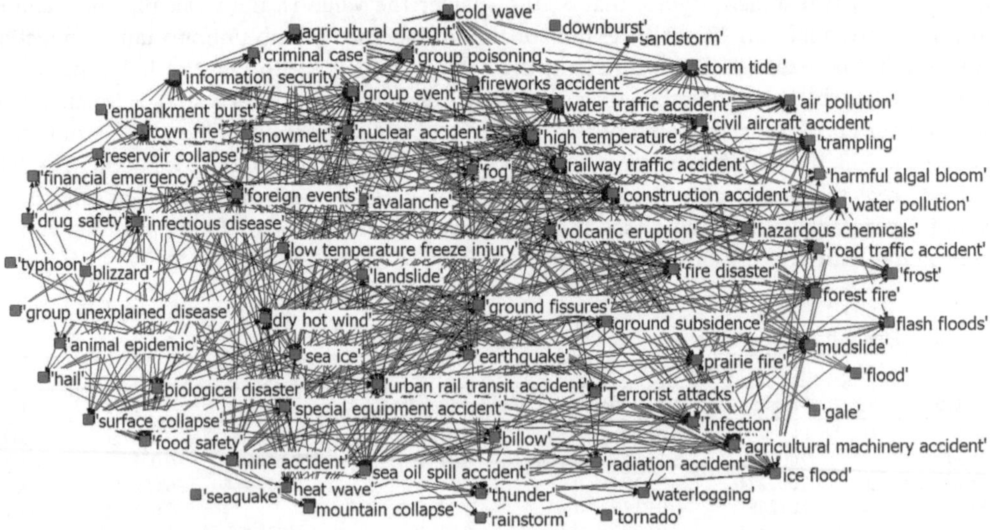

Figure 3. The optimized complex network graph with threshold value 0.03.

Different original disasters may lead to the same secondary one. For example, both typhoon and earthquake are possible to trigger the landslide hazards, which means that when dividing the disaster complex network into different communities, some communities may have common nodes. Therefore, the algorithms for overlapping communities division would better fit the disaster chain extraction task.

There are various algorithms that are adapted to either disjoint communities or overlapping communities. Several algorithms fitting with overlapping communities such as k-Clique-Community Finding Algorithm (Palla et al. 2005), Algorithm Based on Edge Vector Clustering (Evans and Lambiotte 2009; Ahn, Bagrow, and Lehmann 2010), Improved Maximum Flow Algorithm (Ford and Fulkerson 1956), etc., have been tested in our research and the Improved Maximum Flow Algorithm shows the best extraction results. This algorithm has the feature of finding out the cluster around a selected core, which is similar to the process of constructing a disaster chain. Thus, a bipartite network is firstly created based on the disaster complex network, and the improved maximum flow problem of the bipartite network is solved to get the community division results.

Figure 5 shows the process of calculating the flow value for a community. A network G(V,E), that has the node-set V and the edge-set E, is extended to a bipartite network G′’(V,V′,E′). The node-set V′ is equal to V, so if G(V,E) has 6 nodes and 7 edges, G′(V,V′,E′) has 12 nodes and 14 edges. Moreover, we divide G(V,E) into two initial communities: t_1 and t_2. The inter-flow and total-flow of community are respectively defined as Equation (2) and Equation (3). The efficiency of community is calculated by Equation (4). The value of f_{ij} is 1 when there is an edge between node i and node j, else f_{ij} is zero. For community t_1, it has 3 nodes. There are six edges among its internal nodes and one edge between its internal nodes and outer nodes, as the solid lines and dotted line shown in Figure 5 flowing into t_1. Therefore, t_1's inter-flow f is 6 and total-flow k is 7. Its efficiency is 6/7. A node outside could be added into the community only in the case that its entry can promote the efficiency of the community.

$$f = \sum_{i \in t \wedge j \in t} f_{ij} \tag{2}$$

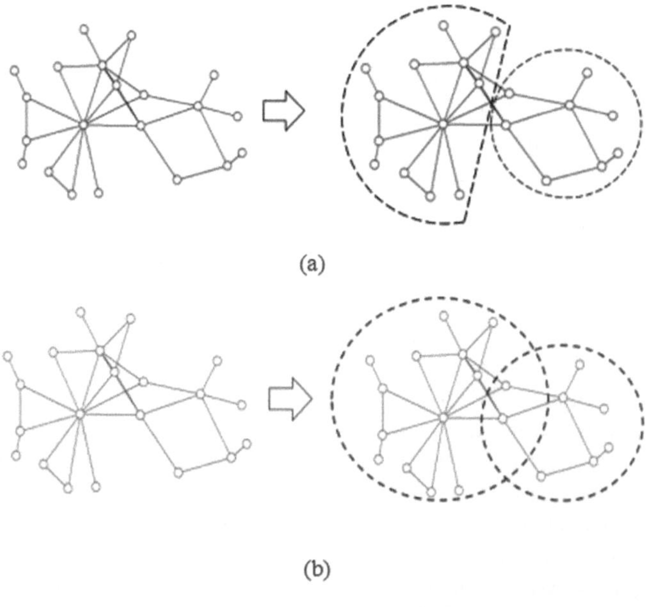

Figure 4. The structures of communities: (a) disjoint communities and (b) overlapping communities.

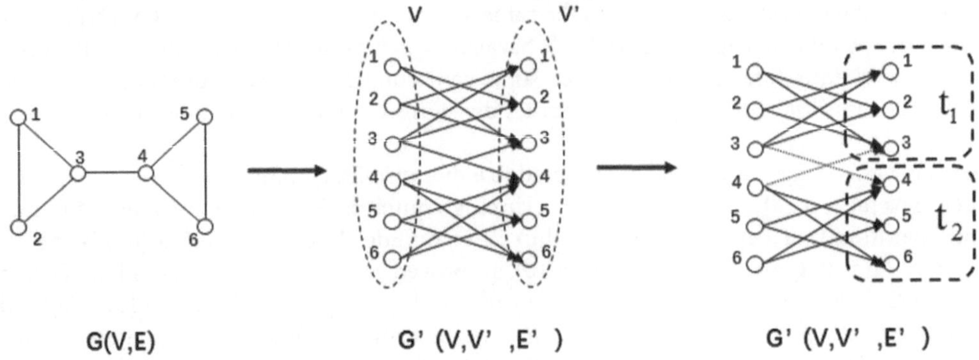

Figure 5. The flow of communities.

$$K = \sum_{i \in V \wedge \in t} f_{ij} \tag{3}$$

$$R_t = \frac{f}{k} \tag{4}$$

When using the Maximum Flow Algorithm to extract a special disaster chain, like the disaster chain of 'typhoon', the following steps are carried out.

(1) Select 'typhoon' node and add any one of its connected nodes to establish an initial community.
(2) Calculate the efficiency of the rest nodes in the whole complex network one by one, supposing it is considered in the initial community.
(3) Find out the node with the maximum efficiency to improve the flow value of the initial community, and add it into this community.
(4) Repeat Step (2) and (3) until the flow efficiency of the community stops increasing anymore.

Based on the above four primary steps, we could acquire the first 'typhoon' community. But it may not be the optimal one. Change the connected node in the Step (1) and repeat Step (1) to (4) to find out all possible typhoon communities. Select the community with the maximum efficiency as the final subnet for typhoon disaster chain.

The result of community division relies on the structure of the disaster complex network and would be altered with the threshold setting of the complex network. Figure 6 provides three different results of the disaster chains of 'flood'. It shows that the smaller threshold leads to a more complex network and more nodes in the community.

When the threshold is zero, the network cannot be divided and all nodes would be included into the community. When the threshold is too big, the community is too simple and some important connections may be eliminated. We explore the threshold decision rule using 'flood' disaster chain and 'typhoon' disaster chain based on Baidu Scholar data. Figure 7 shows the relationship between the number of disaster chain nodes and the setting of threshold. Obviously with the increase of the threshold, there are sharp changes in the number of disaster chain nodes. For example, in 'flood' disaster chain, when the threshold is between 0 and 0.008, the node number is 61 to 65; when the threshold is set 0.008 to 0.03, the node number is 9 to 13; when the threshold is larger than 0.03, the node number remains 1. In the traditional researches (Wang et al. 2016; Liu 2013) about disaster chain, the number of disaster chain nodes are always below 20. Therefore, it is reasonable to believe that the threshold which results in a disaster chain with about 10 to 20 nodes is the optimal one.

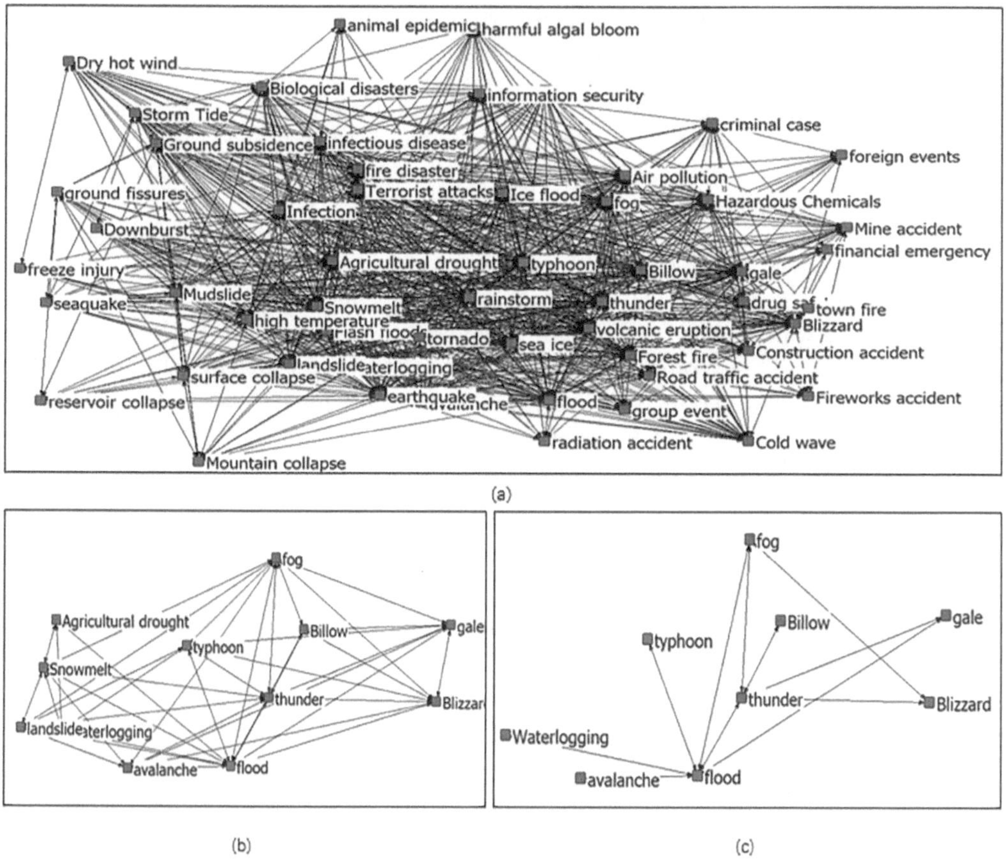

Figure 6. The extracted disaster chains of 'flood' from the complex network with three different threshold settings: (a) threshold is 0.001; (b) threshold is 0.01and (c) threshold is 0.03.

5. The result comparison

5.1. Comparison between Google Scholar data and Baidu Scholar data

Firstly, the search results of Chinese keywords by Baidu Scholar search engine and the search results of English keywords by Google Scholar search engine are adopted to acquire the final disaster chains, and two chains are compared.

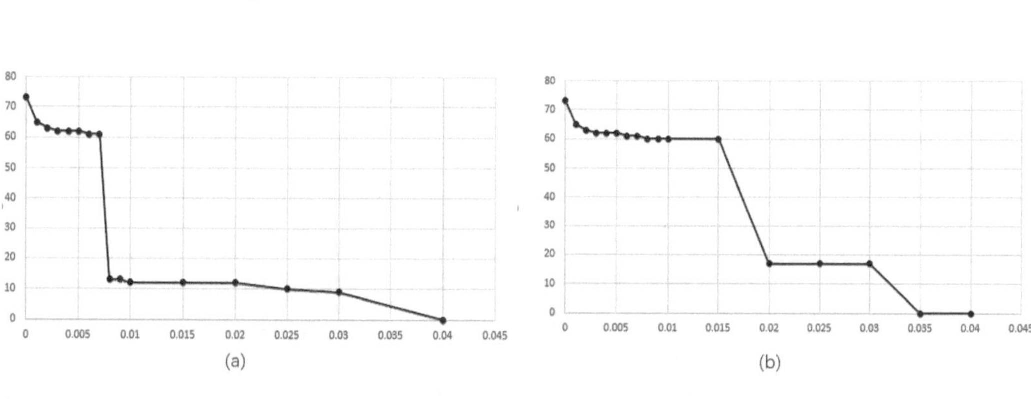

Figure 7. The relationship between the number of disaster chain nodes and the setting of threshold (a) 'flood' disaster chain based on Baidu scholar data and (b) 'typhoon' disaster chain based on Baidu scholar data.

The extracted disaster chain of 'typhoon' based on Baidu Scholar massive data consists of 17 disaster types (including 'typhoon' itself, the same below. The threshold of constructing the complex network is set as 0.02):

Typhoon, flood, waterlogging, storm tide, gale, landslide, rainstorm, blizzard, flash flood, thunder, hail, earthquake, tsunami, freeze injury, mudslide, mountain collapse, nuclear accident.

The extracted disaster chain of 'typhoon' based on Google Scholar massive data consists of 16 disaster types (the threshold is 0.02):

Typhoon, flood, waterlogging, storm tide, gale, landslide, rainstorm, blizzard, flash flood, thunder, hail, earthquake, tsunami, avalanche, volcanic eruption, prairie fire.

The comparison result is illustrated in Figure 8. We can recognize there are 13 common disaster types closely relating with 'typhoon' achieved through both data sources, and the main body of disaster chain is identical with each other. But in these common nodes, the topological relation obtained from two data sources may be different. For example, there is connection between 'tsunami' and 'flash flood' based on Baidu scholar data, but Google denied this link. Therefore, the topological relation among all common nodes should be judged and the edges should be merged. We define that for two common nodes, if the average of their co-occurrence rates in different data sources is larger than the average of thresholds, a connection edge is created between them. For example, in 'typhoon' disaster chain, the threshold for Baidu Scholar and Google Scholar are both 0.02, so the average of thresholds is 0.02 too. In Baidu Scholar, the co-occurrence rate of 'tsunami' and 'flash flood' is 0.022, and in Google Scholar, the co-occurrence rate is 0.006. The average of co-occurrence rates is 0.014, which is less than the average of thresholds. Therefore, the connection edge between 'tsunami' and 'flash flood' would be wiped off. On the contrary, in Baidu Scholar, the co-occurrence rate of 'tsunami' and 'storm tide' is 0.026, and in Google Scholar it is 0.016. The average value is 0.021, which is larger than the average of thresholds. Therefore, the edge between them is reserved.

Using the same approach for extracting the disaster chain of 'fire', 10 disaster types from Baidu Scholar and 8 disaster types from Google Scholar are come out, including 6 identical disaster types. Thus, it shows the results of two data sources have very high similarity.

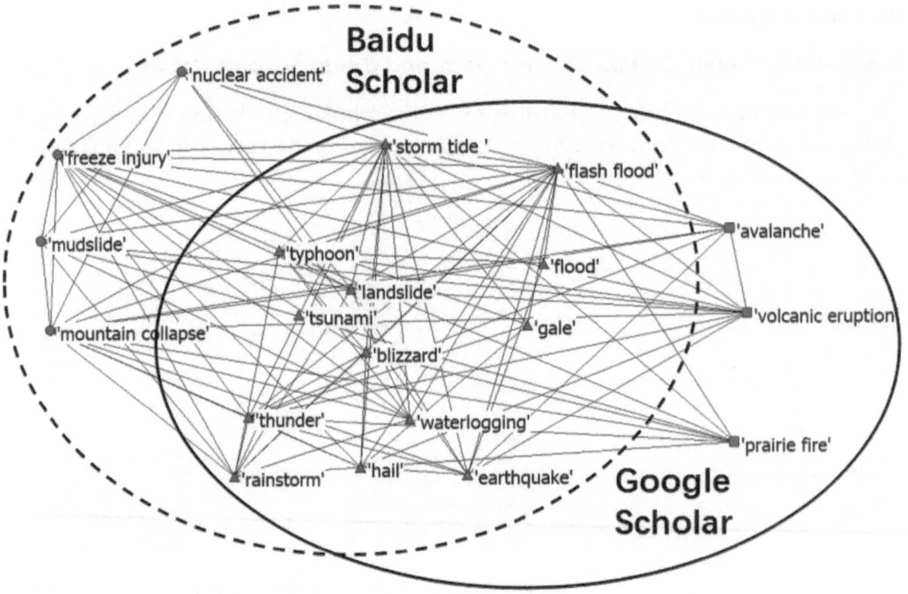

Figure 8. Comparison between Google Scholar data and Baidu Scholar data.

Fire, volcanic eruption, forest fire, mine accident, hazardous chemical, air pollution, water pollution, high temperature, nuclear accident and thunder (Baidu Scholar data source, the threshold is 0.02).

Fire, volcanic eruption, forest fire, mine accident, hazardous chemical, air pollution, prairie fire and elevator accident (Google Scholar data source, the threshold is 0.03).

5.2. Comparison between Baidu Scholar data and Sina News data

The same procedures are applied to Sina News (the threshold of constructing the complex network is set as 0.03), and the disaster chain of 'fire' and its comparison with Baidu Scholar (the threshold is 0.02) is shown in Figure 9. Ten disaster types in the chain are found out including: fire, fireworks accident, hazardous chemical, thunder, high temperature, forest fire, earthquake, traffic accident and criminal case. Six identical disaster types are included both in Sina News and Baidu Scholar. Moreover, social hot news such as 'fireworks accident' and 'criminal case' shows great attention in Sina News massive data, which fits with the features of social Internet news website. Different from scholar data, social news data tends to have a wider coverage, and pays more attention to the influence of human factors and social factors in the development of disasters. Therefore, disasters related to human involved activities appear frequently in disaster chains constructed based on social news data.

5.3. Comprehensive comparison of three data sources and existing methods

In order to verify the rationality of our method, some discussed disaster chains in existing literature works are used to compare with our results. Typhoon and earthquake take place frequently all over the world. The existing researches about them are relatively mature, and researchers have basically reached a consensus on the composition of their disaster chains. Therefore, the union set of typhoon disaster chain nodes in three references (Wang et al. 2016; Shuai, Wei, and Peijun 2012; Pan and Tang 2002) – called typical chain in this paper – is used to compare our results. The nodes outside of the 73 types of disasters involved in this study have been eliminated. Similar comparison is conducted with the earthquake disaster chain obtained from two references (Yin et al. 2012; Cui, Jiming, and Wenping 2012).

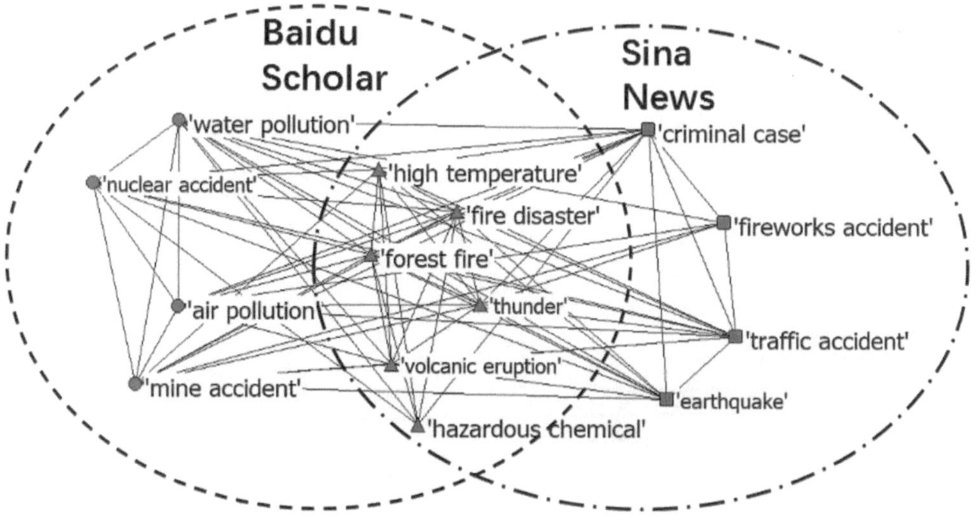

Figure 9. Comparison between Baidu Scholar data and Sina News data.

Figure 10 shows the comparison results of typhoon disaster chains. The thresholds for Baidu, Google and Sina are separately set as 0.02, 0.02, and 0.025. There are 10 common disaster types for all of them, and 6 common disaster types for two of them. The individual disaster types acquired from Sina News have more numbers than Baidu Scholar and Google Scholar, because the focus of news internet website may differ from the focus of scholar web site. Figure 10 also shows an interesting phenomenon. There are 11 common disaster types for three data mining approaches, but comparing with the results from existing literature works, the data mining approaches include two exceptional results of 'earthquake' and 'tsunami'. The exception will be discussed afterwards. Comparison results for the earthquake disaster chain are illustrated in Figure 11. All nodes in a typical chain are included in the results from three data sources.

These two experiments give us a satisfactory conclusion that the disaster chains discovered through data mining approach are consistent with the ones through existing methods. Meanwhile, some disaster types which are less concerned in existing researches have also been excavated, such as 'surface collapse' and 'volcanic eruption' in the earthquake disaster chain. Therefore, the proposed method can be regarded as a useful complement to the existing methods, and may give useful hints to the domain users.

The output of our method is not only these disaster nodes, but also the co-occurrence rate between each disaster-pair, which quantitatively represents the strength of their correlation. For example, in typhoon disaster chain, the co-occurrence rates are respectively 0.044, 0.101 and 0.012 (average value of three data sources) between 'typhoon' and 'flood', 'gale', and 'sandstorm'. It means that the correlation between typhoon and gale is stronger than the other two. This correlation represents the degree of concern in research areas and social domains. There are two factors which could affect the value of co-occurrence rate. First, there is a practical induced relationship between two disasters, so they get a lot of attention in research and social fields. Second, there are some similarities between two disasters, so they are always mentioned at the same time. For example, people always mention some popular disasters like earthquake and typhoon when they talk about

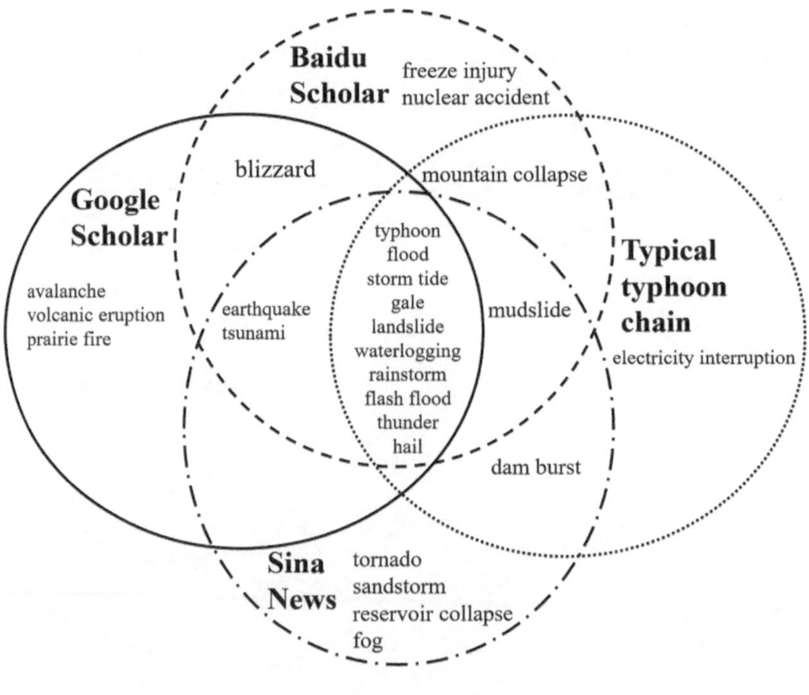

Figure 10. Comprehensive comparison of typhoon disaster chains.

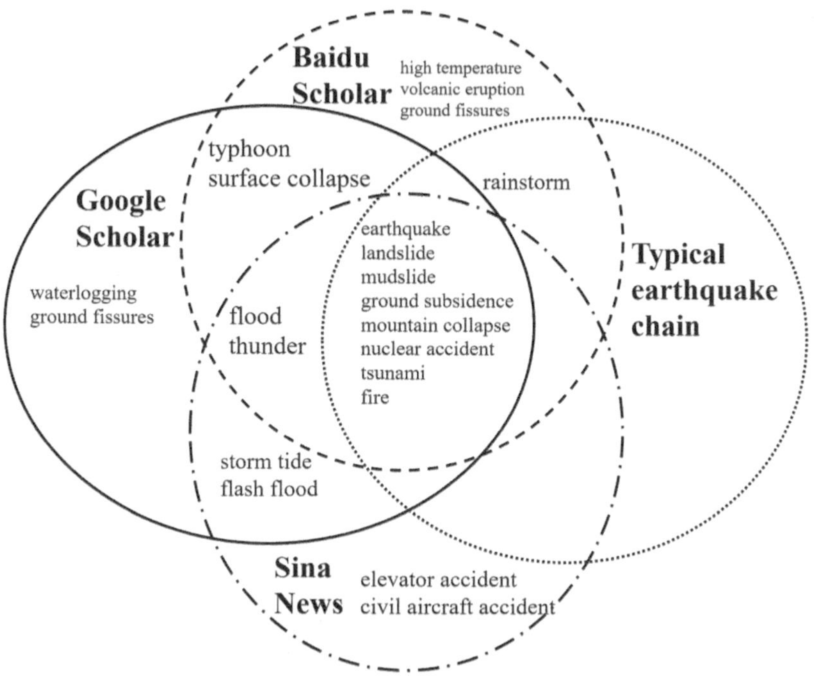

Figure 11. Comprehensive comparison of earthquake disaster chains.

disaster-related topics, even these two disasters are rarely correlated with each other. The research of disaster chain only cares about the first factor. However, filtering the affection of the second factor is out of the scope of this paper, and it is considered as the future work and the last section discusses this issue.

5.4. Comprehensive comparison of three data sources for all 73 disaster types

In this part, we extend our experiments for all 73 disaster types. The comprehensive comparison results of three data sources are visualized in Figure 12.

Comparing the results from Baidu and Google, we calculate the ratio between the number of common nodes and the average number of both nodes. There are 59 types of disasters whose ratios are more than 50%, accounting for 80.8% of all 73 disasters. This shows that in most cases, the results obtained from two scholar databases have high similarity. The same process is applied to compare the results of three data sources, and there are 37 types of disasters whose ratios are more than 50%, accounting for 50.7% of all 73 disasters. Due to the difference concern points of news data and scholar data, the number of common nodes among three data sources is less than that between two scholar data sources, but it is pleasing that more than half of 73 disaster types achieve high similarities.

The superiority of our method is well demonstrated by this extensional experiment. Firstly, it discovers the pattern of disaster chain for all disaster types, including the ones that are rarely involved in existing researches. Meanwhile, the extracted disaster chain contains more latent information. For example, in 'typhoon' disaster chain, most existing research results only contain less than 11 nodes, but our method could dig out 25 relative disasters, including 'storm tide' and 'tornado' which are always ignored in existing researches. Secondly, the whole process relies on pure data analysis, which is not limited by priori knowledge. Therefore, it is flexible to reconstruct the required disaster chain quickly according to the practical needs or the update of databases.

Figure 12. Comprehensive comparison of three data sources for all 73 disaster types; for each disaster type, the first column shows the result from Baidu Scholar, the second one from Google Scholar and the third one from Sina News; the vertical axis indicates the number of nodes in disaster chains.

6. Discussions and conclusions

This paper presents a novel method of constructing the disaster chain complex network based on the massive data. The strength of correlation between each two disaster types is obtained by means of co-occurrence analysis method, based on huge amounts of scholar and social news datasets, through the

Google Scholar, Baidu Scholar and Sina News search engines. This method takes full advantage of the information from scholars and social news media, which can complement existing research results by avoiding experts' experiences and prior knowledge. It explores a new path in the field of disaster management, and can apply in similar applications and may enlighten other researchers.

Moreover, based on the features of complex network, the community division method is employed to extract specific disaster chain from the constructed complex network. This method can discover the pattern of disaster chain for all disaster types. The extracted disaster chain connects different disaster types and well explicitly show their latent relationships.

By comparing our achieved results with the ones from existing literature works, the known secondary and derived disaster types are mostly contained in our results. By comparing of applying our approach to different data sources, even the search keywords in different languages (English and Chinese), highly similar results are achieved to show its robustness. Certainly, the characteristics of different data sources are reflected in the individual results, like the difference between scholar data and new media data.

However, some steps further can be implemented in future work to improve this research. Although we defined the detailed search mode and gave a clear keywords filtering way, such as considering the singular and plural issue in English, and traditional Chinese and simplified Chinese spelling issue in Chinese, still more semantic issues need to be very careful. Same semantic word may be expressed slightly different, for example, 'waterlogging' and 'water-logging' are the identical word and may occur in different scholar papers or news. Synonyms may exist for many words, such as 'mudslide' and 'debris flow', 'tsunami' and 'seaquake'. Different scholar papers or news may adopt either of them to express the same phenomenon. If the above situations are not considered, the search results will be inevitably ignored. Therefore, it is necessary to further study the semantic issues in different linguistic and cultural context (Landauer and Dumais 1997).

In this paper, we compared the final disaster chain results from Baidu Scholar, Google Scholar and Sina News, and each result is individually computed through one single search engine. In future, the data integration and fusion work can be carried out to optimize the final results. There may be several approaches to try in this direction. The first way is to integrate the original massive searched data together, and take the three databases as one database to proceed next steps. Certainly, besides the three mentioned data sources, more disaster-related datasets may be obtained through more engines in various search ways. The second way is to investigate the method of fusing the final three disaster chains together as one chain, based on the particular features of each data source.

Last but not least, the experimental result has shown that there are still abnormal samples occurring in the original filtered search results. As mentioned in Section 5, all of the three disaster chains acquired through three data sources demonstrate the high correlation between 'typhoon' and 'earthquake', 'tsunami', even they are hardly causing each other. Because being as the common catastrophes, these three disasters are always mentioned in the scholar articles or news when talking about the disaster issues. This fact also shows that the data filtering work needs to be updated, and it requires more intelligent way to distinguish appropriate articles. Natural language processing methods may be taken into account to develop suitable algorithms for maximally removing the noisy datasets and assure the quality of data, and thus to achieve a better consequence.

Disclosure statement

No potential conflict of interest was reported by the authors.

Funding

This paper is funded by National Key Research and Development Program of China (Grant No.2016YFC0803107, Grant No.2016YFB0502601) and Shenzhen Science and Technology Innovation Commission (JCYJ20170307152553273).

References

Ahn, Y. Y., J. P. Bagrow, and S. Lehmann. 2010. "Link Communities Reveal Multiscale Complexity in Networks." *Nature* 466 (7307): 761–764.

Cai, Guoray. 2008. "Human–GIS Interaction Issues in Crisis Response." *International Journal of Risk Assessment & Management* 6 (4/5/6): 1–17.

Chen, Xiang, Jing Chen, and Jing Ai Wang. 2007. "Analysis on Typhoon Disaster Chain in Fujian: A Case Study of Typhoon Longwang in 2005." *Journal of Beijing Normal University* 43 (2): 203–208.

Chen, Changkun, Zhi Li, and Yunfeng Sun. 2008. "Study on Characteristics of Disaster Information Transmission Based on Complex Networks." *Journal of Catastrophology* 23 (4): 128–131.

Chen, Quan Liang, L. I. Zhan, Guang Zhou Fan, and Ke Yun Zhu. 2011. "Indications of Stratospheric Anomalies in the Freezing Rain and Snow Disaster in South China, 2008." *SCIENTIA SINICA Terrae* 54 (8): 1248–1256.

China, P. R. 2007. "Emergency Response Law of the People's Republic of China." http://www.gov.cn/flfg/2007-08/30/content_732593.htm.

China Daily. 2011. "China Calls off Alert on Quake Lake." http://www.chinadaily.com.cn/china/2008-06/11/content_6753597.htm.

CNN. 2013. "Hurricane Katrina Statistics Fast Facts." https://edition.cnn.com/2013/08/23/us/hurricane-katrina-statistics-fast-facts/index.html.

CNN. 2017. "The hidden health dangers of flooding." https://edition.cnn.com/2017/08/27/health/health-consequences-flood-waters/index.html.

Cui, Yun, Kong Jiming, and Wu Wenping. 2012. "Cause Characteristics and Prevention /Control Strategies of the Secondary Mountain Disaster Chain of the Wenchuan Earthquake." *Journal of Natural Disasters* 21 (01): 109–116.

Dong, Leilei. 2009. "Modeling Emergency Event Chain Based on Bayesian Networks." Dalian: Dalian University of Technology.

Egghe, Leo. 2009. "New Relations Between Similarity Measures for Vectors Based on Vector Norms." *Journal of the Association for Information Science & Technology* 60 (2): 232–239.

Evans, T. S., and R. Lambiotte. 2009. "Line Graphs, Link Partitions, and Overlapping Communities." *Physical Review E Statistical Nonlinear & Soft Matter Physics* 80 (2): 016105.

Fan, Weicheng. 2007. "Advisement and Suggestion to Scientific Problems of Emergency Management for Public Incidents." *Bulletin of National Natural Science Foundation of China* 21 (2): 71–76.

Fan, Jianrong, Bingwei Tian, Gengwei Cheng, Heping Tao, Jianqiang Zhang, Yan Dong, S. U. Fenghuan, and Bintao Liu. 2008. "Investigation on Damming Object Induced by the Earthquake of Wenchuan on May 12 Based on Multi-Platform Remote Sensing." *Journal of Mountain Science* 26 (3): 257–262.

Ford, L. R. Jr., and D. R. Fulkerson. 1956. "Maximal Flow Through a Network." *Canadian Journal of Mathematics* 8 (3): 399–404.

Forward, J. A. P. A. N. 2017. "Remembering the 2011 Earthquake: Life is Fragile – and Therefore Beautiful." https://japan-forward.com/everyday-life-is-fragile-and-therefore-beautiful/.

Girvan, M., and M. E. J. Newman. 2001. "Community Structure in Social and Biological Networks." *Proceedings of the National Academy of Sciences of the United States of America* 99 (12): 7821–7826.

Gitis, V. G., E. N. Petrova, and S. A. Pirogov. 1994. "Catastrophe Chains: Hazard Assessment." *Natural Hazards* 10 (1–2): 117–121.

Guo, Lirong. 2017. "Python-based Web Crawler Design." *Electronic Technology & Software Engineering* 23: 248–249.

Ha, Si, Jiquan Zhang, Siqin Tong, and Sijia Li. 2016. "Progress and Prospect of the Research on Disaster Chain." *Journal of Catastrophology* 31 (2): 131–138.

Hallegatte, Stéphane. 2008. "An Adaptive Regional Input-Output Model and its Application to the Assessment of the Economic Cost of Katrina." *Risk Analysis An Official Publication of the Society for Risk Analysis* 28 (3): 779–799.

Han, Y. U., Jingai Wang, Mei Chai, Peijun Shi, and School of Geography, and Beijing Normal University. 2014. "Review on Research Methods of Disaster Loss Accumulation and Amplification of Disaster Chains." *Progress in Geography* 33 (11): 1498–1511.

He, Qin. 1999. "Knowledge Discovery Through Co-word Analysis." *Library Trends* 48 (1): 133–159.

Helbing, Dirk, and Christian Kühnert. 2003. "Assessing Interaction Networks with Applications to Catastrophe Dynamics and Disaster Management." *Physica A: Statistical Mechanics and Its Applications* 328 (3–4): 584–606.

Hu, Yingjie, Xinyue Ye, and Shih Lung Shaw. 2017. "Extracting and Analyzing Semantic Relatedness Between Cities Using News Articles." *International Journal of Geographical Information Science* 31 (2): 1–25.

Kothari, R., and V. Jain. 2002. "Learning from Labeled and Unlabeled Data." Paper presented at the annual meeting for International Joint Conference on Neural Networks 2002: 2803–2808.

Kwan, Mei Po, and Jiyeong Lee. 2005. "Emergency Response after 9/11: The Potential of Real-time 3D GIS for Quick Emergency Response in Micro-spatial Environments." *Computers, Environment and Urban Systems* 29 (2): 93–113.

Landauer, T. K., and S. T. Dumais. 1997. "A Solution to Plato's Problem: The Latent Semantic Analysis Theory of Acquisition, Induction, and Representation of Knowledge." *Psychological Review* 104 (2): 211–240.

Li, Husheng. 2017. "Conceptual View and Main Research Content in the Field of Safety and Emergency Management." *Safety* 38 (10): 23–26.

Li, Jian, and Changkun Chen. 2014. "Modeling the Dynamics of Disaster Evolution Along Causality Networks with Cycle Chains." *Physical A Statistical Mechanics & Its Applications* 401 (401): 251–264.

Li, Mao, Jianguo Chen, Tao Chen, and Hongyong Yuan. 2010. "Probability for Disaster Chains in Emergencies." *Journal of Tsinghua University (Science and Technology)* 50 (8): 1173–1177.

Liang, J. T., C. Tang, and J. Wang. 2012. "Exploration- and Analysis-remote Sensing of Geological Hazards Induced by Earthquakes in Typical Areas of Qingchuan, Sichuan, China." *Journal of Chengdu University of Technology* 39 (5): 530–534.

Lin, Dalong, Liang Ming, Shengfang He, and Zhi Li. 2012. "University Fire Derived Disasters Based on Complex Networks." *Fire Science and Technology* 31 (2): 205–206.

Liu, Yang. 2013. "Research on the Typical Debris Flows Chain Based on RS in Palongzangbu Basin of Tibet." *Master diss.*, Chengdu University of Technology.

Liu, Yu, Fahui Wang, Chaogui Kang, Yong Gao, and Yongmei Lu. 2014. "Analyzing Relatedness by Toponym Co-occurrences on Web Pages." *Transactions in GIS* 18 (1): 89–107.

Liu, Aihua, and Chao Wu. 2015. "Research on Risk Assessment Method of Disaster Chain Based on Complex Network." *Systems Engineering -Theory & Practice* 35 (2): 466–472.

Liu, Wenfang, Shengxie Xiao, Yanchun Sui, Jufang Zhou, and Haiwei Gao. 2006. "Analysis of Natural Disaster Chain and Chain-cutting Disaster Mitigation Mode." *Chinese Journal of Rock Mechanics & Engineering* 25: 2675–2681.

Loet. 2008. "On the Normalization and Visualization of Author Co-citation Data: Salton's Cosine Versus the Jaccard Index." *Journal of the Association for Information Science & Technology* 59 (1): 77–85.

Mccarthy, James D., Phil A. Graniero, and Steven M. Rozic. 2008. "An Integrated GIS-expert System Framework for Live Hazard Monitoring and Detection." *Sensors* 8 (2): 830–846.

Men, Ke Pei. 2007. "Time-space Order of Severe Earthquake Disaster Chain and its Prediction." *Progress in Geophysics* 22 (2): 645–651.

Miller, Frederic P., Agnes F. Vandome, and John Mcbrewster. 2010. *Complex Network*. Saarbrücken: Alphascript Publishing.

Palla, Gergely, Imre Derényi, Illés Farkas, and Tamás Vicsek. 2005. "Uncovering the Overlapping Community Structure of Complex Networks in Nature and Society." *Nature* 435 (7043): 814–818.

Pan, A., and X. Tang. 2002. "Preliminary Research Into the Typhoon Disaster Chain in Guangdong Coastal Area." *Journal of Guangzhou University* 1 (3): 55–61.

Sánchez, Cristina Rosales, Massimo Craglia, and Arnold K. Bregt. 2016. "New Data Sources for Social Indicators: The Case Study of Contacting Politicians by Twitter." *International Journal of Digital Earth* 10: 1–17.

Shuai, Jiabing, Xu Wei, and Shi Peijun. 2012. "Characteristic Analysis of Typhoon Disaster Chains in the Yangtze River Delta Region of China." *Journal of Natural Disasters* 21 (03): 36–42.

Tang, Xiaochun. 2008. "Discussion on Disaster Chains of 5.12 Earthquake in Sichuan Province." *Journal of Southwest University for Nationalities* 34 (6): 1091–1095.

Wang, Ran, Lian Fang, Yu Han, Shi Peijun, and Wang Jingai. 2016. "Classification and Regional Features Analysis of Global Typhoon Disaster Chains Based on Hazard-formative Environment." *Geographical Research* 35 (5): 836–850.

Wang, Junjie, Huiying Gao, and Junfeng Xin. 2010. "Application of Artificial Neural Networks and GIS in Urban Earthquake Disaster Mitigation." The paper is presented for International Conference on Intelligent Computation Technology and Automation. Changsha: IEEE.

Wang, Jianxiu, Xueying Gu, and Tianrong Huang. 2013. "Using Bayesian Networks in Analyzing Powerful Earthquake Disaster Chains." *Natural Hazards* 68 (2): 509–527.

Wang, Wei, and Kathleen Stewart. 2015. "Spatiotemporal and Semantic Information Extraction from Web News Reports About Natural Hazards." *Computers, Environment and Urban Systems* 50: 30–40.

Wang, Fei, and Hongyong Yuan. 2010. "Challenges of the Sensor Web for Disaster Management." *International Journal of Digital Earth* 3 (3): 260–279.

Wei, Yi Ming, Ying Fan, F. U. Ji liang, and X. U. Wei xuan. 2000. "Neural Network Based Predicative Method for Flood Disaster." *Chinese Journal of Management Science* 8 (3): 57–62.

Xie, Qi, Jing Guan, and Cuo Yang. 2015. "A New Academic Weapon after Google Scholar - Baidu Scholar." *Journal of Library & Information Sciences in Agriculture* 27 (6): 110–114.

Xu, Mengzhen, Zhaoyin Wang, Le Liu Lijian Qi, and Kang Zhang. 2012. "Disaster Chains Initiated by the Wenchuan Earthquake." *Journal of Mountain Science* 65 (4): 975–985.

Xue, Lan, Zongchao Peng, and Qiang Zhang. 2002. "Public Management and China's Development - Review and Prospect of Public Management Development." *Management World* 2: 43–56.

Yao, Qinglin, and Zuji Qiang. 2006. "Mechanism of Earthquake Disaster Chain and Analysis of Earthquake - Flood Phenomenon." *China's First Disaster Chain Symposium*, 53–58.

Yin, Weixia, Wang Jing'ai, Yu Han, and Shi Qinqing. 2012. "Study of Earthquake Disaster Chains Based on Disaster System Theory." *Journal of Institute of Disaster Prevention* 14 (2): 1–8.

Yin, Yong, and Kazuhiko Yasuda. 2005. "Similarity Coefficient Methods Applied to the Cell Formation Problem: A Comparative Investigation." *Computers & Industrial Engineering* 48 (3): 471–489.

Yu, Shizhou. 2010. "Probability Analysis of Earthquake Disaster Chain and Chain-cutting Disaster Mitigation Method." *China Civil Engineering Journal* 43: 479–483.

Yu, Han, Jing'ai Wang, Jiu Chai, and Peijun Shi. 2014. "Review on Research Methods of Disaster Loss Accumulation and Amplification of Disaster Chains." *Progress in Geography* 33 (11): 1498–1511.

Zheng, Liang, Fei Wang, Xiaocui Zheng, and Binbin Liu. 2017. "A Distinct Approach for Discovering the Relationship of Disasters Using Big Scholar Datasets." Paper presented at the 5th annual 2017 international conference on geo-spatial knowledge and intelligence, Chiang Mai, December 8–10.

Zhou, Hongjian, Xi Wang, Yi Yuan, and Dandan Wang. 2014. "Rapid-assessing Methods of Loss in Extremely Heavy Rainfall Disaster Chain in Semiarid Region – a Case Study on a Flash Flood Debris Flow in Minxian County, Gansu Province." *Arid Zone Research* 31 (3): 440–445.

A cyberGIS-enabled multi-criteria spatial decision support system: A case study on flood emergency management

Zhe Zhang, Hao Hu, Dandong Yin, Shakil Kashem, Ruopu Li, Heng Cai,
Dylan Perkins and Shaowen Wang

ABSTRACT

With the increased frequency of natural hazards and disasters and consequent losses, it is imperative to develop efficient and timely strategies for emergency response and relief operations. In this paper, we propose a cyberGIS-enabled multi-criteria spatial decision support system for supporting rapid decision making during emergency management. It combines a high-performance computing environment (cyberGIS-Jupyter) and multi-criteria decision analysis models (Weighted Sum Model (WSM) and Technique for Order Preference by Similarity to Ideal Solution Model (TOPSIS)) with various types of social vulnerability indicators to solve decision problems that contain conflicting evaluation criteria in a flood emergency situation. Social media data (e.g. Twitter data) was used as an additional tool to support the decision-making process. Our case study involves two decision goals generated based on a past flood event in the city of Austin, Texas, U.S.A. As our result shows, WSM produces more diverse values and higher output category estimations than the TOPSIS model. Finally, the model was validated using an innovative questionnaire. This cyberGIS- enabled spatial decision support system allows collaborative problem solving and efficient knowledge transformation between decision makers, where different emergency responders can formulate their decision objectives, select relevant evaluation criteria, and perform interactive weighting and sensitivity analyses.

1. Introduction

A highly responsive decision-making process based on comprehensive and timely information is critical for emergency managers, planners, first responders, voluntary agency coordinators, and other professionals in the event of hazards and disasters. For example, a well-formed decision-support system can help a community better respond to and recover from a hazard event. Conversely, poor decision making may potentially result in unexpectedly severe casualties and mounting costs (FEMA n.d.a). Therefore, a decision support system plays an essential role in emergency management.

Facilitating the decision-making process during an emergency response can, however, be a challenging task. Disaster responders often need to make quick decisions in complex situations that involve multiple and even conflicting objectives under time stress. The decision goal is usually achieved through an inquiry into a series of spatial questions closely tied to decision objectives and their associated evaluation criteria based on diverse socioeconomic and demographic conditions. In general, the primary goal of emergency management is to protect the health and safety of people (FEMA n.d.a). If a limited amount of rescue resources is available, rescue personnel need to prioritize interventions at the most vulnerable locations ahead of other locations to save more lives and property, which raises the question: where are vulnerable groups of people in a disaster? The second goal of emergency management is to protect property and the environment, as well as to minimize the disruption of community activities (FEMA n.d.a). To meet this goal, governmental organizations would be interested in knowing which area had the greatest economic loss and needs the most significant financial support after a disaster event. Here, risk is defined as the interaction between hazard and vulnerability (Blaikie et al. 2004), where vulnerability is the potential to suffer loss or harm (Cutter 1996).

Geographical Information Systems (GIS) provides the technical capability to help a decision maker in understanding interdependencies between spatial, socioeconomic, and cultural factors in decision-making processes (Sugumaran and Degroote 2011). However, most conventional GIS approaches lack support for taken into account domain experts' knowledge and decision-making capabilities, which weakens their ability to enable decision processes dealing with conflicting objectives (Hendriks and Vriens 2000). Advances in decision-making theories and GIS capabilities have provided opportunities for the development of Multi-Criteria Spatial Decision Support Systems (MCSDSS). Additionally, GIS can provide capabilities for storing, managing, analyzing, and visualizing geospatial data. Multi-criteria analyses offer a rich collection of decision techniques for structuring problems that contain conflicting objectives and enabling the prioritization of decision alternatives and sensitivity analyses.

Various types of MCSDSS has been developed to support decision making in environmental hazard and emergency management. Malczewski (2006) classified GIS-based multiple-criteria decision-making analyses (GIS-MCDM) into two categories: decisions under conditions of certainty and those under uncertainty. Brito and Evers (2016) conducted a literature survey on using MCDM for flood risk management. From 1995 to 2015, there were 180 articles published on this topic, among which 22% of the articles focused on ranking alternatives for flood mitigation, 15% on vulnerability assessment, and 3% on emergency management. The rest of the articles are related to the topics of risk and hazard assessment, susceptibility assessment, coping capacity, and reservoir flood control. Wang et al. (2011) proposed a spatial multi-criteria approach for flood risk assessment in the Dongting Lake region. Karnatak et al. (2007) developed a web-based multicriteria spatial decision analysis application. Mitra et al. (2017) proposed a multi-criteria decision support system that was based on a Bayesian network approach for analysing landslide risk assessment. Zhang et al. (2014) developed a fuzzy multiple-attribute decision-making model to identify the vulnerability of an urban area based on population information. In their model, expert knowledge was collected using a questionnaire combined with a fuzzy modeling approach.

However, most of the aforementioned studies are subject to several limitations. First, decision criteria as well as their weights are often difficult to identify and quantify. Despite the uncertain nature of decision processes, some MCSDSSs do not allow a sensitivity analysis to quantify output uncertainties, and weights are often collected from engineers or system developers rather than domain experts. Second, current MCSDSSs are incapable of facilitating knowledge sharing and communication among different decision makers. In MCSDSS, decisions are often made by a group of decision makers such as domain experts, engineers, system developers, and users with diverse backgrounds, skills, and different preferred terminologies (Turban, Aronson, and Liang 2000). These various users contribute to a synergistic and complex decision space. Thus, how the knowledge can be collected, represented, and transferred between different decision makers promptly under time stress is extremely important to emergency management. Third, knowledge, often derived from geospatial data, comes from various data sources with different formats and scales, leading to tremendous computational challenges (Yang et al. 2017). Therefore, cyberGIS is needed in a MCSDSS to provide high-performance computation support during an emergency, for tasks such as fast data queries and information visualization (Lin et al. 2015).

In this article, we demonstrate a cyberGIS-enabled MCSDSS that can be used to model two decision goals in emergency management mentioned above (FEMA n.d.a). This research considers a flooding case in Austin that happened on 31 October 2013, and combines several data sources such as census data, social media data, and a flood hazard map. In our MCSDSS, domain experts are also users, and a Simple Multi Attribute Rating Techniques (SMART) form is implemented in the user interface to help the domain experts select relevant criteria and generate weights. Weighted Sum Model (WSM) and Technique for Order Preference by Similarity to Ideal Solution (TOPSIS) models are used to combine the relevant criteria along with the criteria weights, and analysis results are interactively visualized on a map. Finally, a sensitivity analysis is performed to evaluate the sensitivity of selected criteria, and a questionnaire was developed to validate the results of the decision analysis. In order to address the above mentioned first and second challenges, our decision support system can facilitate knowledge transformation between different decision makers efficiently, since domain experts (users) can formulate their decision objectives, select relevant evaluation criteria, and perform weighting and sensitivity analysis by using an interactive user interface. The weight of the criteria can be directly generated by a group of domain experts through the user interface. Furthermore, our decision support system is implemented on a cyberGIS-Jupyter platform, which leverages a Jupyter notebook, Docker containers, cloud-based infrastructure provisioning, and high-performance computing. Furthermore, it allows diverse data sources (e.g. census data and social media data) to be integrated into the system and visualized onto an actionable vulnerability map to support spatial decision making. The current implementation of the spatial decision support system, as described in the remainder of this paper, can produce a near real-time vulnerability map depicting the highest overall vulnerability value within a study area during the flooding event.

2. Background and theory

This section provides a theoretical background of key concepts used in this research. It includes various types of vulnerability indicators, spatial decision support systems, multi-criteria decision-making techniques, and decision-making sensitivity analyses.

2.1. Social vulnerability indicators for emergency management

In the context of social sciences, vulnerability often refers to a condition of people as opposed to physical structures, economies, or regions of the earth (Wisner et al. 2004; Juntunen 2006). Inhabitants with higher social vulnerability seem to suffer more significantly from a disaster compared to those with lower vulnerability (Brouwer et al. 2007). In this respect, social vulnerability can be used as an important indicator/tool to evaluate the disaster preparedness of local communities, dispatch emergency responses, and allocate limited resources for disaster relief. Federal agencies, such as Centers for Disease Control and Prevention (CDC) and National Oceanic and Atmospheric Administration (NOAA), have adopted social vulnerability indices to map and identify vulnerable communities that may need support before, during, and after a hazard event. Commonly, an index system, based on a weighted linear combination of relevant variables, is used to develop social vulnerability indicators. For example, the pioneering work by Cutter, Boruff, and Shirley (2003) defines the Social Vulnerability Index (SoVI) using 11 independent factors derived from 42 variables based on a factor analytic approach. Flanagan et al. (2011) developed another social vulnerability index (SVI) based on the percentile ranks of 15 census variables at the census tract level with the goal of improving all 4 phases of the disaster management, including preparedness, response, recovery, and mitigation. Bjarnadottir, Li, and Stewart (2011) proposed a Coastal Community Social Vulnerability Index (CCSVI) to project the social vulnerability of hurricane-prone areas under climate change scenarios by combining both intensities of climate-induced hazard factors and various social characteristics. Various types of social vulnerability indicators for disaster management have been developed in literature, such as family structures (Cutter, Boruff, and Shirley 2003), education (Morrow 1999; Cutter, Boruff, and Shirley 2003), socioeconomic status (Rygel, O'sullivan, and Yarnal 2006; Collins and Bolin 2009), employment (Cutter, Boruff, and Shirley 2003), age (Morrow 1999; Cutter, Boruff, and Shirley 2003; Rygel, O'sullivan, and Yarnal 2006), race and ethnicity (Cutter, Boruff, and Shirley 2003; Rygel, O'sullivan, and Yarnal 2006), and gender (Rygel, O'sullivan, and Yarnal 2006). In addition to the social vulnerability indicators, the availability of motor vehicles will affect the rescue route planning. People with motor vehicles can quickly drive away from the risk area during certain disasters. Social indicators can be developed based on several aspects for disaster management. Fenton and MacGregor (1999) identified five classes of social indicators: informative indicators, predictive indicators, problem-oriented indicators, program evaluation indicators, and target delineation indicators. King and MacGregor (2000) claimed that social vulnerability indicators could be developed based on the aspects of vulnerability and resilience of communities that affected by a disaster. For instance, low income households can be an indicator of vulnerability and high-income households can be an indicator of resilience, and rank trades occupations are more important for resilience than highly paid ones. Community vulnerability and resilience can be further divided into social, economic and demographic characteristics and attributes, behavior and values, which become a separate construct that is indicated by the set of social vulnerability indicators. A review of relevant literature indicates that the commonly used methods for constructing social vulnerability include Analytical Hierarchy Process (AHP), Principal Component Analysis (PCA), and Geographic Information Systems (GIS), which help to identify, manage, and visualize social vulnerability indices in specific geographical areas (Tate 2012; Frigerio et al. 2016; Frigerio and De Amicis 2016).

However, social vulnerability assessment in emergency management currently faces several challenges due to its uncertain and complex nature. First, the social vulnerability index system is difficult to construct, since it is often based on emergency responders' subjective experiences. Emergency management can be considered as a set of collaboration processes among various types of emergency responders, such as fire and rescue departments, emergency medical crew, transportation agencies, and insurance companies, with varying decision criteria and foci. Emergency management requires effective collaboration and coordination among multiple parties and objectives, that may lead to overlapping disaster response efforts, ineffective communication, and poor coordination. The second challenge is to assign weights to the social vulnerability indices for evaluating the vulnerability of an area in a case of any disaster. Emergency responders should create their own decision objectives and assign weights to the indicators according to their professional responsibilities. However, this raises the question of how emergency responders' knowledge can be collected and represented in geospatial modeling processes to efficiently support spatial decisions. Human knowledge exists in a compiled format while computers require information to be represented in a structured way (Turban, Aronson, and Liang 2000).

2.2. Use of social media in emergency management

In addition to social vulnerability indicators, social media data (e.g. Twitter) has been often used to model communication behaviors of disaster responders to improve situational awareness in emergency management. Social media channels allow for quick dissemination of information during a crisis, as well as two-way communication between members of the public and emergency management organizations (Goodchild and Alan Glennon 2010). Chatfield, Scholl, and Brajawidagda (2013) used Twitter as a warning tool in natural disasters. Kryvasheyeu et al. (2016) found that there was a correlation between the number of tweets mentioning hurricane Sandy and the time when the hurricane landed in the continental United States. Schnebele and Cervone (2013) found that the magnitude of the Calgary flood had a similar pattern to the location of flood related tweets. Li et al. (2018) developed a kernel-based flood mapping model to map the flooding possibility for the study area based on the water height points derived from tweets and stream gauges. Kongthon et al. (2012) classified the flood-related tweets into four major categories: situation announcements and alerts, support announcements, requests for assistance, and requests for information. Most of these situational tweets were tweeted by local citizens who knew what had been happening during a disaster situation. This raises the question of how social media could help to improve communications between emergency management organizations and citizens. Therefore, there is a need to establish a disaster-related social media channel and mapping application, so citizens can provide more detailed or accurate information to assist authorized agencies during emergency incidents.

ATXfloods is an online application for collecting real-time flood information and mapping emergency road closures in Austin. Austin city has also Twitter channel, which displays the content of tweets that are related to Austin flood events (@ATXfloods)(Twitter 2018a). Similarly, Federal Emergency Management Agency (FEMA) also provides a Twitter channel, which supports communication between citizens and first responders before, during, and after an emergency (Twitter 2018b). Here, Twitter data is used as a communication tool in applications where the spatial context is ignored. For instance, it is not obvious as to where people might request assistance.

2.3. Spatial decision support systems

Decision support systems (DSSs) have many definitions. Turban, Aronson, and Liang (2000) identified four key components of DSSs: a data management system (e.g. a database management system), a management subsystem (quantitative models that provide the system with analytical capabilities and software management), a knowledge-based management subsystem (an intelligent knowledge base that stores knowledge that can be used to make decision), and an interface subsystem (the user communicates with the DSS through this subsystem). According to above mentioned definitions of DSSs, a spatial decision support system (SDSS) can be defined as a framework for integrating analytical and spatial decision modeling capabilities (management subsystem), spatial and non-spatial data management (data management system), domain knowledge (knowledge acquisition and knowledge-based management system), spatial display capabilities and reporting capabilities (user interface) (Sugumaran and Degroote 2011). The next section introduces two types of decision-making models that were used in our MCSDSS management subsystem.

2.4. Multi-criteria decision making (MCDM)

A variety of GIS analytical methods have been developed to help decision-makers solve spatial decision-making problems with multiple criteria. The WSM is the earliest and probably the most widely used method (Fishburn 1967). The weighted product model (WPM) can be considered a modification of the WSM and has been proposed in order to overcome some of its weaknesses. The analytic hierarchy process (AHP), as proposed by Saaty (Saaty 1980, 2008), is a later development that has recently become increasingly popular. Belton and Gear (1983) suggested a modification to the AHP (which we will call the revised AHP) that appears (as it is demonstrated in later chapters) to be more consistent than the original approach. Some other widely used methods are the ELECTRE and the TOPSIS methods (Hwang and Yoon 1981).

MCDM analysis is a mathematical decision-making framework where many decision criteria are combined to meet one or several objectives to give support for decision making. Given a set of m

alternatives, denoted as $A_1, A_2, A_3, \ldots, A_m$, and a set of n decision criteria, denoted as $C_1, C_2, C_3, \ldots,$ Cn, it is assumed that the decision maker (DM) has to determine the performance value a_{ij} (for $i = 1,$ $2, 3, \ldots, m$ and $j = 1, 2, 3, \ldots, n$) of each alternative in terms of each criterion. A matrix A with a_{ij} values is defined, along with the criteria weight W_j (the weight of the relative performance of the decision criteria). Usually, these weights are normalized to add up to one. Finally, the alternatives are ranked.

2.4.1. Weighted sum model

If there are m alternatives and n criteria, the best alternative is the one that satisfies (in a maximization case) the following expression:

$$A = \max \sum_{j=1}^{n} a_{ij} w_j, \text{ for } i = 1, \ldots, m \text{ and } j = 1, \ldots, n \tag{1}$$

Where A is the WSM score of the best alternative, n is the number of decision criteria, a_{ij} is the actual value of ith alternative in terms of the jth criterion, and w_j is the weight or importance of the jth criterion.

2.4.2. Technique for order preference by similarity to ideal solution (TOPSIS)

The TOPSIS technique is based on the concept that the best alternative is the one that is closest to its ideal solution and farthest from the negative ideal solution (Behzadian et al. 2012). The formal TOPSIS procedure is defined as follows:

Step 1: Calculate normalized rating for each element in the decision matrix.

Step 2: Calculate weighted normalized ratings. The weighted normalized value v_{ij} is calculated through this formula

$$v_{ij} = w_j a_{ij}, \quad i = 1, \ldots m; \; j = 1, \ldots, n. \tag{2}$$

Step 3: Identify the positive ideal A^* and negative ideal A^- solutions. The A^* and A^- are defined in terms of the weighted normalized values, as shown bellow

$$A^* = \{v_1^*, v_2^*, \ldots, v_j^*, \ldots, v_n^*\} = \{(\max_i v_{ij} | j \in J_2) | i = 1, \ldots m\} \tag{3}$$

$$A^- = \{v_1^-, v_2^-, \ldots, v_j^-, \ldots, v_n^-\} = \{(\min_i v_{ij} | j \in J_2) | i = 1, \ldots m\} \tag{4}$$

Where J_1 is a set of benefit attributes and J_2 is a set of cost attributes.

Step 4: Calculate separation measures. The separation between alternatives can be measured by n dimensional Euclidean distance. The separation of each alternative from the positive idea solution is given by

$$S_i^* = \sqrt{\sum_{j=1}^{n} ((v_{ij} - v_j^*)^2}, \; i = 1, \ldots, m \tag{5}$$

Similarly, the separation from the negative ideal solution is given by

$$S_i^- = \sqrt{\sum_{j=1}^{n} ((v_{ij} - v_j^-)^2}, \; i = 1, \ldots, m \tag{6}$$

Step 5: Calculate similarities to ideal solution.

$$C_i^* = \frac{S_i^-}{S_i^* + S_i^-} \quad i = 1, \ldots m \tag{7}$$

$0 \leq C_i^* \leq 1$ where $C_i^* = 0$ when $A^* = A^-$ and $C_i^* = 1$ when $A_i = A^*$

Step 6: Rank preference order. Choose an alternative with maximum C_i^* or rank alternatives according to C_i^* in descending order.

2.5. Sensitivity analysis

Sensitivity analysis is a critical component of MCDM, which evaluates how changes in the entire vector of a decision matrix (or changes in the current weights of the decision criteria) may change the ranking of the alternatives. One important step of sensitivity analysis is to identify critical criteria and then, if needed, to re-evaluate the weights for the criteria. For the current study, we consider critical criteria to be the criterion whose weight change may change the ranking of any alternative. In this case, the *criticality degree of criteria* is the smallest percent amount by which the current value of w_k (weight of criteria C_k) must change, such that the existing ranking of the alternatives will change. An illustrative example of this calculation is available in Triantaphyllou (2000).

3. Methods

This section introduces data collection and processing techniques and methods used in this research. Various types of data have been used in this research such as Austin city census data, a flood hazard map, and Twitter data collected using the Twitter API. MCDM WSM and TOPSIS models are used as core methodologies in the MCSDSS management subsystem to combine different criteria to produce an overall vulnerability map.

3.1. Data collection and processing

The data used in this research comes from three data sources. Austin city census data was downloaded from the TIGER Geodatabase and American Fact Finder from United States Census Bureau (United States Census Bureau 2017). Census data was collected at the block group level for attributes on the age, sex, educational attainment, poverty, income, housing characteristics, and health insurance status. In this study, the following criteria are used to evaluate the vulnerability of people affected by the flood hazard: total population in an area, number of people over 75 years old, number of people without health insurance, number of people in poverty, number of people in a minority group, area median house value, population education attainment, number of children, people without a vehicle, and the number of tweets that are relevant for the flooding event. While most social vulnerability research (as discussed earlier) includes more vulnerability indicators, in this study we considered a smaller set of indicators to demonstrate our MCSDSS. Our proposed system is also flexible enough to incorporate more vulnerability indicators as considered important for particular hazard or geographic context. In addition to census data, a flood hazard map layer was downloaded from FEMA's flood map service center portal (FEMA n.d.b).

We collected Twitter data through the Twitter API over five years (2012–2017) for the continental United States, and the data has been used for many research projects. The collected Twitter data contains 2 billion tweets (about 4TB), which poses significant computational intensive challenges for traditional GIS computing environments. To resolve this challenge, we used the cyberGIS-Jupyter environment that deployed on the first cyberGIS supercomputer ROGER (Wang 2016; Yin et al.

2017). Geographical and geospatial information sciences have been undergoing dramatic digital transformations, which has fueled the development of cyberGIS as a new generation of geospatial information sciences and systems (GIS) based on advanced cyberinfrastructure during the recent decade (Wang 2016). CyberGIS-Jupyter is a new type of cyberGIS gateway platform that provides interactive, flexible, and reproducible GIS interfaces to advanced cyberinfrastructure. It organically synthesized Jupyter notebooks, containerization, cloud computing, data-intensive computing, and high-performance computing, which grants easy access to the power of advanced cyber-infrastructure inside the user-friendly notebook interfaces. To process the Twitter data, we used the Hadoop functionality of cyberGIS-Jupyter to first filter all tweets within the bounding box of Austin (−98.230, −97.279, 29.986, 30.678) from the whole set of US tweets and retrieve 23,802,637 tweets (about 14GB) inside the Austin area. The next step is to develop a strategy in finding suitable keywords that can be used to filter the Twitter data and identify tweets that are related to flooding events, such as reporting the flooding situation of a location or asking for rescue help during flooding events.

First, we selected Tweets tweeted during flooding events that occurred between October 30th and November 6th in 2013 and 2015, and August 25th through 1 September 2017, which contain the key word 'flood'. This query resulted in 806 tweets, a large number of tweets with a variety of relevant and non-relevant content. For instance, the filtered data contained many false flooding related tweets (e.g. 'Can the flood warnings be any more annoying on the iPhones', 'I'm getting really annoyed of these flash flood warnings on my tv and phone … I get it already'), which are not useful for our case. Therefore, 'flood' may not be the best keyword for filtering the data. After that, we combined the keywords 'flood' to filter the data. For the 'flood rescue' keywords, this resulted in 18 filtered tweets where 6 of them are relevant. Many of other tweets that are related to flooding events were taken from a local news channel's Twitter feed (http://www.topix.com/austin), and so are not tweets from local residents.

In the next step, we selected all Tweets in Austin that happened during the same flooding events and sorted words according to their frequency of appearance in the selected Tweets. By doing so, we are able to see which words are most often used during flooding events. The analysis result is illustrated in Figure 1, where the numbers represent number of tweets that contain the corresponding key word. These results indicate most of the frequently appearing key words (with more than 2118 tweets) are not related to flooding. During this time period, some career channels had been

18382	**I**	**5330**	**on**	**2775**	**have**
16581	**to**	5166	this	2697	Austin
15493	**the**	5019	TX	2680	#job
13626	**a**	4859	me	2654	was
11868	**in**	4403	be	2511	#CareerArc
8893	**and**	4231	#Hiring	2505	are
8499	**my**	4209	with	2495	&
8486	**you**	3976	#Austin	2481	#Jobs
8404	**@**	3715	it	2471	out
8399	**for**	3599	that	2430	up
8384	**at**	3314	so	2252	#Job
6776	**I'm**	3183	just	2250	Texas
6725	**of**	2981	Austin	2207	get
6666	**is**	2827	The	2134	all
6592	😄	2827	like	2134	about
6266		2799	our	2118	This
5854	**-**	2789	TX)		

Figure 1. Frequency of keywords in tweets during the flooding events.

actively using Twitter for advertising. Therefore, the key words 'Hiring', '#Jobs', and 'CareerArc' have received high ranks.

The NOAA storm events database contains comprehensive and detailed information on major climatic events from January 1950 to November 2017, as entered by NOAA's National Weather Service (NWS). This database was used to search for flood events that occurred between January 2012 and December 2017 in Travis County (NOAA 2018). Search results indicate 88 flood events were recorded with $122.1 million in total property damages. The City of Austin has created a Twitter channel @ATXflood to support communication before, during and after a flood emergency. Therefore, the keyword @ATXflood might be more reasonable to use for filtering the data than using the above-mentioned methods (e.g. using the keyword 'flood' or 'flood rescue'). The tweets that exactly contained the '@ATXflood' keyword were selected, 305 tweets in total. Each tweet contains geospatial coordinates and a timestamp, which is treated as a user's activity at that point. After that, the normalized number of flood events data were plotted together with the normalized number of tweets to observe the spatiotemporal correlation between them (Figure 2). A positive relationship occurs during 2013 except from June to August, in 2014 from April to December, in 2015 from April to May and from September to November, and most of the time in 2016. After that, we computed the correlation coefficient between number of the @ATXflood tweets and number of flood events. The overall correlation coefficient is 0.52, which indicates a moderate positive correlation, meaning people use @ATXflood channel to tweet more frequently during flooding events than other times. Therefore, it is reasonable to include @ATXflood Twitter data in the flood emergency decision process as we have implemented in our proposed MCSDSS.

3.2. Case study

Austin city has experienced serious flooding problems throughout its history. Since the late 1970s, the federal government has paid out $54 million for flood claims in Austin and $43 million more in the rest of Travis County (Price 2016). To present the MCSDSS, this research considers a flooding case in Austin that happed on 31 October 2013. The neighborhoods near Austin-Bergstrom International Airport were overwhelmed when water from the nearby Onion Creek surged past its banks and rolled into their streets, filling homes within minutes. The flood killed four people and damaged

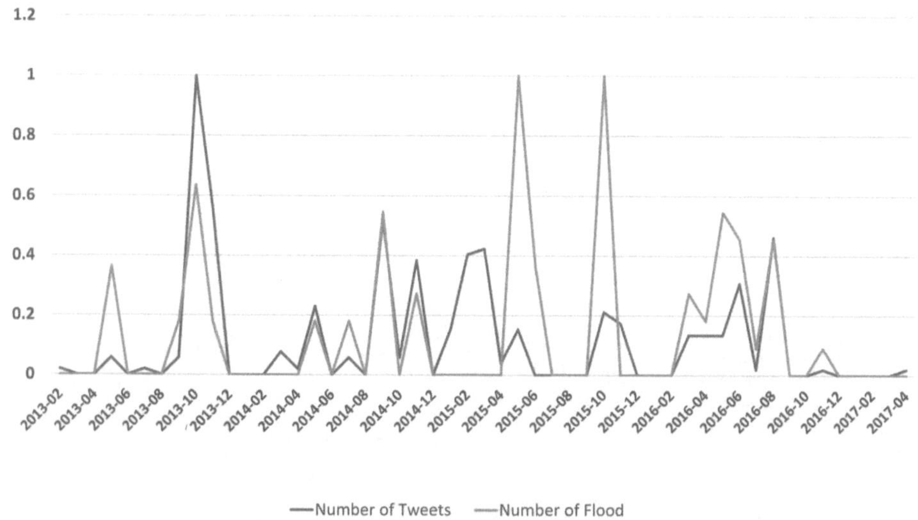

Figure 2. Relationship between the number of Tweets from the @ATXflood channel and the number of flood events in Travis County.

more than 1,200 homes (Jervis 2013). During this flooding event, there were 52 @ATXflood tweets, 178 tweets containing the 'flood' keyword, and 11 tweets with both the 'flood' and 'rescue' keywords captured. In the introduction section, we introduced two decision goals based on FEMA's literature from the perspectives of emergency response and recovery (FEMA n.d.a).

Decision goal 1: which area should rescue personnel go to first in order to save more lives? If the disaster occurs over a large area and the government has limited resources, rescue personnel need to prioritize interventions at the most vulnerable locations ahead of other locations in order to save more lives.

Decision goal 2: which area had the most significant economic loss and needs the greatest financial support to recover from a flooding event?

The structure of our cyberGIS-enabled MCSDSS is presented in Figure 3. In this framework, knowledge for making decisions is collected through interactions between domain experts (users) and the user interface. The collected expert knowledge and geospatial data are stored in the knowledge base, which can be used as input for the model management system. The model management system includes two decision-making models that are used to combine multiple criteria values (social vulnerability indicators) for creating an overall vulnerability score for the decision goals. Each census block is considered as one alternative. For instance, the value for the a_{ij} variable in Equations (1) and (2) were generated using normalized social vulnerability indicator values attached to each block. A weight was simulated using the Simple Multi Attribute Rating Techniques (SMART), which assigned an importance of 100 points to the most important criterion and the next-most-important criteria is assigned a number of points reflecting the ratio of relative importance to the most important dimension (Olson 1996). Furthermore, our decision support system application is designed to enable group decision-making capabilities, so that a group of decision-makers (DMs) can participate in the decision-making process and assign weights to the criteria simultaneously. In this research work, six GIS experts participated in simulating the weight for each criterion based on two decision goals. A weight matrix W was generated where d refers to the number of DMs and n refers to the number of evaluation criteria (Figure 3). The final weights were calculated by averaging the value from each row in matrix W. Simulated weighting results are illustrated in Table 1. The weights were normalized to represent variable w_j in Equations (1) and (2). The area that has the highest (or maximum) overall value is considered as the recommended place for earliest intervention.

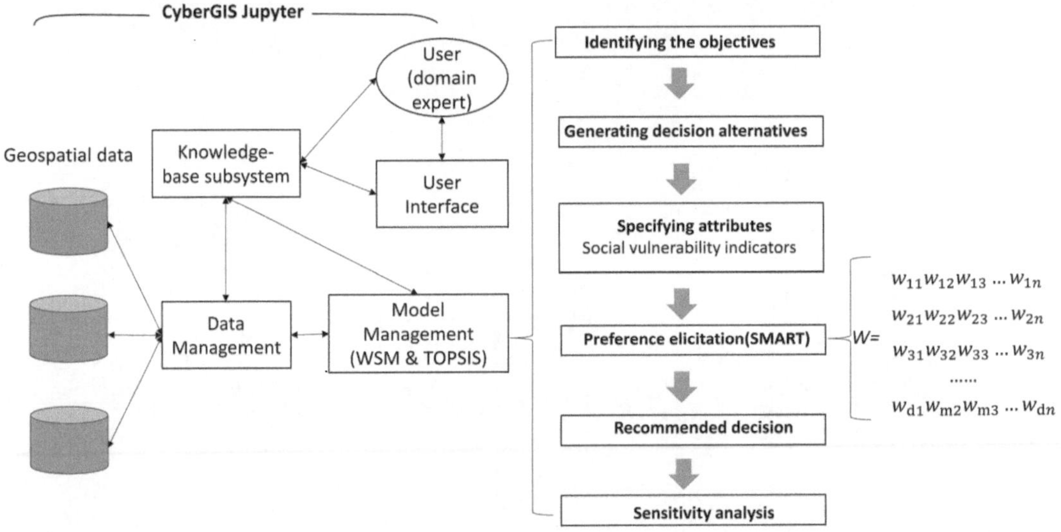

Figure 3. An overview of cyberGIS-enabled MCSDSS.

Table 1. Simulated weights for the evaluation criteria in perspectives of two decision goals.

Evaluation Criteria	Decision goal 1	Decision goal 2
Total population	74	83
People over 75 years old	87	70
People without health insurance	60	82
People in poverty	63	83
People in a minority group	60	68
Area median house value	47	66
Population education attainment	41	56
Number of children	93	79
People without a vehicle	78	58

4. Results

This section presents the results of the case study. Our cyberGIS-enabled MCSDSS user interface was implemented using the cyberGIS Jupyter computing platform. Figure 4 illustrates the user interface of the implemented system. It contains several layers, such as a flood hazard layer and a Twitter layer that the DM can switch on or off. In this article, the Twitter layer is used as an additional decision

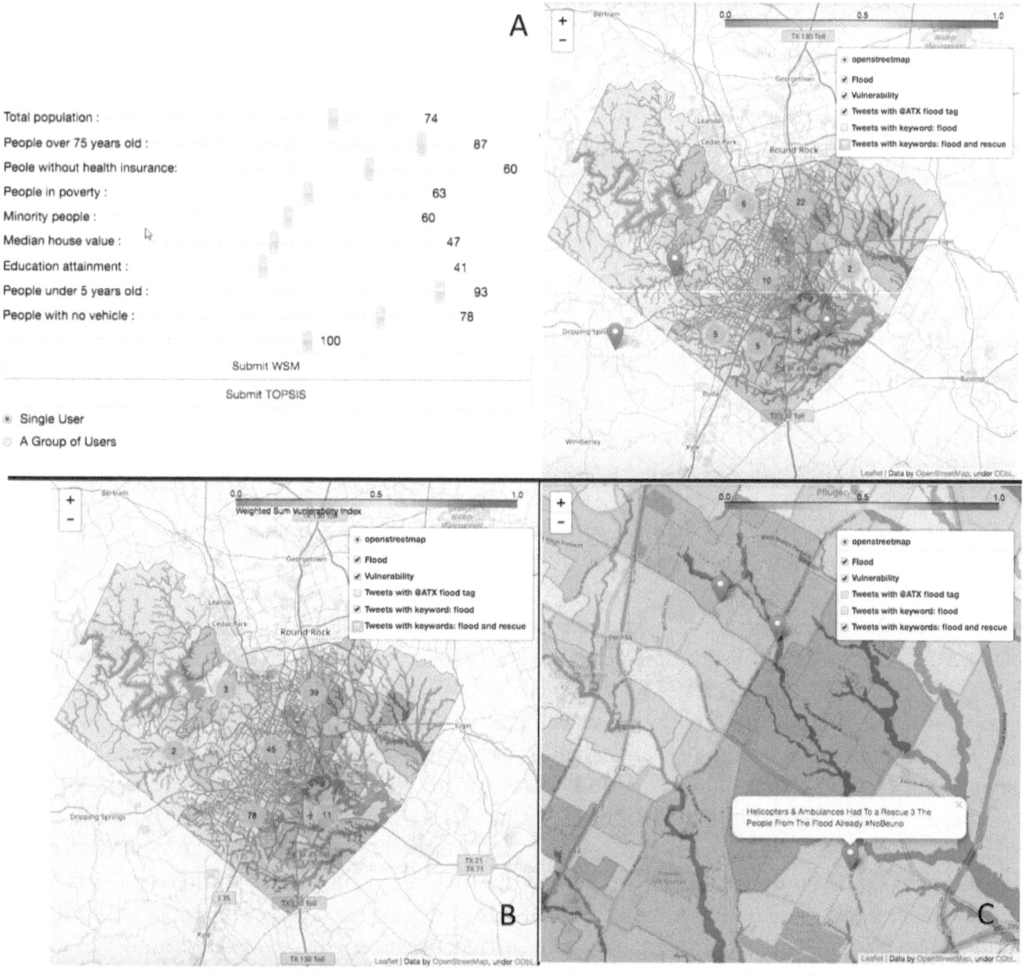

Figure 4. The cyberGIS-enabled MCSDSS user interface.

support tool added on the top of the decision model. The decision maker can use this tool to filter the Twitter data using different keywords. In this article we used three keywords sets: '@ATXflood', 'flood', and 'flood' combined with the keyword 'rescue'. The filtered results are illustrated in Figure 4(A–C) respectively. In the future, this decision support tool can be developed further so the decision maker can enter any keyword of interest and interactively observe the filtered results on the map. In Figure 4(C), the blue pinpoint on the map represents a single tweet with content mentioning 'flood' and 'rescue' during the flooding event. If multiple tweets are geographically close to each other, they are grouped and visualized as clusters on the map with the number indicating the count of tweets in that region (see Figures 4(A and B)). The user can zoom in and out on the map and click on the blue pinpoint to see the tweet contents. This application is capable of visualizing flood-related tweets in real time, and thus help emergency management organizations identify areas and locations of people that are at risk.

The left side of the Figure 4(A) illustrates the weighting slider, which can be used to assign weights to the criteria that are relevant to the decision goals. If the DM selects the option 'using weight from a single user', the vulnerability map will be generated based on the single submitted list of weights. If the DM selects the option 'using weight from a group of users', the assigned weights will be first combined with a weight matrix stored in a common accessible file managed by the CyberGIS-Jupyter notebook. This weight matrix has the weight information previously submitted by other DMs in the same group, and it will be updated based on the latest submitted weight. After that, the weight will be averaged and normalized to generate the final weight to produce an overall vulnerability map.

The weights illustrated in Table 1 resulted in a vulnerability index map shown in Figure 5. For the first decision goal, both models show block numbers 1, 4, and 5 received high vulnerability values, and rescue personnel should go there first in order to save more lives in an emergency situation. Most of the blocks that received middle-high category overall values in the TOPSIS model received the highest overall values in the WSM. Block numbers 1, 2, 17, and 6 received high overall values in the TOPSIS model but lower vulnerability values in the WSM model. In addition, block number 15 received the low category for both maps. For the second decision goal, block numbers 3, 4, 5, 13 and 14 received the highest overall values for both maps, meaning those blocks may suffer the most economic loss and need the most financial support for post event recovery. Most blocks that received middle-high overall values in the TOPSIS model had high vulnerability value in the WSM. The opposite situation applied to block numbers 1 and 2. Generally speaking, WSM produces a wider range of values and higher output category estimations than the TOPSIS model. Block number 15 represents the Austin-Bergstrom international airport, which received a low vulnerability value for all the decision models. This can be caused by fewer people living in that area compared to the other blocks. However, there are 11 tweets contain 'flood' keyword near block number 15, and most of them are relevant tweets. In this case, the decision maker should not ignore this block due to the low vulnerability score produced by the decision models.

Sensitivity analysis is performed to observe the sensitivity of the criteria, and the results are illustrated in Table 2. The values in Table 2 represent the rank of criticality degrees of the criteria that are explained in section 2.3. For the first decision goal, the number of children and people without vehicles are the most sensitive criteria and median house value is the least sensitive criterion. It means that if the weight of people without vehicles is changed, even by a small value, the census block with the highest vulnerability values will most likely also change. In other words, the weight of people without vehicles should be carefully determined by experts in a real-world decision- making scenario. On the other hand, median house value is the least sensitive criterion so that the census block with the highest vulnerability index is not likely to be changed even if the weight of median house value is significantly increased or decreased. However, a big ranking difference between the two decision models appears for some of the criteria, such as the criteria of people without health insurance and the number of children. For the second decision goal, people without vehicles and insurance are the most sensitive criteria, and population education attainment and the area median house value are the least sensitive criteria. The number of children and people without health

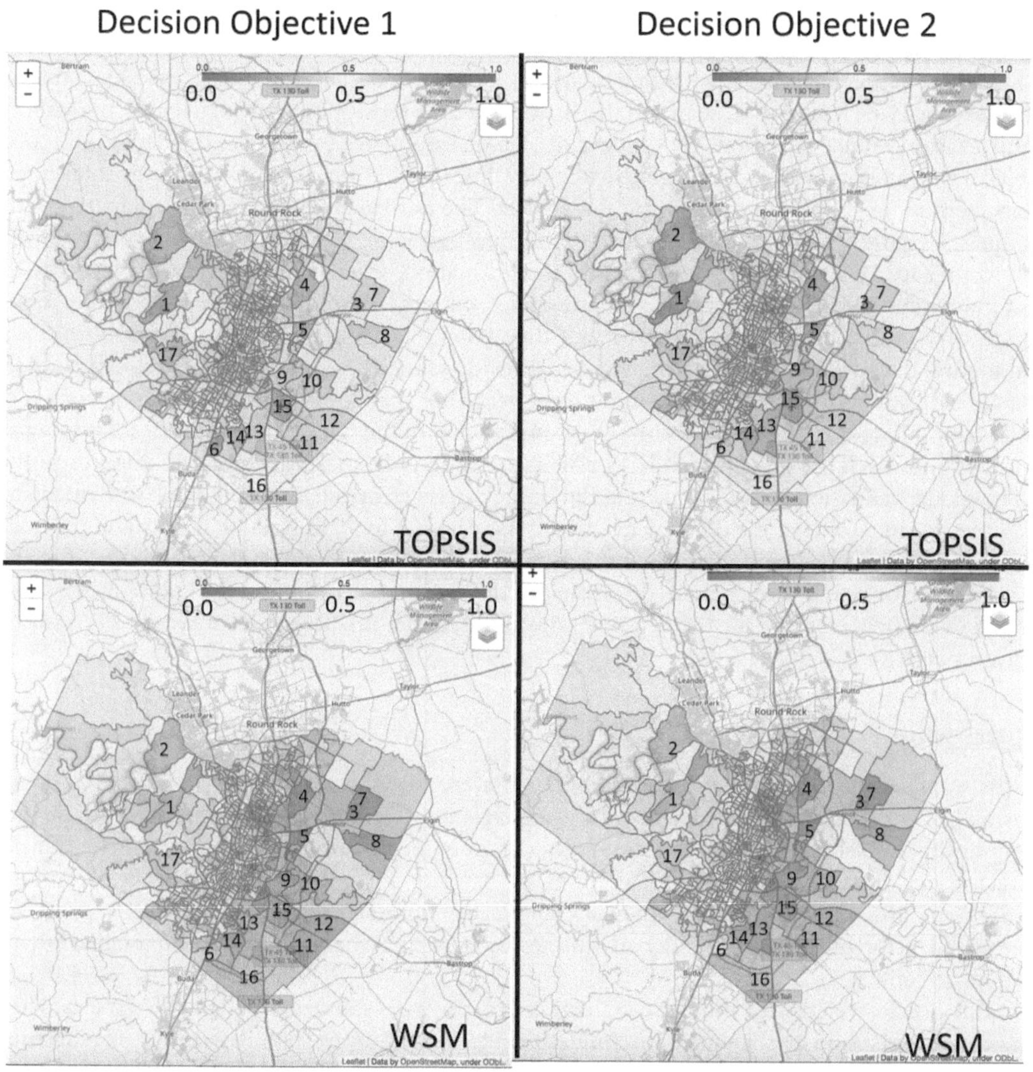

Figure 5. Vulnerability maps based on two decision-making models (WSM and TOPSIS).

Table 2. Sensitivity analysis results.

Criteria	Objective 1 (Rank of Criticality Degrees of the Criteria)		Objective 2 (Rank of Criticality Degrees of the Criteria)	
	TOPSIS	WSM	TOPSIS	WSM
Total population	3	5	2	5
People over 75 years old	2	4	5	6
People without health insurance	7	3	6	1
People in poverty	5	6	3	4
People in a minority group	6	7	4	7
Area median house value	9	9	7	9
Population education attainment	8	8	9	8
Number of children	4	1	8	2
People without a vehicle	1	2	1	3

insurance criteria received the greatest and the second greatest ranking difference between the two decision models.

Model validation has been challenging task for multicriteria decision analysis (Qureshi, Harrison, and Wegener 1999; Meyer et al. 2014). In this article, we used sensitivity analysis to check the stability of the decision model. Additionally, we developed a questionnaire to evaluate the decision model's results. The questionnaire contains all the attribute information for 10 blocks, such as total population, number of children, people without a vehicle. The participants who have participated in the first questionnaire was asked to fill in the second questionnaire. The decision maker has to give a rank from 1 to 10, based on the given attribute information, where rank number 1 represents the most vulnerable block. This is similar to a real-life emergency situation, in that rescue personnel have to make the decision based on how well they know the area. All the participants said it was hard to rank the 10 blocks based on many given attributes, and there is a need to develop a multicriteria decision support tool to make the ranking easier. After that, these blocks were ranked based on the decision model's vulnerability results and compared with the questionnaire results. By doing so, we can see how well the decision model's results match decision makers' rapid thinking in real life.

The model validation result is illustrated in Table 3. The error rate for each block was calculated using the absolute value of the difference between the decision model's ranking results and the questionnaire's ranking result and divided by the number of 10 (maximum value of the rank). After that the average error rate was calculated. According to the results, both WSM and TOPSIS models produced a over 40% error rate for the decision objective 1, and a smaller error rate (over 20%) for the decision objective 2.

5. Discussion

In this study, MCDM was used to combine different social vulnerability indicators to reach decision goals that are related to emergency management. While we used a limited set of indicators in this case study, more indicators, such as family income status, family structures, housing characteristics, and flood severity (e.g. rainfall intensity), can be included in our proposed MCSDSS in order to make this framework applicable for different kinds of disaster scenarios. The flexible structure and portability of this technology also make it quickly adaptable for any geographic context.

In this article, weights for different evaluation criteria were simulated by six GIS experts. However, the experts who participated in simulating the weights were not emergency management domain experts, therefore the results of the study cannot be used to answer the decision goals correctly in a real-world situation. In the future, real domain experts such as rescue operation officers or public agency professionals should use the application to generate weights in order to make the analysis results more accurate and realistic for an actual disaster scenario.

Table 3. Model validation results.

Blocks	WSM Model		TOPSIS Model	
	Objective 1	Objective 2	Objective 1	Objective 2
1	0.1	0	0	0.2
2	0.8	0.8	0.8	0.8
3	0.2	0.1	0.2	0.1
4	0.4	0.5	0.4	0.5
5	0.6	0	0.5	0.1
6	0.2	0.3	0.4	0.4
7	0.8	0.1	0.9	0
8	0.3	0	0.4	0.1
9	0.4	0.1	0.4	0.1
10	0.6	0.5	0.4	0.3
Average Error	0.44 (44%)	0.24 (24%)	0.48 (48%)	0.26 (26%)

One objective of this work is to develop a software application that enhances the communication between citizens and emergency management organizations. Our application provides fine-grained location (or exact location) information for rescue personnel to evacuate people who need help during the disaster event using a Twitter data layer. For instance, rescue personnel can identify the areas and people at risk by observing the content of a tweet and its corresponding location on the map. Currently, we used three keyword sets to filter the tweets. In the future, text mining techniques can be embedded in the tool to categorize tweets (e.g. according to Kongthon et al. (2012)'s work) in order to support different types of decision problems. For instance, this tool can be developed further so the decision maker can enter any keywords that they think are useful and interactively observe the filtered results on the map. Besides Twitter, this framework can be further expanded by incorporating other social media postings (e.g. Facebook, Instagram, Snapchat/Snap Map, etc.) to make it more useful for emergency responders and more accessible to wider audience.

In this application, our analysis is at the block group level, which is the finest-grained geographic level in U.S. Census where socio-economic data is reported. It would have been better if we could use the parcel level but enumerating data at that granular level will be a separate research project by itself. In this article, we are proposing a computational framework, and it can be implemented later by rescue organizations using more detailed data (e.g. parcel level data or their own rescue data, which might not be publicly available) to provide fine-grained location information.

Other important aspects of this application are ensuring consistent results or facilitating different actors in quickly reaching an agreement. WSM produces more diverse values and a higher output category estimation than the TOPSIS model. Simanaviciene and Ustinovichius (2010) found that if the values of criteria differ by 10% from the average criterion value, the TOPSIS method is more sensitive than the WSM. In our case, for both decision objectives, both models produced consistent answer for some places that have high vulnerability. Both models provide interfaces (through weights) for trials and errors from decision makers. We used sensitivity analysis to check the stability of the decision model. Additionally, we developed a questionnaire to validate decision model's results. As a result, both WSM and TOPSIS produced similar validation results for each decision objective. However, clear error differences appear between the two objectives. For instance, both models produced a higher error rate for the first decision objective, and a lower error rate for the second decision objective. Finally, the parameterization and evaluation of the models should depend on a panel of experts and decision makers following Delphi style approaches. For instance, if there are extremely limited rescue resources available, a decision maker might want to consider using the results that are produced by the TOPSIS model first since it produces fewer hot spots than WSM results.

6. Conclusion

In this article, we described a cyberGIS enabled decision support framework, which combines a high-performance geospatial computing environment (cyberGIS Jupyter), advanced multi-criteria decision analysis models, various types of social vulnerability indicators, and Twitter data to efficiently support spatial decision-making processes in emergency management. The advantages of this MCSDSS come from several perspectives. First, cyberGIS Jupyter enables collaborative big data analytics so various types of data sources can be integrated into the decision-making processes. Second, multiple and conflicting objectives are common decision problems in emergency management. In this study, two types of MCDM models (WSM and TOPSIS) were constructed, where many different criteria can be combined to meet one or several objectives and give support in decision making. Third, this framework supports group decision making and collaborative problem solving, which is important since emergency management requires effective collaboration and coordination among emergency responders. In this framework, emergency responders can work together to formulate decision objectives, select relevant evaluation criteria, and perform interactive weighting and sensitivity analysis. In addition, we developed an innovative way to use Twitter data to enhance communications between citizen and emergency management organizations, and therefore to improve the awareness of various spatial situations. For instance, this framework can support geo-visualization of near real-time Twitter data and help emergency organizations identify the areas and people that are at risk. Finally, the interactive user interface enables the expert knowledge of decision makers to be collected and incorporated to efficiently support spatial decision making for emergency management.

Acknowledgement

Any opinions, findings, and conclusions or recommendations expressed in this material are those of the authors and do not necessarily reflect the views of the National Science Foundation. We would like to acknowledge Mr Johnathan Rush and other members of the CyberGIS Center for Advanced Digital and Spatial Studies at the University of Illinois at Urbana-Champaign for their kind support and helpful inputs.

Disclosure statement

No potential conflict of interest was reported by the authors.

Funding

This material is based in part upon work supported by the U.S. National Science Foundation under [grant numbers: 1047916, 1429699, 1443080, 1551492, and 1664119].

ORCID

Zhe Zhang http://orcid.org/0000-0001-7108-182X
Heng Cai http://orcid.org/0000-0001-6068-5150

References

Behzadian, Majid, S. Khanmohammadi Otaghsara, Morteza Yazdani, and Joshua Ignatius. 2012. "A State-of-the-Art Survey of TOPSIS Applications." *Expert Systems with Applications* 39 (17): 13051–13069. doi:10.1016/j.eswa.2012.05.056.

Belton, Valerie, and Tony Gear. 1983. "On a Short-Coming of Saaty's Method of Analytic Hierarchies." *Omega* 11 (3): 228–230. doi:10.1016/0305-0483(83)90047-6.

Bjarnadottir, Sigridur, Yue Li, and Mark G. Stewart. 2011. "Social Vulnerability Index for Coastal Communities at Risk to Hurricane Hazard and a Changing Climate." *Natural Hazards* 59 (2): 1055–1075. doi:10.1007/s11069-011-9817-5.

Blaikie, Piers, Terry Cannon, Ian Davis, and Ben Wisner. 2004. *At Risk: Natural Hazards, People's Vulnerability and Disasters*. Oxford: Routledge.

Brito, M. M., and M. Evers. 2016. "Multi-Criteria Decision-Making for Flood Risk Management: A Survey of the Current State of the Art." *Natural Hazards and Earth System Sciences* 16 (4): 1019–1033. doi:10.5194/nhess-16-1019-2016.

Brouwer, R., S. Akter, L. Brander, and E. Haque. 2007. "Socioeconomic Vulnerability and Adaptation to Environmental Risk: a Case Study of Climate Change and Flooding in Bangladesh." *Risk Analysis* 27: 313–326.

Chatfield, Akemi Takeoka, Hans J. (Jochen) Scholl, and Uuf Brajawidagda. 2013. "Tsunami Early Warnings via Twitter in Government: Net-Savvy Citizens' Co-Production of Time-Critical Public Information Services." *Government Information Quarterly* 30 (4): 377–386. doi:10.1016/j.giq.2013.05.021.

Collins, Timothy W., and Bob Bolin. 2009. "Situating Hazard Vulnerability: People's Negotiations with Wildfire Environments in the U.S. Southwest." *Environmental Management* 44 (3): 441–455. doi:10.1007/s00267-009-9333-5.

Cutter, Susan L. 1996. "Vulnerability to Environmental Hazards." *Progress in Human Geography* 20 (4): 529–539. doi:10.1177/030913259602000407.

Cutter, Susan L., B. J. Boruff, and W. L. Shirley. 2003. "Social Vulnerability to Environmental Hazards." *Social Science Quarterly-Wiley Online Library* 84: 242–261. https://onlinelibrary.wiley.com/doi/full/10.1111/1540-6237.8402002.

FEMA. n.d.a. "Preparedness for emergency response, FEMA training." https://training.fema.gov/hiedu/docs/fem/chapter 9 - preparedness for emergency response.doc.

FEMA. n.d.b. "Flood Map Service Center: Welcome!" https://msc.fema.gov/portal.
Fenton, D. M., and C. MacGregor. 1999. *Framework and Review of Capacity and Motivation for Change to Sustainable Management Practices: Theme 6: Project 6.2.1., Social Science Centre*. Canberra: Bureau of Rural Sciences.
Fishburn, Peter C. 1967. "Methods of Estimating Additive Utilities." *Management Science* 13 (7): 435–453. doi:10.1287/mnsc.13.7.435.
Flanagan, Barry E., Edward W. Gregory, Elaine J Hallisey, Janet L. Heitgerd, and Brian Lewis. 2011. "A Social Vulnerability Index for Disaster Management." *Journal of Homeland Security and Emergency Management* 8 (1). doi:10.2202/1547-7355.1792.
Frigerio, Ivan, and Mattia De Amicis. 2016. "Mapping Social Vulnerability to Natural Hazards in Italy: A Suitable Tool for Risk Mitigation Strategies." *Environmental Science & Policy* 63 (September): 187–196. doi:10.1016/j.envsci.2016.06.001.
Frigerio, Ivan, Stefania Ventura, Daniele Strigaro, Matteo Mattavelli, Mattia De Amicis, Silvia Mugnano, and Mario Boffi. 2016. "A GIS-Based Approach to Identify the Spatial Variability of Social Vulnerability to Seismic Hazard in Italy." *Applied Geography* 74 (September): 12–22. doi:10.1016/j.apgeog.2016.06.014.
Goodchild, Michael F., and J. Alan Glennon. 2010. "Crowdsourcing Geographic Information for Disaster Response: A Research Frontier." *International Journal of Digital Earth* 3 (3): 231–241. doi:10.1080/17538941003759255.
Hendriks, Paul, and Dirk Vriens. 2000. "From Geographical Information Systems to Spatial Group Decision Support Systems: A Complex Itinerary." *Geographical and Environmental Modelling* 4 (1): 83–104. doi:10.1080/136159300111388.
Hwang, C. L., and K. Yoon. 1981. *Multiple Attribute Decision Making: Methods and Applications*. New York: Springer-Verlag.
Jervis, Rick. 2013. "Historic flash flood leaves devastation in Austin." USA Today. November 7. Accessed March 8 2018. https://www.usatoday.com/story/news/nation/2013/11/07/flash-flood-austin-residents-killed-displaced/3459181/.
Juntunen, Lorelei. 2006. "Addressing Social Vulnerability to Hazards." *Disaster Saftey Review*, TsuInfo Alert 4 (2): 3–10.
Karnatak, Harish Chandra, Sameer Saran, Karamjit Bhatia, and P. S. Roy. 2007. "Multicriteria Spatial Decision Analysis in Web GIS Environment." *GeoInformatica* 11 (4): 407–429. doi:10.1007/s10707-006-0014-8.
King, David, and Colin MacGregor. 2000. "Using Social Indicators to Measure Community Vulnerability to Natural Hazards." *Australian Journal of Emergency Management* 15 (3): 52–57.
Kongthon, A., C. Haruechaiyasak, J. Pailai, and S. Kongyoung. 2012. "The Role of Twitter during a Natural Disaster: Case Study of 2011 Thai Flood." In 2012 *proceedings of PICMET '12: Technology Management for Emerging Technologies*, 2227–2232.
Kryvasheyeu, Yury, Haohui Chen, Nick Obradovich, Esteban Moro, Pascal Van Hentenryck, James Fowler, and Manuel Cebrian. 2016. "Rapid Assessment of Disaster Damage Using Social Media Activity." *Science Advances* 2 (3): e1500779. doi:10.1126/sciadv.1500779.
Li, Zhenlong, Cuizhen Wang, Christopher T. Emrich, and Diansheng Guo. 2018. "A Novel Approach to Leveraging Social Media for Rapid Flood Mapping: A Case Study of the 2015 South Carolina Floods." *Cartography and Geographic Information Science* 45 (2): 97–110. doi:10.1080/15230406.2016.1271356.
Lin, Tao, Shaowen Wang, Luis F. Rodríguez, Hao Hu, and Yan Liu. 2015. "CyberGIS-Enabled Decision Support Platform for Biomass Supply Chain Optimization." *Environmental Modelling & Software* 70 (August): 138–148. doi:10.1016/j.envsoft.2015.03.018.
Malczewski, Jacek. 2006. "GIS-Based Multicriteria Decision Analysis: A Survey of the Literature." *International Journal of Geographical Information Science* 20 (7): 703–726. doi:10.1080/13658810600661508.
Meyer, Spencer, Michelle Johnson, Robert Lilieholm, and Christopher Cronan. 2014. "Development of a Stakeholder-Driven Spatial Modeling Framework for Strategic Landscape Planning Using Bayesian Networks Across Two Urban-Rural Gradients in Maine, U.S.A." *Ecological Modelling* 291 (10): 42–57.
Mitra, Debashis, Chandrani Bhandery, Anirban Mukhopadhyay, Abhra Chanda, and Sugata Hazra. 2017. "Landslide Risk Assessment in Darjeeling Hills Using Multi-Criteria Decision Support System: A Bayesian Network Approach." In *Disaster Risk Governance in India and Cross Cutting Issues*, Disaster Risk Reduction, 361–386. Singapore: Springer. doi:10.1007/978-981-10-3310-0_18.
Morrow, Betty Hearn. 1999. "Identifying and Mapping Community Vulnerability." *Disasters* 23 (1): 1–18. doi:10.1111/1467-7717.00102.
NOAA. 2018. "Storm Events Database." https://www.ncdc.noaa.gov/stormevents/details.jsp.
Olson, David L. 1996. *Decision Aids for Selection Problems*. Springer Science & Business Media.
Price, Asher. 2016. "In Texas Flash Flood Alley, problems extend beyond terrain." MyStatesman, May 28. Accessed Feb 13 2018. https://www.mystatesman.com/news/local/texas-flash-flood-alley-problems-extend-beyond-terrain/qTACYqmPbxCURBhCHUJ0JM/.
Qureshi, M. E., S. R. Harrison, and M. K. Wegener. 1999. "Validation of Multicriteria Analysis Models." *Agricultural Systems* 62 (2): 105–116.

Rygel, Lisa, David O'sullivan, and Brent Yarnal. 2006. "A Method for Constructing a Social Vulnerability Index: An Application to Hurricane Storm Surges in a Developed Country." *Mitigation and Adaptation Strategies for Global Change* 11 (3): 741–764. doi:10.1007/s11027-006-0265-6.

Saaty, Thomas L. 1980. *The Analytic Hierarchy Process. Vol. VI. The AHP Series*. New York, NY: McGraw-Hill.

Saaty, Thomas L. 2008. "Decision Making with the Analytic Hierarchy Process." *International Journal of Services Sciences* 1 (1): 83–98. doi:10.1504/IJSSci.2008.01759.

Schnebele, E., and G. Cervone. 2013. "Improving Remote Sensing Flood Assessment Using Volunteered Geographical Data." *Natural Hazards and Earth System Sciences; Katlenburg-Lindau* 13 (3): 669–677. doi:10.5194/nhess-13-669-2013.

Simanaviciene, Ruta, and Leonas Ustinovichius. 2010. "Sensitivity Analysis for Multiple Criteria Decision Making Methods: TOPSIS and SAW." *Procedia Social and Behavioral Sciences* 2: 7743–7744.

Sugumaran, R., and J. Degroote. 2011. *Spatial Decision Support Systems, Principles and Practices*. UK: Taylor & Francis Group.

Tate, Eric. 2012. "Social Vulnerability Indices: A Comparative Assessment Using Uncertainty and Sensitivity Analysis." *Natural Hazards* 63 (2): 325–347. doi:10.1007/s11069-012-0152-2.

Triantaphyllou, Evangelos. 2000. "Multi-Criteria Decision Making Methods." In *Multi-Criteria Decision Making Methods: A Comparative Study*, Applied Optimization, 5–21. Boston, MA: Springer. doi:10.1007/978-1-4757-3157-6_2.

Turban, Efraim, Jay Aronson, and Ting-Peng Liang. 2000. *Decision Support Systems and Intelligent Systems. Seventh*. New Jersey: Prentice Hall.

Twitter. 2018a. "ATXfloods@ATXfloods." https://twitter.com/atxfloods?lang=en.

Twitter. 2018b. "FEMA @ fema." https://Twitter.com/fema.

United States Census Bureau. 2017. "Tiger Geodatabases." https://www.census.gov/geo/maps-data/data/tiger-geodatabases.html.

Wang, Shaowen. 2016. *CyberGIS. In International Encyclopedia of Geography: People, the Earth, Environment and Technology*. Hoboken, NJ: John Wiley & Sons.

Wang, Yamei, Zhongwu Li, Zhenghong Tang, and Guangming Zeng. 2011. "A GIS Based Spatial Multi-Criteria Approach for Flood Risk Assessment in the Dongting Lake Region, Hunan, Central China." *Water Resources Management* 25 (13): 3465–3484.

Wisner, Ben, Piers Blaikie, Terry Cannon, and Ian Davis. 2004. *At Risk: Natural Hazards. People's Vulnerability and Disasters*. London: Routledge.

Yang, Chaowei, Qunying Huang, Zhenlong Li, Kai Liu, and Fei Hu. 2017. "Big Data and Cloud Computing: Innovation Opportunities and Challenges." *International Journal of Digital Earth* 10 (1): 13–53.

Yin, Dandong, Yan Liu, Anand Padmanabhan, Jeff Terstriep, Johnathan Rush, and Shaowen Wang. 2017. "A CyberGIS-Jupyter Framework for Geospatial Analytics at Scale." In *Proceedings of the Practice and Experience in Advanced Research Computing 2017 on Sustainability, Success and Impact, 18:1–18:8. PEARC17*. New York, NY: ACM. doi:10.1145/3093338.3093378.

Zhang, Zhe, Urška Demšar, Jaakko Rantala, and Kirsi Virrantaus. 2014. "A Fuzzy Multiple-Attribute Decision Making Modelling for Vulnerability Analysis on the Basis of Population Information for Disaster Management." *International Journal of Geographic Information Science* 28 (9): 1922–1939. doi:10.1080/13658816.2014.908472.

Index

Note: Page numbers in *italics* and **bold** denote figures and tables, respectively.